TIMEWALKERS

TIMEWALKERS

The Prehistory of Global Colonization

CLIVE GAMBLE

Harvard University Press
Cambridge, Massachusetts

Published by arrangement with Penguin Books Limited

Typesetting and origination by Alan Sutton Limited

Printed in the United States of America

Second printing, 1996

This book is printed on acid-free paper, and its binding materials have been chosen for strength and durability.

Library of Congress Cataloging in Publication Data

Gamble, Clive.
Timewalkers: the prehistory of global colonization/Clive Gamble.
p. cm.
Includes bibliographical references and index.
ISBN 0–674–89202–X (cloth)
ISBN 0–674–89203–8 (pbk.)
1. Man, Prehistoric. 2. Anthropology, Prehistoric. 3. Man-Migrations.
4. Archaeology. I. Title.
GN740. G63 1994
930. 1—dc20 93–28825
CIP

Contents

Acknowledgements vii

Preface ix

1 Introduction: "To turn the hero" 1

2 "With wand'ring steps and slow:" progress toward a world prehistory 13

3 Past cradles and ice age time 29

4 The modern cradle for human origins 47

5 Why Africa? 74

6 Social climbers and migrant workers on the early African savannahs 96

7 800,000 years down the Old World track 117

8 Ancients and Moderns: what happened to the Neanderthals? 144

9 Pioneers and diehards in the new lands of the Old World 179

10 Humans almost everywhere 203

11 Why people were everywhere 241

Notes 249

Bibliography 270

Index 303

For Elaine

Acknowledgements

This book has benefited from good advice from many people. I am grateful to the following for the time and care they spent on reading earlier drafts and answering questions: Alaisdair Whittle, Chris Stringer, Robin Torrence, Peter White, Colin Ridler, John Pfeiffer, Elaine Morris, John Gowlett, Pat Carter, Ian McNiven, Richard Cosgrove, Olga Soffer, Graham Richards, Mark Roberts, Chris Gosden, Paul Mellars, Francis Wenban-Smith, Rhys Jones, Robin Dennell, Derek Sturdy, Phillip Edwards, Steve Shennan, Iain Davidson, Wil Roebroeks, Andrew Chamberlain, Roger Powers, Steven Mithen, Robin Derricourt, James Steele, Michael Fisher, Bob Whallon, David Yesner, Rupert Harding, Brian Molyneaux, John Speth, and Peter Ucko. Particular thanks are due to Andy Garrard, Nikolai Praslov, Janusz Kozlowski, Jim Allen, Rob Foley, Tim Murray, Elisabeth Vrba, Clark Howell, Geoff Bailey, Lew Binford, K. Paddayya, and Todd Whitelaw for showing me either the field evidence or the errors of my discussions on evolutionary ecology and hunters. The challenges from students on several courses and many lectures in England and Australia have, as always, been invaluable and I hope they enjoy the results.

Any attempt at a world prehistory must also acknowledge the example provided by Sir Grahame Clark who showed thirty years ago that it was both possible and profitable. Field work has been supported by the British Academy, the University of Southampton, and the Sir Robert Menzies Australian Studies Centre. I should also especially like to thank Paul Graves for many stimulating discussions as this book grew. Although I know he will scratch his head over many of the arguments he will, I hope, find some things of value.

All the original artwork was designed and produced by Rob Read. I am grateful to Mike Smith (pl. 10.4), Rudy Frank (pl. 10.7), Bjarne Grønnow (pl. 10.2), Morten Meldgaard (pl. 10.3), Nikolai Praslov (pl. 9.1), Nick Bradford (pl. 10.8) and Gustavo Politis (pl. 9.4) for the photographs they supplied. Mary Iles compiled the index. To all of them, many thanks.

"This farther land has never yet been trodden by civilized Man; and if he ever does reach it, he will thus probably find it occupied by men who may have forgotten how and whence their fathers came."

The Duke of Argyll, 1869. *Primeval Man.*

"To think is to forget differences, generalize, make abstractions."

Jorge Luis Borges, 1964. *Funes the Memorious.*

Preface

Five hundred years ago a single voyage ushered in a new world. With every further exploration an astonishing variety of peoples and places was uncovered. These discoveries about the world and its inhabitants formed the basis for many new orders of science to be tested in the global laboratory.

The subject of this book, the human prehistory of how and why the world was colonized, is part of Columbus's legacy, although neither he nor his contemporaries could ever have anticipated it. While he is credited with dispelling myths about a flat earth, it is sometimes forgotten that he sailed with a full complement of equally implausible ideas about what else he should find. He expected to find lands populated, if not by humans, then by fabulous creatures straight from the medieval imagination. On his return he wrote to his patrons, "I have not found the human monsters which many people expected."[1]

The fact that people were almost everywhere raised no eyebrows, even when discovered on remote islands, across inhospitable deserts, and up rugged mountains. While the great voyages of discovery that began with Columbus gradually uncovered the near totality of prehistoric colonization, this was rarely seen as an issue requiring the attention of the new science of human prehistory which appeared in the middle of the nineteenth century when the adventure of global empires had reached maturity. Prehistoric archeology, with its methods that chart the course of human mental, social, and cultural development, documented the paths toward civilization. The how and why of the first global colonizers evoked little curiosity even if it was established that entire continents had been settled without the basic advantages of European civilization such as iron, wheels, or writing. Explanations, as I shall show, were always ready to hand.

This lack of interest in such a fundamental issue about our global prehistory continues. When colonization is considered, the obvious is often stated, as in this very recent archeological example: "the configuration of islands in the Pacific . . . . rewarded Pacific peoples, more than peoples of other oceans, for developing maritime skills."[2]

The purpose of this book is to remedy such missed opportunities. I will show how prehistory contributes new knowledge about ourselves through the study of a shared past, the end result of which was global colonization. Rather than this being an *adaptation*, as the quotation above implies, I will argue that this process is an *exaptation*—an unforeseen consequence of fundamental changes in human behavior rather than the reason for them. Five hundred years after Columbus, our concerns should be not only with the rise of civilizations but also with understanding the commonality of the mosaic of humanity. The global village needs its own prehistory. To achieve this we need a question to investigate and the one I have chosen for this book is, Why were people everywhere?

In order to answer it I shall look at both the history of prehistoric archeology as well as its recent discoveries. It is sometimes forgotten that the great success of 150 years of systematic prehistory has been not only to extend the span of human history to include deep time but also to alter our perception of the past by changing the way in which such longterm records of action are preserved and presented to us. Prehistory is not a poor person's history because it deals with stones, bones, and landscapes rather than documents and pictures. We are dealing with a singular reality, another form of self-knowledge which ultimately may serve, as all history serves, several contemporary purposes. The timewalkers in this book are therefore both ancient and modern. They are the peoples of prehistory as well as the archeologists who reconstruct them. At the end of a century when the concept of geographical remoteness has shrunk so that the idea of "getting away from it all" has little hope of fulfillment, the sense of remoteness in time has become the substitute. Archeologists have replaced those earlier explorers. We investigate landscapes expecting to find prehistoric occupation even though the area may be uninhabited today.

In 1990 I joined a team from Melbourne's LaTrobe University digging a small limestone cave by the Maxwell River in the World Heritage Wilderness Area of South West Tasmania. Warreen, the cave of the wombat, had last been used 18,000 years ago when its roof collapsed during the height of the last global ice age. No Aborigines lived in the southwest when Europeans first made contact. Animals and plant foods are scattered while snakes abound in the stands of ti-tree and sun themselves on the tussocks of button grass. This is difficult country where it can take a day to walk a kilometer. Helicopters have changed that, but no one had been to this valley with its small cave for over 12,000 years. What stands between the present and the past in the cave is a thin layer of moon milk, a platey stalagmite covering the floor. Remove this and the deposits begin, super-rich in butchered animal bone and chipped stone flakes and tools.

Levering himself out of the narrow trench one day, the team's leader Professor Jim Allen called me over.

"There's timewalking for you," he said. "One foot at 25,000 B.C. and the other on A.D. 1990. That's why prehistory is special. Put that in your book."

1
Introduction: "To turn the hero"

What made us human and how did our humanity evolve? My aim in this book is to examine not only the evidence for human origins and evolution but also the framework for explaining why change took place—when, where, and how. To obtain the answers I believe we need to investigate the neglected question of why we are the only animal with a near-global distribution. This topic is the overlooked subject matter of human prehistory.

Let me start a long story with an example of recent exploration. It is Christmas 1776. Captain Cook, whose discoveries unveiled the shape of the world, is on his third and final voyage. He is anchored at Kerguelen Island in the southern Indian Ocean. Although it is summer the place lives up to its other name, the Desolate Isle. By sailing round it Cook has just punctured the optimistic reports of Kerguelen who discovered the island four years earlier and named it after himself. Back in Paris, Kerguelen has magnified his glimpse of land into part of the sought-after southern continent. He described it favorably and inaccurately as like "Southern France" to potential backers of a return voyage, but he never tried to go ashore.[1]

Plate 1.1 Resolution *and* Discovery *in Christmas Harbour, Kerguelen Island 1776,* by John Webber.

This was one of the few places visited by Cook where neither natives nor their fires greeted him as he sailed into the bay. Remoteness and desolation provided a ready explanation.[2] Even so, John Webber, the ship's artist, painted emperor penguins as a welcoming party on the shore.

I began this book out of curiosity. Seeing Webber's picture in an exhibition at the British Museum, it struck me forcefully that people were expected everywhere. Like penguins, they were another bipedal part of the fauna. Their presence proved so universal that when found in the normal course of discovery they merited no comment in the ship's log. If natives were absent, the nearest representative was painted in their place.[3]

But why did the early European explorers show so little surprise at their discovery of the unrecorded colonization of the globe? Their journals never raise the presence of these natives either as an issue for speculation or later, during the scientific voyages of men like Cook and Banks (1768–71), as a circumstance requiring systematic investigation.[4] Nor did this fact depend upon religious or scientific convictions. The journals from the voyage of the *Beagle* (1831–6), which saw Darwin sharing a cabin with an ardent fundamentalist, Captain Fitz-Roy, contain many comments about the indigenous peoples they both saw but few about why or how they got there.[5] They could have been placed by God as part of the natural fauna and flora of a region, or migrated there in some slow manner from a center of origin. The nearly universal distribution of humanity, which so exceeded the geographical range of any other mammal, never drew their attention. One of the great Victorian naturalists, Alfred Wallace, who arrived at the principle of natural selection independently of Darwin, expressed this indifference perfectly. In his preface to *Island life* in 1880, a work which founded the study of biogeography, he commented that, while there was a great deal to understand and describe about the geographical distribution of all other species, with *Homo* all that could be said was "the bare statement – 'universally distributed' . . . and this would inevitably have provoked the criticism that it conveyed no information."[6] So he left us out.

This lack of curiosity about a signal fact concerning our own species now strikes me as one of the central issues in the development of archeology and particularly the study of prehistory. During prehistory, which comprises 99.9 percent of first hominid and then human evolution, we developed all the essential characteristics of our humanity—language, culture (including art and technology), society, intelligence, religious, ethical, and moral codes—as well as exploring most of the ways of using and abusing them through cooperation, economic arrangements, bureaucracies, warfare, invention, competition for status, rank and prestige, and colonization. We have inherited the view from the Enlightenment of the eighteenth century, supported by the scientific advances of the last century, that these characteristics of our humanity developed piecemeal, like receiving gifts, and over time have been combined into ever more complex patterns and relationships. Working on this assumption archeologists have rejected the alternative proposition, common in the last century[7] and still found outside of science,[8] that mankind emerged fully civilized and then in some parts of the world and among some societies degenerated to lower states of barbarism and savagery.

The reasons are not difficult to find. The indifference, as I shall show in chapter 2, stemmed from the fact that they expected to find people. They were only surprised, like Columbus, that they did not have two heads and that they were not ten feet tall. Long before there was a scientific record of human fossils and archeological finds, or a theory of evolution to account for changes, firmly held beliefs existed about our earliest geographical origins, physical appearance, mental abilities, economies, and political systems. Deeply embedded in these stories were the moves from simple to complex, primitive to civilized, crude to sophisticated, camps to cities, hunting to agriculture; in short, the celebration of progress.

Nowhere is this better illustrated than by the theme of this book, global colonization. The arrogance of empire still insists that the West discovered the world. The mission of progress was to colonize and accelerate the benefits of civilization. However, the effortless sense of superiority in much of the history of Western exploration sits uneasily next to the realization that wherever anchors were dropped, mountains climbed or deserts crossed, people were there to be discovered.

What also sustained my curiosity was the historical role of archeology in this portrayal. Prehistoric archeology appeared in the middle of the nineteenth century. As I show in chapter 2, it was dedicated to the demonstration of progress and hence the inexorable rise of civilization. At its worst it formed a branch of Herbert Spencer's Social Darwinism.[9] Here antiquity conferred status on change and on such spurious social laws as "survival of the fittest." Today's prehistorians reject progress as a guiding principle, but continue to follow the agenda into human origins that was set over 150 years ago. Identifying a center, or cradle, and then watching the drift of new populations to other parts of the world in its time-honored method of investigation. I expand on why this was and continues to be the case in chapter 3, where I shall pay attention to the importance of climate and time as explanations. Here I will only note that the question about our universal distribution, which Wallace consigned to obscurity, is not often investigated. One of the aims of this book is to justify a more prominent position for it on the agenda of research into human origins.

Instead of asking "why were humans everywhere?" archeologists have stuck instead to "why was civilization not universal?" The choice of question made prehistory a subject with a narrow regional focus and a bias toward material remains which would reveal the heroes of the present's past. Archeological textbooks are often sermons on the inevitable. All developments are seen as a consequence of elapsed time during which the human species achieved full humanity, like a child becoming an adult, and there are frequent references to the "infancy of the human race" and "the cradle of mankind."

The problems with stories

Another suggestion for the familiar shape of research into human origins can be found in one of the most entertaining papers to appear in recent years. In it Misia Landau tackles our origins with a literary model.[10]

Her point is that irrespective of what data the pioneers of human origins had to hand, the explanations they put forward for human evolution were akin to a good folktale or heroic myth. The familiar story opens to reveal the hero, a fossil ancestor, in a state of calm and ease which is then disrupted by outside forces. The hero then sets out, is tested, receives gifts and as a result changes, is tested again and then triumphs. Much of the debate between different versions of human evolution comes down to disagreements over the gifts and challenges but not the structure of the story itself. These gifts continue to provide the traditional answers to my opening question. Bipedalism, intelligence, technology interact with climatic and environmental challenges. Progress realizes potential. As Landau puts it,

> Here, in the narrative of evolution, man's struggle often takes a turn away from nature and toward men. In any case, the function of these tests is to develop civilization and thus *to turn the hero into a modern human* (my emphasis).[11]

Her analysis rightly draws attention to the way we present our arguments. The history of science is full of similar examples where the plot which achieves the happy ending, in this case civilization, can accommodate any number of different characters. Chapter 4 presents the current, albeit simplified, cast list of early fossil ancestors from Africa and chapter 5 their various gifts and challenges.

Fit to or fit for?

Having identified the storyline, what next? Well, literary forms other than fairy stories do exist. Landau's simplification of research into human origins is only partially helpful because she too has backed away from considering alternatives by telling us only what is familiar.

These alternatives are not about the characters or even the ending to the human story, but about understanding the process of evolution as applied to ourselves. I have set out the main positions in chapter 5. By posing the simple question, Why Africa?, attention is drawn to both the specific components of hominid evolution in that continent as well as the more general principles of evolutionary theory.

The tests or challenges in Landau's storyline can of course be fitted very well to the data and theories. They mesh nicely with the notion of adaptation. Derived from the latin *ad* (to) and *aptus* (fit) the term is commonly used as a means to explain change in form and behavior. This is seen as a result of selection for such changes coming from outside forces such as climate. The implication is that all aspects of an organism have been developed for a present purpose. Goodness of fit between form and function is assumed. Feet are optimally designed for walking, brains for making tools, pelvises for giving birth to larger headed children, and so on. Each element is therefore adapted to a specific function because they are all under relentless selection for improvement. The pay off comes from the increased biological fitness and therefore survival. This applies to both the organism and, more importantly, to its offspring. Here is a mainspring of Darwinian evolutionary theory with its mechanism to explain change as descent with modification.

The problem with adaptation is that it can become a tautology. Just as frogs, being adapted to princesses, always get kissed and so evolve into princes, everything, if we try hard enough, becomes adaptive. The huge beetling brow ridges so characteristic of Neanderthal skulls are a good case in point. Every few years a new adaptive function is put forward. This bar of bone becomes a sunshade, a way of reducing snow glare, a threat display, or a means of absorbing stresses and strains from the grinding action of their large teeth. Whatever the answer, the ingenious suggestions that continue to appear point to our scientist's dislike of a feature without an accepted function.

An alternative, as Stephen J. Gould and Elisabeth Vrba have argued, is to acknowledge that not all features are currently adaptive.[12] They are instead the source for future change. They have not been designed during evolutionary history to fill the particular role they may now be serving. They propose exaptation (fit by reason of their form) as the missing term. Exaptations are coopted for use rather than designed. They cite feathers as an example. The fossil record shows that these did not originally appear as adaptations for flight, but instead as a means of keeping warm. Only later were they coopted—or exapted—as the materials for natural selection to work on. After further adaptations the materials for flight were produced.

Stone tools provide an archeological example. Their shape might be determined by functions such as cutting, chopping, and scraping as well as by the requirement for such visible items to carry information, via their style, about group affiliation or personal identity—lithic "badges" which carry messages about being and belonging conveyed in stone. Let us suppose that this stylistic component of arrowhead design emerged late in prehistory as changes in social behavior then required, for the first time, the support of this means of visual communication. Any changes in shape would therefore be exaptations, coopted to meet a current need. The many styles of later arrowheads merely draw on the potential of stone to be chipped in a variety of ways. Their shapes are not explained by the historical circumstance that for 2 million years previously people had used stone for tools, even though they were under selection to do so since tool use, we assume, contributed to longterm evolutionary fitness.

Independence from the adaptive drive of evolutionary history is, as Gould and Vrba point out, important for our understanding of the effect of evolutionary changes. Nowhere is this more so than for human behavior generated by that infinitely flexible and subtle organ, the brain. Their conclusion is that most, if not all behavior, is exapted rather than adapted.[13]

Why is such an alternative concept important? To return to Landau, the notion of exaptation contradicts her preferred form of narrative for arguments about human evolution. The teleology in a story, with the inevitable happy ending of civilization, is exposed. The plot that depends on adaptive concepts no longer holds up.

Migration and colonization

But you might counter that my chosen questions about the origins of global humanity are also teleological; they presuppose a story based on the wisdom of historical hindsight—an earth colonized in prehistory. This would indeed be the case if I were to argue that such an ending was what we were, and still are, adapted for. However, as I shall endeavor to show in chapter 5, the fact of global colonization is as likely to be an exaptation as an adaptation forged from a pair of feet and a large brain. The result will be a different sense of history rather than the adaptive *Just so stories* identified by Landau or the conclusion to Bruce Chatwin's thoughtful *Songlines*:

> Natural selection has designed us—from the structure of our brain-cells to the structure of our big toe—for a career of seasonal journeys on foot through a blistering land of thorn-scrub or desert.[14]

This places particular importance on the words we use to analyze the global journey.

Colonization, migration, dispersal are widely but often imprecisely used terms. They embody political as well as biological notions and deal with operations that take, and took place, at different scales of time and space. They can also refer to individuals, groups, or entire populations, the movement of peoples, genes or, in the case of behavior and material culture, ideas. There is also the difficult area of intention and motivation implicit in the shortterm historical usage, but often denied in longterm, evolutionary treatments. My use of these terms is set out in Table 1.1 and discussed further in chapter 5 along with the question of speciation. Adaptive radiation is a familiar term that describes a pulse of new species in the evolutionary record.[15] I have balanced this with exaptive radiation as an alternative means of accounting for the same pattern. It places more emphasis on behavior as the framework for change among species. This behavior provides purpose; it produces a different picture of colonization to what is usually considered a random process.

Adaptive and exaptive approaches are examined in chapter 6 with a reconstruction of African hominids and their colonizing history. This illustrates the variety of scales and processes which are at work. These need to be separated before turning to the archeological evidence presented in chapters 7 through 10 for the last million years of the process.

These chapters are guided by the proposition that the evolution of humans and the acquisition of humanity came through the process of global colonization. In other words, what was important about becoming human is represented by the very fact of our "getting there." The fact of arrival is more important than the technical means which made it possible and the additional cultural baggage this has left for the archeologist to catalog and pigeon-hole. This stance is consistent with a view of evolutionary effects as exapted rather than adapted. While technological aspects provide important details, I shall argue that they are not always the best way of establishing either how or, indeed, whether we have changed.

Table 1.1 Terms and definitions.

Migration
A discrete event, involving directed or intentional, though not necessarily calculated, movement from one type of place to another; between areas that are distinct environmentally, geographically, or seasonally. Short timescales or even single events, though not necessarily short geographical distances. One-off events that may or may not have lasting consequences in terms of colonization, as defined below.

Dispersal
A more general process but still fairly small-scale in time and space (referred to here as ecological time). It has the connotation of spreading out—of individuals or groups of the same species filling up the available habitat and reaching an area not previously inhabited by them. It is the basis of the spreading or colonizing process.* Dispersal may or may not have lasting consequences, depending on whether allopatric speciation (in another place or geographically isolated) occurs.

Colonization
A process occurring on a larger scale both temporally and geographically. Major extension of species habitat or range to include established occupation of areas previously unoccupied and use of new ecological niches. This may occasionally be due to the removal of environmental barriers but more likely to behavioral and biological changes. If the latter, then adaptive and exaptive explanations need to be investigated.

Adaptive radiation
A larger scale version of dispersal, in the sense of longer time spans (geological or evolutionary time), but not necessarily larger geographical areas. The diversification of species from a common ancestor to fill up the available ecological space. Natural selection adapts them to new niches that appear either locally or through colonization.

Exaptive radiation
Another large-scale and longterm process taking place in evolutionary time. Diversification of new species through the effect of phenotypic and behavioral potential being coopted in colonizing situations.

* Udvardy 1969: 11.

There, then, is the outline for my story. The journey will form the narrative and there will be plenty of heroes, gifts, and challenges along the way. The twist in the tale lies in a new opportunity for interpreting the legacy of our humanity. Archeologists have usually dwelt upon achievements such as the first use of fire, stone tools, or art as signposts which can be arranged chronologically to see when we drew closer to the full modern package, within spitting distance of civilization. While presenting this material I will emphasize instead those aspects we cannot see in the remains we recover but which we need to infer. Consequently, what I believe matters most is the process of the past—the behavior which led to the manufacture of artifacts, the alteration of landscapes and hence to archeology—a process which, contrary to the conventional methods of scientific research, we cannot directly observe.

My solution to this problem is to argue that the prehistory of global colonization is the prehistory of humanity. We will learn more about human evolution by studying the history and process of world colonization than by any other means. Through colonization we observe the evolution of our humanity.

The story so far

In my story world colonization is neither a goal nor a triumph. After all, who would have thought that by filling up the earth the opportunity would arise to rediscover it? The process of colonization has its own framework of *space* in the form of the arrangement of major habitats across the continents and islands, and *time*, as shown by the archeological evidence for getting there.

To emphasize these points I shall now briefly set out the archeological section of the book, chapters 4 and 7 through 10. The divisions are not based on the shape of fossil skulls or stone tools. Neither are they closely related to geological periods. The three main divisions, based on the pattern of colonization, are set out in table 1.2. I have also included some of the actors to help later in sorting out the inevitable tangle of names.

Each of these timewalking divisions saw major new lands or habitats colonized for the first time. The time taken varies enormously between each period. By far the most complicated, as we shall see in chapters 7 and 8, covers a million years of the Ancients. At the end of this period is a brief Pioneer phase when further colonization begins. It is only visible for a short time, some 20,000 years, in some continents, such as Europe, where the data are richest. The detail from this transition phase can be used to investigate how modern humans

Table 1.2 A guide to the prehistory of world colonization. Three major timewalking divisions are identified (B.P. = before present).

5 million–1 million	**Early hominids** Apiths (Australopithecines), early *Homo* *sub-Saharan Africa*
1 million–200,000 B.P.	**Ancients** *Homo erectus*, "archaic" *Homo sapiens*
200,000–60,000 B.P.	early Neanderthals, anatomically Modern humans *Africa, mid-latitude Asia and Europe*
60,000–40,000 B.P.	Pioneer or transition phase classic Neanderthals, "archaic" *Homo sapiens* anatomically Modern humans *continental Eurasia*
50,000–10,000 B.P.	**Moderns** anatomically Modern humans *Australia, Eastern Siberia, Pacific margins, Japan, Americas, unglaciated mountain chains*
10,000 B.P.–1,500 A.D.	*Arctic, Indian Ocean, deep Pacific, tropical rain forests, great sand deserts*
1,500 A.D.–present	*central and southern Atlantic Ocean*
unoccupied	*Antarctica*

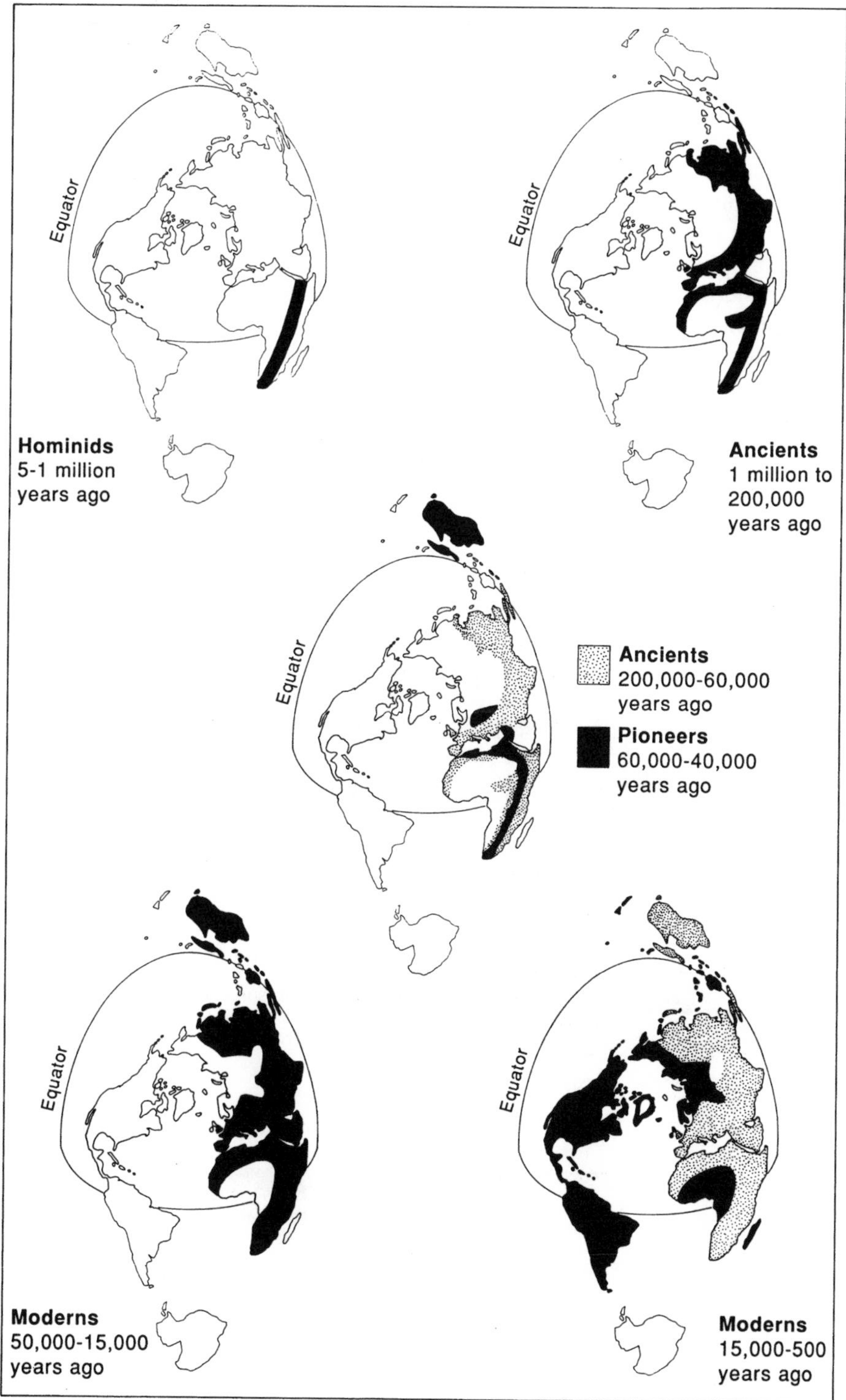

Figure 1.1 Steps toward global colonization.

evolved, either through local developments or by replacement from outside of genes, anatomy, and culture.[16] The contrast between the Ancients, with only small geographical gains once timewalkers had moved out of sub-Saharan Africa, and the rapid expansion of the Moderns in the third major phase, marks the great pivotal point in human prehistory (fig. 1.1). Understanding how and why is my goal in chapters 9 and 10. Table 1.2 also shows how comparatively little was discovered about the world in the voyages since Columbus. This involved mostly a few small islands in the central and southern Atlantic, from Bermuda in the north to Bouvet in the south.

Another way to look at this staccato pattern of global colonization is by the different properties of the environments first settled in each period. This has been done in fig. 1.2 where I have compared ten of the earth's major terrestrial habitats. The measure I have chosen is significant for longterm survival and

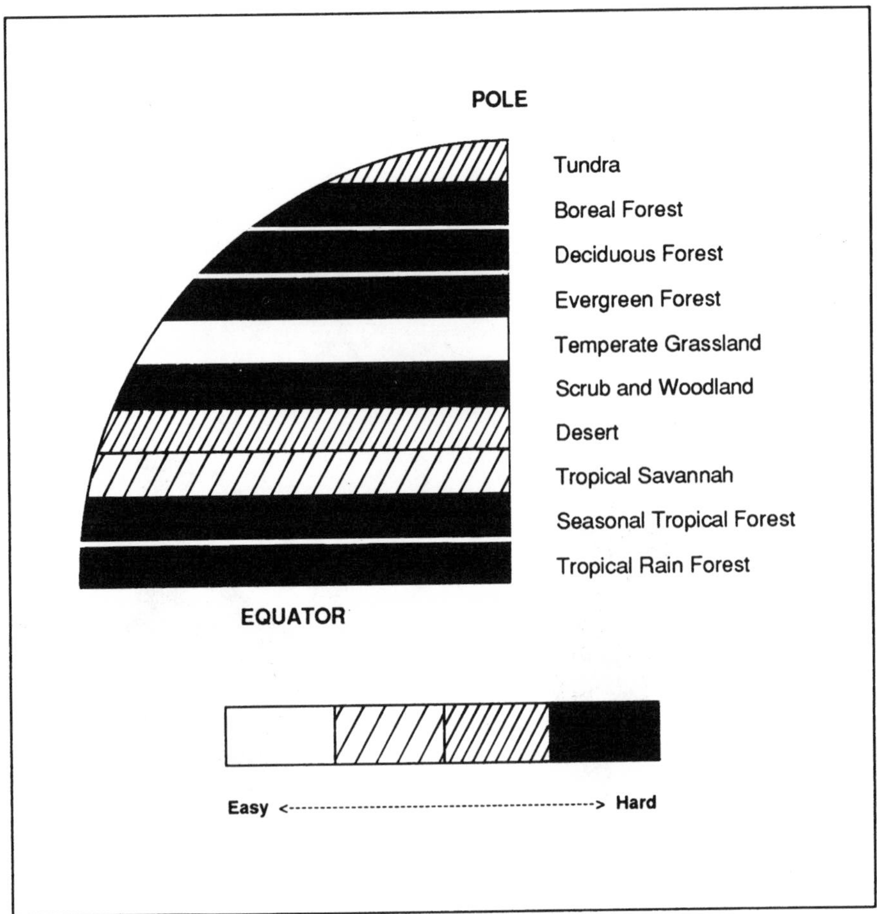

Figure 1.2 Ease of making a living as measured by the accessibility of food for a foraging economy in ten major habitats from the Equator to the pole.

evolution: the accessibility of food resources, plant and animal. This measure was first proposed by Robert Kelly,[17] who calculated it from estimates about the amount of primary (plant) and secondary (animal) biomass and the net primary productivity (the amount of new growth every year) in each habitat. Figure 1.1 gives a simplified measure of accessibility in order to rank the habitats as they change from Equator to pole, the timewalkers' route. Four divisions are obviously an oversimplification but worth presenting for all that. When combined with the timescale of world colonization offered in table 1.2 a clear pattern emerges. The two easiest habitats—tropical savannahs and temperate grasslands—are colonized first, with the difference in time apparently due to the barrier created by the intervening Sahara Desert. It is interesting that the highly productive temperate grasslands of southern Africa are not separated from the savannahs of East Africa by any such obstacle. They not only have evidence from the early hominid phase but may yet prove to have the earliest timewalkers of them all.

Once away from these habitats with their plant and animal resources life becomes harder. There may be huge growth and productivity, as in a tropical rain forest. But diversity and abundance are not everything. The distribution and comparatively lesser usefulness for timewalkers of many of the forest resources result in much later colonization. Only those tundras which appeared in mid-latitude Europe and Asia during the Pleistocene ice ages came anywhere near the productivity of the temperate grasslands. Even so, these tundras were only rich in animals and not plant foods.

The estimates of accessibility focus our attention on two patterns to which I shall return repeatedly. In the first place, to what extent did the environment shape human evolution? How were the barriers to colonization in these habitats overcome? Was it a matter of waiting for the right adaptation to gradually evolve or was it instead the opportunistic cooption of existing physical structures and behavior?

The second pattern deals with time. Instead of a slow continuous spread, as many once believed (chap. 2), we have a highly punctuated chronology measured in Myr (million years) and Kyr (thousand years) B.P. (before the present). These staccato bursts of colonization are not associated exclusively with any single fossil ancestor as recognized either by skull shape or indeed by any significant step forward in technology. We shall see time and again that the major pattern of world colonization was the result of changes in behavior which did not necessarily march in time with anatomical or cultural developments.

I also need to explain my use of some vernacular terms for the fossil hominids. First, there is hominid itself, which strictly speaking includes all living humans as well as their fossil ancestors. We are hominids; so too were the Australopithecines, the southern apes, although they are often referred to as "early" hominids. The various species in this genus are sometimes lumped together as apemen. This is an unfortunate term. It conjures up an image of early hominids that is either all too red-in-tooth-and-claw or overstresses the involvement of one sex rather than the other in evolution and prehistory. To avoid the Yeti-ization of prehistory I have settled for an abbreviation of the Australopithecine mouthful, to Apith.

I shall term the next major group the Ancients. This covers a wide range of head shapes and sizes throughout the Old World and includes *Homo erectus*, archaic *Homo sapiens* and *Homo sapiens neanderthalensis* (or Neanderthal people). The current controversy over what to call *Homo erectus* (chap. 4) highlights the need for a catch-all term such as Ancients. The final group are the Moderns who need no anatomical introduction.

These shorthands are only used to ease the text, where possible, of cumbersome jargon. While they refer to the fossil evidence they do not, as already stated, imply corresponding levels of behavior. Indeed, we shall see that modern behavior was not exclusive or essential to the Moderns; table 1.2 illustrates that anatomically modern humans are to be found in the Ancient and Pioneer phases.

We are supposed to restrict the term human to ourselves, *Homo sapiens sapiens*, on the grounds that until we know what the differences are between ourselves and extinct taxa only confusion will arise if the distinction is dropped.[18] This ignores the fact that the confusion may stem from the way we classify ourselves and the expectations this brings to human evolution. I will look at this problem in later chapters. It is because of such taxonomic procedures that I have invented timewalkers to convey, in as neutral terms as I can muster, the agents involved in the process of global colonization. Freed from anatomy and expectations of which type of Paleolithic (Old Stone Age) tools they should be making, they serve to remind us of the all-important measures of time and space in understanding colonization as the evolution of behavior, rather than the bestowing of gifts on heroes.

2
"With wand'ring steps and slow:" progress toward a world prehistory

Humans everywhere: a non-issue?

The reasons for ignoring world colonization as something needing explanation and later as a framework for studying human evolution are embedded deep in Western culture. From at least Classical times the world was never thought of as an empty place. Pliny's famous catalog of monstrous races and peoples, primarily from Africa or India, mixes travelers' anecdotes with fantasy and imagination.[1] The Blemmyae of the Libyan Desert lacked necks and heads and so their faces were attached to their chests. Then there were horned men, hairy men and women, fish eaters, one-legged Sciopods, hole-creeping Troglodytes, Amazons and Antipodes (who walked "opposite-footed" or upside down on the underside of the world), dog-headed men and one-eyed races as well as giants, pygmies, and people with horses' hooves. Pliny's fabricated races were still evident in medieval texts where they were portrayed in bestiaries or hidden in the corners of illuminated manuscripts. They were painted and carved on churches and could often be found lurking at the edges of maps of the world. They emerged from the Classical world as early as the seventh century A.D. when St Isidore of Seville spoke of a land in the south "within the bounds of which the Antipodes are fabulously said to dwell."[2] And powerful images of these strange creatures found their way into Shakespeare:

> And of the Cannibals that each other eat,
> The Anthropophagi, and men whose heads
> Do grow beneath their shoulders . . .
>
> *Othello*, I, iii, 128ff.

The presence of these monstrous if picturesque races was undisputed, although proof of their existence tended to be as elusive as that today for the abominable snowman. On returning from his discovery of the West Indies, Columbus apologized for having found only men and not monstrosities as "many people expected."[3] For many years the Straits of Magellan, the gateway to the Pacific, were thought to be guarded by a race of giants, a belief which symbolized the fact that Europeans were entering an alien world where their conventional standards did not apply.[4] But it was part of a world that was inhabited by potential competitors even if, like Pliny's Panotii, they had ears which reached to their feet like blankets.[5]

These early explorers set out fully expecting to find other peoples as well as monsters and dog-headed cannibals. Their expectations were well summed up in the fictitious journeys of Sir John Mandeville from the late fourteenth century who claimed that wherever a traveler went in the world, "And always he should find men, lands and isles, as well as in this country."[6]

The tradition continues today with the widely held view that intelligent life, usually anthropomorphic, will be found in the galaxy. Before Neil Armstrong set foot on the moon he fielded sincere questions at a press conference about what he would do if small green creatures popped up from behind a rock. Deep space probes such as *Voyager 2* carry messages of goodwill in the expectation that they will be translated. Science fiction has added immeasurably to the number of extra-terrestrial Plinian races.

As the explorers failed to encounter the monsters which travelers like Marco Polo in 1292 had claimed were waiting just over the next horizon, so attention settled upon the indigenous peoples. Since their discovery had been expected, the interlopers' main expressions were those of distaste for unfamiliar customs rather than surprise at human presence. Practical information was sought and a useful rule-of-thumb for explorers, always interested in precious metals, lay in whether newly discovered peoples were clothed, in which case civilization and gold might be present, or naked and hence savage and best ignored.[7] Future voyages were planned on the successful sighting of shirts.

Indeed, naked savages were sometimes regarded as invisible, as shown by the infamous Terra Nullius Doctrine applied to New South Wales. This stated that the Australian Aborigine had no political or social existence, and that therefore the land which they occupied was open to occupation and colonization by Britain. Cook's observations, as he sailed the coast of New South Wales for the first time in 1770, that "We saw on all the Adjacent Lands and islands a great number of smooks, a certain sign they are inhabited, and we have dayly seen smooks on every part of the coast we have lately been upon"[8] proved no hindrance to the Terra Nullius Doctrine. People were expected everywhere. What mattered was finding society.

The human drift

The effortless Western sense of superiority was also nurtured by a belief that the primitive world moved at a different pace. Given enough time, everything was possible, even global colonization. The pace and direction set by Milton in the closing lines to *Paradise lost* (1667),

> The World was all before them, where to choose
> Their place of rest, and Providence their guide:
> They hand in hand with wand'ring steps and slow,
> Through Eden took their solitary way.

would find support two hundred years later from one of Darwin's close friends and neighbors, Sir John Lubbock. He wrote in his seminal book *Pre-historic times* (1865), "There can be no doubt that man originally crept over the earth's surface, little by little, year by year, just for instance as the weeds of Europe are now gradually but surely creeping over the surface of Australia,"[9] since this suited the Darwinian view of slow, gradual change. The plodding steps were matched by aimless wanderings. The purposeful voyages of discovery by the Europeans from the fifteenth century onward were triumphs of the mind, unlike the unrecorded human drift, as Jack London put it in 1919, fueled by animal passions:

> Dominated by fear, and by their very fear accelerating their development, these early ancestors of ours, suffering hunger pangs very like the ones we experience today, drifted on, hunting and being hunted, eating and being eaten, wandering through thousand-year-long odysseys of screaming primordial savagery.[10]

And still the tradition continues. Writing for an issue of *National Geographic* devoted to the peopling of the earth, Wilbur E. Garrett summed up the position in 1988:

> Whether driven by fear, hunger, curiosity, sense of destiny, or a mix of all, mankind—then as now—has continually migrated, as if to fill every void on this planet. Scientists generally agree that man drifted north out of Africa to Europe and eastward to Asia. Millenia later, with the appearance of modern Homo sapiens, the great migration pushed on to Australia and the Americas.[11]

Desert islands

Against this background it comes as little surprise that the question of why people had come to distant lands provoked scant interest. When Roggeveen found Easter Island in 1722, while searching for the great southern continent, he only commented that it was inhabited although it looked sandy and barren.[12] Perhaps he did not appreciate the remoteness of the tiny speck in the Pacific they had stumbled upon. But even if he had, the fact was that whether anchors were dropped by continents or islands, native canoes would soon be clustering around the ships.

Indeed, very few islands were uninhabited at first western contact. Among these were the Galapagos in the Pacific, Bermuda, Ascension, St. Helena, and probably the Malvinas/Falklands in the Atlantic, the Seychelles and Mascarenes in the Indian Ocean. There was always a commonsense reason such as too little water, poor soils, or the lack of a landbridge from a nearby continent. Remoteness did not always mean a lack of occupation as Hawai'i and Easter Island demonstrate; sometimes it did, as Cook found at Kerguelen.

These true desert islands must be distinguished from vacant Pacific islands such as Norfolk, Christmas, and Pitcairn, the best known of all since it was found by the *Bounty* mutineers in 1789. We now know all three have evidence of prehistoric occupation. In chapter 10 I shall look in more detail at these and consider the apparently surprising late colonization of islands as large as Madagascar, Iceland, and New Zealand.

The desert island is an uninhabited island. Because so few of them were found the adjective now has an ironic twist—a state not found in nature but which has come to represent an ideal state of nature we would like to find ourselves in! "Blow me to Bermuda!" as Merlin cries in *The once and future king*.

Robinson Crusoe is the most famous inhabitant of a desert island. Namesake of the earliest novel in the English language, his story was first published in 1719 as the journal of a man shipwrecked for twenty-eight years.[13] In all this long period of time Crusoe expresses no curiosity about why his island, which was large and well stocked with goats, turtles, birds, fruits, and water should have no one on it but himself. He is there for over fifteen years before finding out that one of the beaches he had never visited, but which was close to his "castle," had been regularly used as a human barbecue by "savages" from a nearby island. Crusoe is relieved, as well he might be, rather than surprised that for so many years he did not have to contend with wild beasts, "savages" or the Conquistadors. The novel makes it very clear that humans were expected everywhere. When they were not it was generally cause for celebration rather than curiosity as to why that should be.

Few human deserts have been found on the continents although, as we shall see in chapter 9, occupation in some environments is more recent than in others. King Solomon's Mines never lacked inhabitants and the lost world of the Kalahari has its people. The southwest corner of Tasmania was vacant at first contact. In chapter 10 I will outline the recent archeological work which has revealed a long sequence of prehistoric cave occupation fringing its rivers. High mountains also held surprises as shown by the Leahy brothers discovery in 1933 of one and a half million people living in the Highlands of Papua New Guinea, an area until then believed to be deserted.[14] Popular tradition has mountains as the hiding places of fabulous beasts, shy retiring almas, bigfoots, yetis, and other monsters[15] which, if finally contacted, turn out to be no different biologically than their discoverers and usually fugitives from the law.

Living prehistory

Western exploration led to the plotting and precise charting of the world. The catalogs of fauna, flora, landforms, and peoples provided the data for new sciences which sought to explain them, and a resource to be used or plundered by those nations which did the discovering.

One of the new sciences to emerge was prehistory, which made particular use of the ethnographic accounts of Native Americans, Africans, Australians, and peoples of the Pacific as shown by Lubbock's subtitle to *Pre-historic times, Illustrated by ancient remains and the manners and customs of savage peoples*. Indeed, the nineteenth-century folk would probably have reversed L.P. Hartley's famous opening line to his novel *The go-between*[16] to read, "A foreign country is the past, they do things differently there," and, much earlier, this was definitely the opinion of the French political economist Turgot in his discourse "On universal history" delivered in 1750:

> A glance over the earth puts before our eyes, even today, the whole history of the human race, showing us traces of all the steps and monuments of all the stages through which it has passed from the barbarism, still in existence, of the American peoples to the civilization of the most enlightened nations of Europe.[17]

One very important development was the use of space as a metaphor for time. Remoteness in the present world became synonymous with remoteness in the past. The voyages of discovery that produced an accurate map of the world and accounts of its peoples now became prehistoric investigations of the societies which initiated such timewalking. The French explorer Baudin, while planning his voyage to the South Pacific in 1800, received the following advice from Joseph-Marie Degérando about the observation of "savage" peoples:

> We shall in a way be taken back to the first periods of our own history; we shall be able to set up secure experiments on the origin and generation of ideas, on the formation and development of language, and on the relations between these two processes. *The philosophical traveller, sailing to the ends of the earth, is in fact travelling in time; he is exploring the past; every step he makes is the passage of an age. Those unknown islands that he reaches are for him the cradle of human society* (my emphasis).[18]

In this foundation text for anthropology the peoples of the world became the living representatives of the ancestors of Western society. The journey might have been in space but it very clearly came to the shores of living prehistory and earlier ages. Drawing a geological analogy Lubbock later spelt out the relationship to living fossils:

> If we wish clearly to understand the antiquities of Europe, we must compare them with the rude implements and weapons still, or until lately, used by savage races in other parts of the world. In fact, the Van Diemaner [Tasmanian] and South American [Fuegian] are to the antiquary, what the opossum and the sloth are to the geologist.[19]

The history of these newly discovered peoples was unimportant. Their contribution lay in what light they could shed on earlier conditions in regions now enjoying industrial civilization. Furthermore, how could these peoples have any history if they showed no evidence for development? And how could they have developed significantly toward civilization if they were the ones who had been discovered? Thus, they came to be regarded as the lesser races, a living prehistory, many of whom were still "infants" when compared to their paternalistic discoverers. They deserved pity rather than praise, as Darwin recalled in *The descent of man* (1871) forty years after the *Beagle* visited Tierra del Fuego,

> The astonishment which I felt on first seeing a party of Fuegians on a wild and broken shore will never be forgotten by me, for the reflection at once rushed into my mind—such were our ancestors. He who has seen a savage in his native land will not feel much shame, if forced to acknowledge that the blood of some more humble creature flows in his veins.[20]

Enough has now been said about why finding humans everywhere failed to prompt the question—Why? The early voyagers and explorers had their lockers and saddlebags full of folk traditions and knowledge. The hard evidence about indigenous peoples which replaced them were put into new schemes, where although living in the modern world by virtue of discovery, these people nonetheless lived in its past. Participation in the present and possession of a history were denied to many of them and, as we shall see, their qualifications to be fully human questioned. We must now turn to the society which created our ancestors in order to understand why this happened and how it has colored prehistoric studies ever since.

Uses of the past

All societies have a concept of the past and the means of building it are varied.[21] None has elaborated it quite as much as industrial societies for whom the past serves two main uses.

In the first instance the past is pressed into service to sanction the use of power and authorize action. This is primarily achieved through the use of symbols embodied in objects as different in size, shape, and style as the Stone of Scone, Gothic railway stations and chocolates made to look like pyramids. The objects act as touchstones for action and contain information about how to behave in everyday life. This may involve accepting an abstract concept such as monarchy by appreciating the historical associations of a lump of rock, celebrating technology in a shrine to the gods of speed and steam, or simply seeking a competitive marketing edge. The metaphors of the past now pervade literature, television, and movies to such an extent that we rarely stop to analyze why the fiction of the past is so important, unless it is in the guise of the Heritage industry, where its wheels turn a handsome profit.

The power of our past stems from its link with the changes to life wrought by the Industrial Revolution which, as Melvyn Bragg observes, "dug into antiquity for help and respectability and dynamic metaphors."[22] Appeals to the past were a sure sign that new orders were struggling to emerge, as shown in this rallying call from the antiquarian J.J.A. Worsaae during the formation of the Danish state in 1849:

> The remains of antiquity thus bind us more firmly to our native land; hills and vales, fields and meadows, become connected with us, in a more intimate degree; for by the barrows, which rise on their surface, and the antiquities, which they have preserved for centuries in their bosom, they constantly recal [*sic*] to our recollection, that our forefathers lived in this country, from time immemorial, a free and independent people, and so call on us to defend our territories with energy, that no foreigner may ever rule over that soil, which contains the bones of our ancestors, and with which our most sacred and reverential recollections are associated.[23]

Prehistory was intimately connected with both the rise of nation-states and the middle classes which derived ascendancy as these were formed. Ancestors, as long as they were ours, were now important.

The second use of the past lay in the development of these new societies through the application of science rather than the appeal of emotion. I define science very broadly as the discipline which conducts a dialog about nature, the purpose of which is to understand variety in these human and natural worlds. The fact that a dialog can take place at all is due to the human ability to construct the world and then live in it. Animals, we believe, just live in the world. Humans obviously create the social and cultural worlds. They also modify, classify, and give meaning to the natural world. Nature, in that sense, never exists independently. Hunting societies and industrial states have both created their versions of nature.

The many worlds which humans build depend on ideology which summarizes the beliefs and ideas which justify why society exists and how it came into existence. One purpose of ideology is to explain the world, usually from the perspective of a single society. Such explanations are never straightforward and often have to try and reconcile many conundrums thrown up by cultural practices and traditions as well as by conflicting interests and changing circumstances. Ideology is a means to deny the logic and importance of many of these contradictions. The beliefs and actions which are approved through a society's ideology serve to resolve those contradictions in the interests of individuals, groups, classes, and even such a fuzzy unit as "society itself." In the last three hundred years, science has been increasingly important in filling this role and producing the means to support the ideas. An example of ideology in practice would be the pursuit of scientific goals, among which must be numbered the investigation of human origins.

The changing ideology of the nineteenth century, fostered by the growth of science, helped turn us all into regular users and consumers of the past. These changes drew in turn upon the development of ideas about humanity and the world known collectively as the Enlightenment of the eighteenth century. The hallmarks of science that emerged during this time, including accurate measurement and repeatable experiments, had their impact on historical disciplines. Reliable history, rather than folktales and origin myths, became the order of the day with great emphasis placed on dates and genealogy. The concern with accuracy and corroborating the facts was closely allied to the rise of non-hereditary forms of government supported by new ideologies of the past on which they based claims of legitimacy to govern.[24]

The role of prehistory in the development of science was therefore to reduce ambiguity about the past and turn it from myth into fact. It became a social weapon, as solid as geological explanations and demonstrations of high antiquity from which so many of prehistory's methods came. A very striking example can be seen in the English Act of Parliament, sponsored by Lubbock and passed in 1882, that made provision for ancient monuments to be brought under the care and protection of the state.[25] This was only possible because speculation and ambiguity concerning the prehistoric monuments of that nation's past had been replaced by agreed definitions and a schedule of values relevant to the late nineteenth century. The benefits to at least one class are nicely summed up by General Pitt-Rivers, father-in-law to Sir John Lubbock, pioneer of scientific, accurate excavation, ardent evolutionist, and first inspector of Ancient Monuments for England: "The law that Nature makes no jumps can be taught . . . in such a way as at least to make men cautious how they listen to scatter-brained revolutionary suggestions" (1891).[26]

The stages of the past

We no longer live in the industrial world of the nineteenth century but some Victorian values have trickled down. These concern the "natural" order of different societies according to evolutionary stages. Since world colonization was largely achieved by peoples with a lifestyle based on hunting and gathering we need to find out where they were put in the world order and why.

The philosopher John Locke declared in 1690 that "In the beginning all the world was America."[27] The Native North Americans were the model for the original society from which others had progressed. Peter Heylyn's *Microcosmos*, published in 1636, had compared those parts of North America, inhabited but as yet uncolonized, to Europe about three hundred years after Noah's flood.[28] These views were endorsed by Adam Ferguson in his "Essay on the history of civil society" of 1767 where he declared that "It is in their [the Native North Americans] present condition that we are to behold, as in a mirror, the features of our own progenitors."[29]

Turgot, in common with Ferguson and Millar from the Scottish Enlightenment,[30] stressed the importance of basing divisions between societies on how they obtained subsistence. Further differences in laws and policies could be understood by concentrating on these aspects of behavior. In 1725 Vico had described in his *New science* the order of human history as determined by habitat, settlement, and society, "First the forests, after that the huts, thence the villages, next the cities and finally the academies" (Axiom 239).[31] Adam Smith (1763) later put it in economic terms to describe the "progressive" nature of social development in this classic stadial model: "There are four distinct states which mankind pass thro:—1st, the age of hunters, 2dly, the age of shepherds, 3dly, the age of agriculture; and 4thly, the age of commerce."[32]

The upshot of these ideas was that subsistence livelihood came to explain differences between people.[33] Nations and classes were judged not only by whether or how they held their knife and fork, but by their recipes and the means by which they provisioned their tables. The discoveries of peoples and cultures on the many voyages of discovery of the eighteenth century were reduced, back home, to a simple staircase of economic development. Savages, or hunters, lay on the bottom step, because of their apparent lack of political organization, while barbarians, capable of agriculture and some temporary political institutions, were to be found one riser up.

By the end of the eighteenth century the foundations of an anthropological system had been laid. The first great syntheses arrived much later with the evolutionary schemes of Morgan (1877) in America and Tylor (1881) in England. These provided a place in the development of ancient society for all present and prehistoric cultures. Morgan's seven major ethnic periods, from savagery through barbarism to civilization, turned to modern tribes for examples of the condition of his own remote ancestors.[34] Progress towards civilization was a goal of history that could not be attained by all the tribes and nations running in the race.

For Tylor the great principle that supported his descriptions of prehistoric and contemporary primitives was that "no stage of civilisation comes into existence spontaneously, but grows or is developed out of the stage before it."[35] Following Morgan he identified three great stages of culture—savage, barbaric, and civilized. The first included all hunters and gatherers irrespective of whether they were found in tropical forests or the colder regions of the globe. In Morgan's threefold division of savagery, the earliest or "lower status" had no living survivors among the world's hunting peoples. Language, however, did commence in this period. Middle savagery saw the development of fishing and fire and so the Australians and Polynesians were cited as examples, while upper savagery was marked by the invention of the bow and many groups in the Americas were held to still represent such a stage.

Prehistory made its own significant contribution to these stages. The three-age system where C.J. Thomsen classified the collections in the Copenhagen museum into stone, bronze, and iron set the trend in 1836, and has been elaborated ever since. The job was tackled with a model of prehistoric societies, past and present, based on technology. Lubbock's great synthesis of 1865 used technological similarities as the bridge from the present to investigate the mental and moral state of past societies. As we have seen there was no hesitation in casting modern peoples as examples of prehistoric societies.

Those who suffered most from this matching of the past with the present were the hunters and gatherers. They were judged to lie at the bottom of the scale due to primitive technologies and uncivilized habits. This is well illustrated in the frontispiece to the Reverend Wood's *Natural history of man* first published in 1870 (fig. 2.1). In the center is the pinnacle of evolution, male and W.A.-S.P. (white, Anglo-Saxon, Protestant), surrounded by the "lesser" races and cultures who fade away into the distance until at the edge of the picture we can just make out some hunters and gatherers from Australia and the Arctic. These would have been widely regarded at the time as living representatives of our ancestors who in earliest prehistory made stone tools and hunted animals in the Old World.

Their position relative to the central civilized figure and against whom all others are judged is determined by lifestyle, technology, and geographical location at the uttermost ends of the earth. It is a long way from the image promoted in the modern world where all peoples could apparently be joined together in peace and harmony by the shared culture of a soft drink. Just think; the Victorians might have suggested that because everyone, everywhere can bash rocks together and make some sort of stone tool, perhaps even the real thing, then world fellowship could also follow. This was very definitely not the use to which prehistory was put.

Thus the mold was fixed. During the decades that followed there were various elaborations but no substantial disagreements or alterations. Of these the most influential for early prehistory was by Sollas who, in 1911, took a threefold division of the Paleolithic, or Old Stone Age, and matched, respectively, the periods of the

Figure 2.1 The natural history of man according to the Reverend Wood in 1870.

Lower and Middle Paleolithic to the Tasmanians and Australian Aborigines and the Upper Paleolithic to the Eskimo and Kalahari bushmen. Interestingly, he neither acknowledged Morgan's scheme nor discussed why he now saw fit to include a living representative in lower savagery. No longer was there a comment on why such a scheme should be followed. There was just a scheme. These views were well entrenched by the time Boas, in 1928, forcefully argued for the study of artifacts by their geographical and historical distribution and not by their association with races.[36]

In the years that followed, Lubbock's term Paleolithic gradually replaced savagery, although some prehistorians have continued up to the present day to follow this division and its nomenclature when discussing wider issues in prehistoric cultures.[37] A great deal of effort has been expended on collecting data to fill out the three divisions of Sollas and others. This has involved the study of stone technologies, fossil remains, environmental reconstruction, and more accurate means of dating. Change among stone tools was treated as a form of descent with modification analogous to the evolution of plants and animals; except that modifications always led to progressive technological complexity in a positively upwardly mobile direction, as figure 2.2 shows. For many years the discussion of change did little more than follow Tylor's great principle of development.

Unity of mankind, disunity of humanity

Race is a term, as Latham observed in 1851, that "if it conceals our ignorance, proclaims our openness to conviction."[38] The founding fathers of prehistory had many firmly held convictions on this important matter. We need to understand them in order to appreciate how research into human origins became channeled on a well-worn track. In chapter 3 I shall look in more detail at the practical outcomes but here we must consider the basic issue of classification and membership in humanity.

The great systems of classification are another example of how science reduced ambiguity. This was not achieved without considerable disagreement over what a species was and how many varieties, or subspecies, it contained. In the English translation of the 1758 edition of Linnaeus' *Systema naturae*, the genus *Homo* was divided into two species *sapiens* and *monstrosus* [*sic*].[39] Five varieties of *Homo sapiens* were listed—Wild Man, American, European, Asiatic, and African—using skin, eye and hair color and cultural criteria; Europeans, for example, are governed by laws and Asiatics by opinions. The varieties of *Homo monstrosus* were conditioned by environment and culture as with the head-flattened Canadians and the small, active but timid Mountaineers.

Blumenbach in 1776 abandoned the use of cultural criteria and defined five principal varieties—Caucasian, Mongolian, Ethiopian, American, and Malay—on the basis of anatomy, using skull shape, as well as skin, hair color, and other superficial features.[40] His rules for defining new species were widely followed. Classification through measurement came to be accepted and many other divisions were subsequently put forward so that a century later Darwin, in his discussion of the human races, listed fourteen schemes ranging in the number of varieties recognized from one to sixty-three.[41]

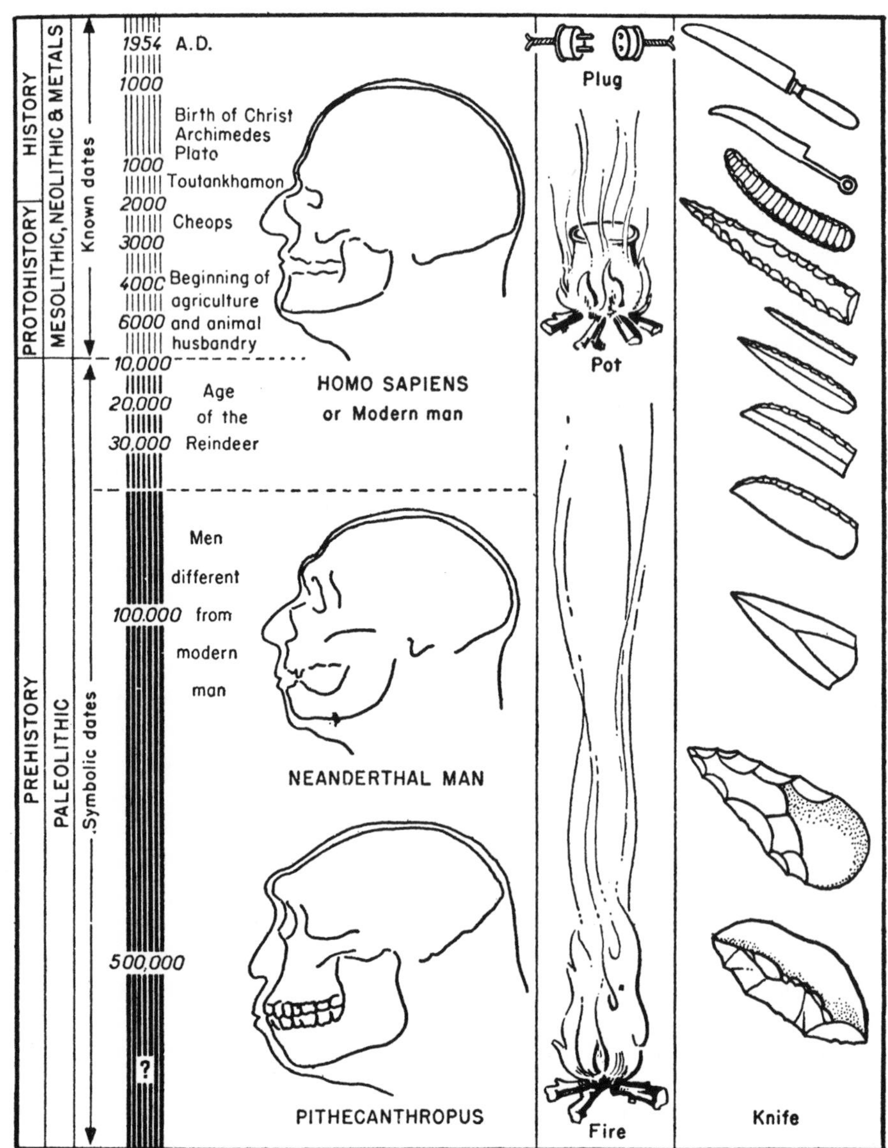

Figure 2.2 An example of progressive evolution. Human history as a celebration of technology made possible by increasing intelligence. In his caption to this figure the archeologist Leroi-Gourhan (1957) noted the "continuity through time that marks the gradual development toward *perfection* of a tool such as the knife."

The importance of these schemes rested on the fierce debate over whether the races of mankind constituted a single species. On the one hand the *monogenists* fervently believed that all races were descended from a common ancestor. Opposed to them were the *polygenists* who argued, with no less heat, that the varieties of mankind were original.[42] The idea of monogenism can be found in the foundation texts of natural history by Buffon (1749) and Blumenbach (1776). The nineteenth-century advance was to marry up the basic principles of taxonomic classification with a mechanism to account for descent. A species could be accounted for in several ways and monogenists were quite prepared to defend their position by first pointing to solutions such as Lamarck's, based on the transmission of acquired characteristics and later to Darwin's mechanism of natural selection leading to descent with modification.[43]

Polygenist explanations, such as those of Lord Kames (1774–5), were often, but by no means exclusively, based on Genesis, with the major races traced back to the wanderings of Noah's sons after the flood. An influential scientist such as Agassiz argued up to his death in 1873 for the separate creation of races within a divine plan but on a long timescale. The man who was instrumental in proving the existence of former ice ages was apparently converted from his monogenist views when he moved to Boston in 1846 and came in contact with Americans of African descent.[44] While he did not claim that the races were separate species, he did make it abundantly clear that he believed these varieties were original. Racist opinions and polygenist doctrines often went together and, as Stocking pointed out in his masterly discussion (1968), formed the legacy for the twentieth century.

Monogenists and polygenists could still find themselves on the same side of an argument. Blumenbach and Captain Fitz-Roy held different views on whether species were original but both agreed that degeneration was the mechanism which explained racial differences. Fitz-Roy claimed that modern savages, although descendants of Abraham, had degenerated as they moved farther away from the center of creation. During this process the civilizing trademarks of clothing and cultivation dropped away. He scornfully rejected the alternative, that mankind started from a simple state and evolved in some areas towards civilization. Who, he asked, would look after the "infant" in its cradle until it could fend for itself?[45]

Accounting for the condition of primitive peoples by a process of degeneration had a wide appeal. It often went hand in hand with a creationist view of how the world and mankind were formed rather than with any strong association with either the mono- or polygenists. Among the latter was the Duke of Argyll who, when writing in 1869 about primeval man, criticized Lubbock's view of progressive evolution and argued instead that man's capability for degradation was an indisputable fact. Pressure of population numbers would force weaker groups to the peripheries where climate would exert its influence over their physical and cultural degeneration.

The idea that species were not immutable was championed by Thomas Huxley in his influential essays on *Man's place in nature* (1863) and eventually won the day. Sir Charles Lyell's conversion to the view that species arose through transmutation, set out in his influential *Antiquity of man* (1863), was one of the turning points in the acceptance of natural selection. But equally significant was Alfred Wallace's U-turn in 1864 about the ability of the mechanism of natural selection to account for man's moral faculties as well as other highly characteristic traits—the hand, upright posture, speech, beautiful symmetry, and expressive features. Instead he argued for a power which had guided these developments, a Creator's will as the anti-Darwinians were quick to claim.[46]

This was a debate which, when won, had little practical influence on general opinion. As Edmund Leach has remarked, the implied unity of mankind seemed to stem from indifference.[47] Prichard's popular *Researches into the physical history of man* (1813) advocated the monogenist hypothesis. Later he conjured up in his writings a visitor from another planet to look at customs and color and decided that on face value, and without the benefit of tracing the history of the diversified races, the Martian would mistakenly side with the polygenists and interpret varieties as original species.[48] Stocking reminds us that Darwin was prepared to accept that a naturalist would, on first sight, probably regard Africans and Europeans as "good and true species,"[49] but whether human races were to be classed as species or subspecies was a matter of little interest to him.

The biological unity of mankind did not imply full membership of humanity. For some, color was like the rungs on a ladder, providing an unambiguous guide to absolute social, moral, and political order. This was White's view in 1799 enshrined in a gradation model that went from the white European at the top to the black "savage" at the bottom and, like any rigid caste system, allowed no contact between the various grades. Absolute values were naturally "fixed" between societies and there was never any doubt how the pyramid should be drawn up. What began as a dispassionate, "objective" exercise in describing patterns of diversity acquired a significance through the ideology which endorsed the activity. Either way some races and societies, particularly hunters and gatherers, always lost out.

Others like Prichard in 1855 believed in the "psychic unity of mankind," which is probably best translated as human nature. He concluded that "all mankind sympathise in deeply impressed feelings and sentiments, which are as mysterious in their nature as in their origin."[50] But these mental endowments were also widely regarded as subject to change and improvement and so became part of the comparative method for studying societies—normally by comparing everyone else to the values and achievements of the industrial West (as figure 2.1 so convincingly shows). Tylor and Morgan argued that all peoples began with the same mental apparatus but with civilization there had been mental evolution to a superior intellect. Customs and culture, and the mental states these implied, became the criteria for comparison and classification. When biological and mental unity were compared with behavioral diversity as manifested in the different cultures of the world, past and present, the fragmentation of humanity as a "natural" fact was confirmed. The burden of proof was left firmly with those wishing to show that it was otherwise.

The confusion of classification and interpretation has continued despite the best efforts of anthropologists, like Boas in the twentieth century, to separate race from culture. This involved repudiating the conclusions based on the vast bodies of measurement of humans that began with physicians such as Morton (1839), were elaborated by Broca in the 1860s and which became rampant in the eugenics of Galton (1883). The approaches of these head measurers are thoroughly discussed by Stephen Jay Gould in *The mismeasure of man* (1984). Their "objectively" derived conclusions supported racist creeds. They were not just confined to measuring primitive society but also to demonstrating the intellectual inferiority of the "weaker" sex and the "lower" classes through such means. Equality, freedom, humanity were, and presumably always will be, qualified terms. Republicans could be slave owners, Christians preach the right of limited franchise, and abolitionists urge the resettlement, elsewhere, of freed slaves. Science was marshaled to support the logic of these contradictions and so preserve the interests these beliefs defined. Ambiguity was buried under a mountain of spurious statistics.

Allied with this belief in progress and perfection was the conviction that race was only a temporary complication. In a paper read to the Anthropological Society of London and acknowledging the influence of Herbert Spencer's ideas on progressive evolution, Alfred Wallace argued that under the operation of natural selection

> It must inevitably follow that the higher—the more intellectual and moral—must replace the lower and more degraded races . . . till the world is again inhabited by a single homogeneous race, no individual of which will be inferior to the noblest specimens of existing humanity.[51]

These would of course be the "wonderful intellects" of the Germanic races.

A more recent example of the confusion was provided by Carleton Coon in *The origin of races* (1962). He used the term race as a loose synonym for geographical populations of various sizes and equated it with the taxonomic rank of subspecies. He recognized five modern races—Caucasoid, Mongoloid, Australoid, Congoid, and Capoid—and explained the differences between these five distinct regional or "natural" races by climate and historical continuity over the past half million years. There were five independent developments from *Homo erectus* with, according to Coon, the Negroids having a shorter evolutionary history than the Caucasoids which would affect the level of civilization attained by some of the former populations. His intellectual baggage bears many familiar, faded, peeling stamps: "[Africa] was only an indifferent kindergarten. Europe and Asia were our principal schools."[52]

Declarations

The tree of progress has long roots feeding into our past. Evolutionary science continues to be cited as an unbiased standard with which to measure change rather than to understand variability.[53] Put in the context of the human species we can see that nineteenth-century science was interested in demonstrating taxonomic unity but that was as far as it went. It never saw, as part of its brief, the possibility that taxonomy, based on biological principles, and behavior assessed by culture, could be unified to contribute towards a study of humanity in general and prehistory in particular. Instead biology, in the guise of mental powers or intelligence, was used to account for the state of behavior as measured against technological achievement and cultural status—savage, barbaric, or civilized. Either way, God or science was very definitely a thinking "Euromerican."

It would be a long time before the biological unity of mankind was also interpreted as an intellectual and behavioral unity with differences between cultures attributed to history and not to ability or potential. It can eventually be found in the *Universal declaration of human rights* of 1948 and UNESCO's *First statement on race* in 1950:

> The unity of mankind while based firmly in the biological history of man rests not upon the demonstration of biological unity, but upon the ethical principle of humanity, which is . . . the right of every human being to the fulfilment of his [*sic*] potentialities as a human being."[54]

We should surely expect such a fundamental change in attitude to be met by a shift in the study of the origins of this unity. This implies a move away from the material, technological, cultural, and anatomical and toward the behavioral as we try to penetrate the ambiguity surrounding the emergence of our common humanity.

3
Past cradles and ice age time

The three great problems for nineteenth-century ethnology and prehistory were identified by Latham in *Man and his migrations* (1851) as

1. the unity or non-unity of the human species;
2. its antiquity;
3. its geographical origin.[1]

This short list has formed the basis for research into human origins ever since. The ambiguity surrounding each question has been reduced to every generation's satisfaction, then thrown open again as changes in opinion about the world and its peoples have led to revisions. This cyclical process has provided the spur to fieldwork and the development of new techniques of classification, analysis, and dating.

Latham was writing at an interesting time, eight years before the *Origin of species* was published. This was the foundation text for the biogeography of Darwin and Wallace which accounted for the distribution of life on the planet. The importance of these studies was their contribution to the scientific investigation of variation via the principle of natural selection. Individuals were the units under selection with the evolutionary results measured by their differential reproductive contribution to the next generation. The explanation of variation in the world was

> grounded on the belief that each new variety, and ultimately each new species, is produced and maintained by having some advantage over those with which it comes into competition; and the consequent extinction of less-favoured forms inevitably follows.[2]

Darwin's insistence on the mechanism of natural selection as a "belief" held out a brief hope that, by setting out the underpinning assumptions, value judgments might be explicitly recognized in science. However, the vocabulary of "advantage", "competition" and "less-favored" was too easily transferred to the study of human societies. The goal of progress, so forcefully put by Spencer in his *Principles of sociology* (1876–96), was now a consequence of evolutionary principles applied to the ladder of societies in this version of Social Darwinism. Science had proved the direction of progress, which, according to Condorcet writing during the French Revolution, never went in reverse.

This was not Darwin's view. In a telling passage in *The descent of man* he pointed out that the ancient and oriental civilizations neither expected society to change nor believed that change, when it did occur, was directional.[3] He was insistent in his biological writings that evolution had no goal and followed no great design. The mechanisms which led to descent with modification passed no judgments in terms of success or failure and only provided a way of comprehending variability. He appreciated, however, that the process of supplantation continued in his day largely through the skills of civilization—a product of the intellect. "The intellectual faculties," he suggested, "have been gradually perfected through natural selection."[4]

What effect did this new synthesis have for answers to Latham's three questions? In this chapter I will look at three key issues: geographical centers, time, and climate. These have been important elements in any discussions of human origins. Their significance, as we shall see, has for a host of reasons been open to modification and reinterpretation.

Geographical centers

Cradles and crèches

The notion of a cradle for mankind, a discrete geographical center for human origins, is an ancient idea. The Garden of Eden is the best known example. Adam and Eve might be replaced, as they were in the last century, by the first formed people or *protoplasts*,[5] but the idea of an ancestral homeland continued. So too did the idea of expulsion and dispersal with the fall replaced by environmental change and the evolution of upright walking.

The scriptwriters for this new version were Darwin and Wallace in their biogeographical writings. There were two main propositions. Firstly, following Lyell and all other geologists, the pattern and position of continents were regarded as constant. The only significant changes were the temporary appearance of landbridges due to the rise and fall of oceans as a result of glaciations. Secondly, against this fixed geographical background was the restlessness of life and geology on the continents and in the oceans. Through natural selection organisms adapted to niches as they appeared, and in turn this led to reproductive isolation and so to speciation. These new species then dispersed from their centers of origin across the continents and landbridges and, as a result, the varied realms of plants and animals came into being through this constant pulse of life.

The Darwin/Wallace model did not bring consensus or clarify where the human cradle would be. In the eighteenth and nineteenth centuries three main areas were favored. The first of these was a fertile tropical island. Linnaeus argued in 1781 for Paradise as a sole island existing at a time when the continents lay under the sea. Humans, as well as all animals and plant species, originated here by divine creation then dispersed around the world.[6] Obvious benefits followed "where the warmth of the air was such, that no clothing was necessary, and where there were no wild beasts to endanger their safety," as Lyell once commented.[7] The tropical island found its strongest advocate with the German biologist Haeckel who writing in 1876 favored Lemuria, a sunken continent in the Indian Ocean. From here his twelve races of man migrated to their present homes. In a later edition he acknowledged the force of geological arguments against Lemuria and moved Paradise into southern Asia in the region of modern Afghanistan and Pakistan.[8]

The second region, Asia, had many supporters. In 1749 Buffon proposed that the cradle contained whites and lay in central Asia,[9] while a century later Latham speculated on somewhere in tropical Asia for a single locality involving a single protoplast pair.[10] In 1776 Blumenbach was swayed by the beauty of the modern Georgians and, being a firm believer in degeneration, made them the primal nation and put the cradle in Asia north of the Caucasus.[11] Haeckel used his repositioned Paradise to stress the importance of the southern Himalayas in the transformation of the mammalia and especially the primates. He also pointed to the Todas and Dravidas who live in its foothills and whom he regarded as lesser races "most closely related to the long extinct Primaeval Men."[12] "Somewhere in Asia" was also Haddon's view, with a preference for western Asia and within that area a more southerly locality where he believed primitive characteristics could still be found.[13]

The most comprehensive case for Asia was contained in W.D. Matthew's geographical study of speciation *Climate and evolution* (1915). His global perspective on this problem took as its starting point the view that the center of origin could be identified by the distribution of the most advanced forms of a species since "At any one time . . . the most advanced stages should be nearest the centre of dispersal, the most conservative stages farthest from it. It is not in Australia that we should look for the ancestry of man, but in Asia."[14]

He went on to locate his cradle for animal species more precisely in, or about, the great plateau of central Asia.[15] He was not the first to plump for Tibet. Quatrefages had made a similar detailed case in 1879. This may seem a strange choice since the roof of the world is, and always has been, very sparsely populated. Matthew's "empty" cradle was defined by the ring of early civilizations surrounding it in the Middle East, India, and China which pointed inward to the original home of the "higher" races. There seems little advance over Kant who could write in 1775 that the Old World "seems to earn the name of Old World even from the standpoint of peopling"[16] without a shred of fossil or archeological evidence.

Following Matthew's lead Asia's claim was vigorously taken up by the primatologist, Davidson Black writing in 1925, and by geographers, such as Griffith Taylor in 1927, who favored eastern Asia on the borders of the warm steppes and woodlands. The center was revealed through racial zones and strata. The former were successive spreadings of racial stocks where each new wave covered less area than the last. If these waves were sectioned then strata were revealed, giving the age and evolutionary position of each race. Those races at the farthest ends of the earth—Bushmen, Tasmanians, and Fuegians, yet again—were the oldest and of course the lowest.[17]

The third cradle has been Africa. This was Darwin's choice because our closest living animal relatives, the African great apes, lived there and had not migrated to other parts of the world.[18] Applying his theory that species arose, then dispersed from centers, he concluded that they still existed where they had evolved alongside humans.[19] Matthew and Black on the other hand had all primates migrating to Africa, southeast Asia, and America from a cradle in central Asia.[20] This was not Osborn's view in *The age of mammals*, published in 1910, where he concentrated upon the Ethiopian region as an important center for mammalian evolution.[21] This was in contrast to earlier schemes, Wallace's published in 1876, Lydekker's in 1896, and Stratz den Haag's in 1904, that saw mammals originating in the northern latitudes and then diversifying once they reached Africa and the Southern Hemisphere generally. However, in 1915, by the time he had compiled *Men of the old stone age*, Osborn inclined to Asia as the chief theater.

This of course by no means exhausts the list of places where the cradle has been put. By 1901 the eminent paleontologist Otto Schötensack saw Australia as the scene of the first development on account of its "primitive" animals and aboriginal races.[22] America, North and South has had its supporters as have both poles and a host of extra-terrestrial cradles.[23] Europe, however, has not. The Piltdown forgery of 1912, which gave Englishmen the ancestor they wanted with a big brain and a weak chin, did not shift the cradle.[24] The American anthropologist Ales Hrdlicka writing in 1922 got close by putting southeast Asia as the center for hominids but southwest Europe as the cradle for humanity. Grafton Elliot Smith, a principal in the Piltdown affair, declared in 1916 that the arrival of anatomically modern people in Europe should be regarded as "the greatest event in history,"[25] presumably because he believed, like Coon after him, that this continent brought out the best in our ancestors and set the direction for progress into the higher grades of social life.

European claims were probably in the back of Latham's mind when he argued in 1851 that

> The conventional, provisional, or hypothetical cradle of the human species is, of course, the most central point of the inhabited world; inasmuch as this gives us the greatest amount of distribution with the least amount of migration; but, of course, such a centre is wholly unhistorical.[26]

This had been Kant's feeling as long ago as 1775 when he pointed out that between latitudes 31° and 52° in the Old World (roughly Shanghai–Delhi–Cairo and London–Berlin–Kiev), there was a happy mixture of climate and a perfect jumping off spot for all transplantations.[27] However, no one can rival the precision of Thiselton-Dyer, writing in the Darwin centenary of 1909 where he declared,

> If we accept the general configuration of the earth's surface as permanent a continuous and progressive dispersal of species from the centre to the circumference, i.e. southwards, seems inevitable. If an observer were placed above a point in St. George's Channel . . . he would see the greatest possible quantity of land spread out in a sort of stellate figure. The maritime supremacy of the English race has perhaps flowed from the central position of its home. That such a disposition would facilitate a centrifugal migration of land organisms is at any rate obvious, and fluctuating conditions of climate operating from the pole would supply an effective means of propulsion.[28]

More recent biogeographers such as Darlington continued this imperial tradition albeit from a different but no less central location around the Equator. He argued in 1957 that the more competitive vertebrate animals, including humans, repeatedly emerged in the Old World tropics and then spread north and south.[29] This seems to combine all three cradles into one giant crèche. He applied the model to the data in Coon's *The story of man*, published in 1954, for both fossil and modern races. Africa is selected as "a good guess" for our cradle with initial dispersal spilling into the warm parts of the Old World. He concluded that the geographical history of humans closely conforms to the main principles of zoogeography for all other vertebrates. Barriers to migration are noted as important but not insurmountable for humans since they have the advantage of culture. Therefore the near-universal distribution of the species comes as no surprise given the timescales involved. The language of Social Darwinism is strong: "Ability to spread is one of the attributes of dominance. . . Dominant animals spread. . . and replace other animals . . . animals spread to obtain advantages, not to escape disadvantages."[30]

Leaving home

The theme of dominance allied to a fixed pattern of continents can be found in many works on the origin and spread of species. Darwin put it particularly forcefully when he claimed that the extinction of tropical and southern floras and faunas was inevitable given "the more dominant forms, generated in the larger areas and more efficient workshops of the north,"[31] even though he advocated Africa as the human cradle.

The shape of the world was considered decisive for the production of life and the potential for its enhancement. Darlington emphasized how according to his division movement went from larger to smaller areas and from favorable to less favorable climates. The dominant migration in the Old World was, according to Matthew,[32] east and west with the mountains restricting southern movement so preserving earlier speciation cycles. Great store was set by the unequal distribution of land in the Northern and Southern Hemispheres and its physical arrangement. This led in Ratzel's opinion, expressed in his monumental *Völkerkunde* (translated in 1885–8 as the *History of mankind*), to interdependence and separation respectively. The southern margins of continents served as refuges for less successful races. The uttermost ends of the earth—Tasmania, the Cape, Tierra del Fuego—were defined from a Western perspective geographically, socially, culturally, and ethnically.[33] As we saw in chapter 2, distance in space became remoteness in time to provide the logic for a living prehistory. Although these regions were regarded, along with Europe, as cul-de-sacs,[34] ideology in the guise of science was quick to explain any apparent contradiction by emphasizing how the bracing *northern* climates had led to civilization.

The fixed arrangement of continents created barriers to dispersal. The main problem was presented by oceans and islands when identical animal species could be found on either side. This dichotomy was accounted for in two main ways. Chance dispersal was much favored. Plants and animals were put on logs, vegetation mats, and icebergs and floated around the ocean currents until they reached their destination aided by long timescales. This drift model was complemented by landbridges at times of low sea level when the oceans were locked up in ice caps.

G.G. Simpson succinctly summarized the paleontological position in his classic paper of 1940 on *Mammals and land bridges*. A cradle was not an exact spot but a single biotic district or province with populations in an expansion phase moving along the periphery. These provinces were determined from the distribution of animals within the continents. Landbridges and barriers acted either as corridors or filters and were critical for understanding how distributions had come about. The rules of the sweepstake often governed the history of island occupation as shown by the idiosyncratic range of animals that reached them. Random elements were therefore important. All three mechanisms had to be judged by the important consideration that "ability to expand and ability to survive are two different faunal characteristics . . . with no apparent relationship to each other."[35]

The examples of colonization I have used in this chapter are all based on identifying centers of creation by a variety of means. Then arrows are drawn to where descendants are found. Archeologists, very familiar with these means for explaining distribution patterns, term them diffusionism. It is a perfectly respectable model that traces the movement of peoples and the transmission of ideas through material remains. It received a bad press due to the excesses of the hyper-diffusionist works of Elliot-Smith in the late 1920s and early 1930s. His central doctrine stated that inventions occurred but once and were then diffused around the world. The center of development was always Egypt and it was from here that the rest of the world received the various benefits of urban civilization.

What Elliot-Smith's hyper-diffusionism denied was any spark of independence or originality to the areas that received such bounty. The world had a few active centers and a great passive hinterland. The parallels between his prehistory and the world of fading empires needs no further emphasis.

Blowing with the wind

What caused human migration during world colonization? What tipped the baby out of the cradle? Not surprisingly the human drift is usually pushed by external forces. Writing in 1589 José de Acosta suggested that "men came to the [West] Indies driven unwittingly by the wind,"[36] as well as taking the overland route through Asia where starvation and other hardships drove them onward. Almost three hundred years later hunger was put forward by T.H. Huxley as an adequate reason for dispersal. It propelled primitive families around the globe during which time natural selection got to work on family variation.[37]

Population was expected to multiply, causing problems with food supplies, and spread to escape a self-inflicted woe.[38] Zimmermann's *Geographic history of man and the universally distributed animals* appeared in 1778. He saw the primal nation of white-skinned brunettes leaving central Asia, because of pressure from increased population, in four great columns headed in succession toward the Caucasus, the area north of the Altai mountains, Arabia/India/the eastern islands (perhaps Africa), and southeast Asia.[39]

Against this background Malthus' *Essay on population*, first published in 1798, was very influential. With great clarity he argued how population growth caused by the unavoidable passion between the sexes will outstrip the available food supply and continue to do so even with improvements in yields. However, he was well aware that the phenomenal growth rate in the Europe of his day was not true of all times and places. He showed how population growth among hunters was limited by their mobility and subsistence. He suggested that a reduced passion existed among hunters which, combined with the means of subsistence and the extra burdens that fell on the women because of this mobile lifestyle, served to limit numbers.[40] Growth rates rose dramatically with shepherding and more sedentary lifestyles.

Malthus showed that the potential for increasing numbers was always present and, as interpreted by others, could start from a center with a single pair and proceed at a geometrical rate.[41] The description by Quatrefages for the Tibetan plateau describes the process perfectly: "The first human beings appeared and multiplied till the populations overflowed as from a bowl and spread themselves in human waves in every direction."[42]

This law of increase always caused problems.[43] As Lyell supposed, the first races might linger on their fertile island "but as soon as their numbers increased, they would be forced to migrate into regions less secure and blest with a less genial climate."[44] For the Duke of Argyll, it would be the weakest tribes that went first and those driven farthest would be "the most engrossed in the pursuits of mere animal existence,"[45] as was the case, he claimed, with the Eskimos and Fuegians at either end of the Americas. Warfare was inevitable as population outstripped resources.

Technology was considered a minor component of the dispersal pattern. According to Lubbock, "The lowest races of existing savages must, always assuming the common origin of the human race, be at least as far advanced as were our ancestors when they spread over the earth's surface."[46] From this proposition he deduced the original technology of his ancestors by cataloging what present-day groups lacked. Since pottery, bows, clothes, and spinning were not found in some hunting groups, he concluded, correctly in many respects, that these items were unknown to our earliest ancestors as they colonized the world.[47]

In 1877 Morgan assigned the migration from an original habitat to the greater part of the globe to middle savagery when fire and fishing were introduced. Moreover, these were not migrations in the historical sense, describing movements of Nomads around the steppes or those of the Angles and Saxons, but rather "dim impulses"[48] and vague wanderings whether by boat or on foot. Ratzel claimed that "restless movement is the stamp of mankind,"[49] while Marett in a popular book wrote that "The leading characteristic of our species is a taste for adventure. . . . Something was there in the mind of the race to cry "Forward", though no man knew exactly whither."[50]

Time rings the changes

The key notion of centers and movement from them, be it diffusion, migration, or just plain drift, is intertwined with concepts of time. These can only be understood when time is conceived as a process. As the philosopher Bergson put it, time only has meaning when it concerns invention.[51] Time creates nothing by itself and only stands as a proxy for those processes and mechanisms, such as natural selection, which do the work. To argue that "enough time" was all that was needed to complete global colonization tells us nothing.

This is exactly what happened in the nineteenth century when the scientific establishment gave its unequivocal support to the view of the great antiquity of mankind as part of a widespread revolution in the earth sciences. In the last decade of the eighteenth century James Hutton and later, in the next century, Sir Charles Lyell provided the methodology to explain former changes of the earth's surface by reference to causes now in operation. This was called the principle of uniformitarianism by which it was accepted that the route to understanding the past on geological, prehistoric, or historic timescales lay in the present. They also established a concept of time as a cycle, undirected and purposeless in the sense that Darwin would later insist was the pattern of evolution. The cycles were measured in the inexorable roll of repeated mountain building, decay, and sedimentation.

The principle of uniformity required a long timescale in order for demonstrably slow processes such as ice action and mountain building to produce present and past landforms. The slow, gradual tempo was also applied to the formation of species. Whether it was explaining the formation of coal in the Carboniferous or discussing how species arose through natural selection, one thing was certain: the date of the creation set at 4004 B.C. by Archbishop Ussher, who added up peoples' ages in the Bible, fell well short of requirements.[52]

Establishing human antiquity in deep time was without doubt a triumph for the method of uniformitarian reasoning and unlocked the past to systematic analysis. But it was not the case that antiquity and theories of species change went hand in hand. Some reviewers of Lyell's *Antiquity of man* (1863) criticized his muddying of the waters by pointing out that theories of development needed great antiquity but that demonstrating great antiquity did not prove the reverse. Donald Grayson is quite right when he points out that demonstrating a great human antiquity did not provide a crucial test for any explanation of human origins, even though he sees the establishment of such antiquity for humans as *the* crucial *event* in the attempt to understand such origins.[53] In other words, it paved the way for the demise of the short chronology creationists whether of a polygenist or monogenist persuasion. The long chronology polygenists, such as Agassiz, who also believed in divine creation, took longer to disprove.

The consensus view of evolutionary change for all species was that it formed a slow, gradual process with new forms periodically radiating out from centers. I regard this as the classic concept of *geological* or *evolutionary* time. This should not be interpreted as a belief that given enough time change is bound to occur. Far from it, and the presence of so-called "living fossils" such as aardvarks and coelocanths which have not changed over many millions of years shows that this is the case. But set beside the working of natural selection in geological and evolutionary time is the widely expressed belief, set out in chapter 2, that world colonization by humans required just that—enough time. G.G. Simpson was very explicit on this point: "Any event that is not absolutely impossible (and absolute impossibility is surely rare in problems of dispersal) becomes probable *if enough time elapses*."[54]

As Geoff Bailey has pointed out, it is strange that archeology, which spends so much of its resources on dating and chronology, has had so little to say about time concepts especially when it has such long time spans to examine.[55] The reason is not hard to find. Archeology has continued to follow an agenda, such as Latham's, and accepted, rather than developed, the accompanying concepts about time and space which gave meaning to its discoveries.

Alternative concepts and tempos are possible. Evolution, it is claimed, proceeds in bursts separated by long periods of stasis or no change. This interpretation of evolutionary time by Stephen Jay Gould and Niles Eldredge as *punctuated equilibria*[56] is not without its critics. The main principle which they challenge in the classic concept of evolutionary time is that longterm events can be explained in terms of the sum of continuous, small-scale selection processes. Most important for prehistory, as Bailey argues,[57] is the lesson that differing timescales bring into focus different features of behavior requiring different sorts of explanation. Such time perspectivism puts great store on the duration of time as a key variable in understanding past behavior. The alternative view, which archeologists have inherited from their nineteenth-century roots, is that the past is like the present and so the passage of time is irrelevant to understanding human evolution and only useful for classifying material. Figure 2.2 is a good example of the use of time in a linear manner. An arrow shot through the ages. An unrepeatable process since its passage is marked by unique events and achievements.

Natural selection operates through local adaptation and not through length of time. As the circumstances of survival alter, so selection through differential survival of offspring can lead to change. Here the concept of alternative selection strategies developed by MacArthur and Wilson leads us into *ecological* time.[58] Animals have developed various ways of coping with shortterm selection pressures from their environment. They either opt for investment in few offspring and match their number to the resources available, a process known as "K" selection (think of constant or stable), or else they aim to produce as many progeny as possible, and flood the habitat in the knowledge that some will survive (see chap. 5). This second tactic is known as "r" selection, or the rapid rate of intrinsic increase characteristic of opportunistic species (table 3.1). Animal species conforming to either strategy can be found living side by side in the same habitat.

"K" and "r" are not simply different strategies because in different environments an organism's behavior is only considered in as much as it helps the genes in their mission to get to the next generation. The genetic route may explain the

Table 3.1 Two types of selection (after Pianka 1978).

	r selection	K selection
Climate and environment	Variable and/or unpredictable, uncertain	Fairly constant and/or predictable, more certain
Mortality	Often catastrophic, boom and bust, density independent	More directed, in equilibrium, density dependent
Survival	High juvenile mortality	More constant mortality
Population size	Variable in time, well below carrying capacity, recolonization each year of empty territory	Fairly constant in time at or near carrying capacity, "saturated" habitats, recolonization not needed
Regional occupation	Migratory, ephemeral. Ranges	Resident, permanent. Territories
Competition within and between species	Variable, often lax	Usually keen
Selection favors	Rapid development High maximal rate of increase Early reproduction Small body size Single reproduction Many small offspring	Slower development Greater competitive ability Delayed reproduction Larger body size Repeated reproduction Fewer, larger progeny
Length of life	Short, usually less than 1 year	Longer, usually more than 1 year
Leads to	Productivity	Efficiency

opportunistic "r" strategy but "K" combines this with an ecological route where animals invest in behavior to buffer the results of environmental pressures. They not only select and alter the environments they and their offspring inhabit, but they also have the ability to learn. In this way the effect of environmental capriciousness becomes relative to knowledge about the environment. The material for exaptive rather than adaptive solutions to survival is present in such behavior. The importance of these measures of ecological time will emerge in chapter 5 when I discuss the role of behavior in species evolution and its contribution to stasis and change.

Finally, these concepts of time are not necessarily dependent on its absolute measurement. The antiquity of man and the age of the earth and its changes were all established with a relative framework. All time concepts, however they are measured, are social artifacts, part of the world which humans construct. We tend to regard the estimates which Lyell and others put forward as guesses, and forget that their lack of modern scientific methods, such as measuring the decay rate of isotopes, to come up with absolute dates did not stop them from applying concepts about the tempo and pulse of evolutionary time. The advent of absolute dates is yet another example of reducing possible ambiguities, replacing myth, or in this case guesses, with scientific fact.[59] The pattern of human evolution will not, however, emerge from an accurate chronology alone. While extremely welcome and nowadays indispensable such advances bring us no closer to an objective study of the past. More attention needs to be paid to critical concepts such as time and space.

The climate for progress

Climate and environment were regarded as potent forces in shoving along the human drift, upward toward civilization and sideways to global colonization. For example, those like Quatrefages, who favored northern latitudes as a cradle, pointed to the onset of ice ages which pushed humans south.[60]

But more important was the belief in the active role of climate for shaping society. This view had been championed in 1748 by Montesquieu in *De l'Esprit des lois* where he made the link between temperate climates and progress and the climatically easy life of the tropics and stagnation.[61]

The contrary argument was advanced by Tylor for the warm regions of the world since they favored existence without culture in the form of houses and clothes and were thus highly suitable for the human race in its infancy. They did not foster progress but did provide a safe environment. Consequently, cooler latitudes were settled later and, in a classic extension of his correlation between race and ability, the white races which achieved this were also "gifted with the powers of knowing and ruling which gave them sway over the world."[62] Just as Sir Arthur Keith earnestly believed that the Englishman's judgment in matters of race was adequately safeguarded by his "sportsmanlike qualities,"[63] it would seem that cold showers did wonders for the nurturing of national character.

Explaining race by climate was commonplace. Buffon in 1749 had argued for a single species which once distributed across the earth underwent various changes as the result of climate, food, and mode of living and that through time individual varieties arose that later became, in his opinion, specific. This did not impress the philosopher Kant who pointed out in 1775 that "similar stretches of land and atmosphere do not contain the same race,"[64] and suggested that earth history might have something to do with this—a suggestion I will follow up in chapters 5 and 7.

Ice ages and time

Ever since the acceptance of former widespread glaciations, the ice ages have been both a means to measure time and a potent external force for change. Since our understanding of the ice ages has grown so considerably in the last 150 years, they serve as an illustration of how frameworks change and thus how concepts need to be reworked to accommodate such changes. This is science in action; adequate and authoritative for every era but temporary in the long term. The astronomical theory of the ice ages illustrates the competing claims of the twin concepts, time's arrow and time's cycle, particularly well.[65] The importance of these concepts for archeology is that they established the way data would be collected and organized (e.g. fig. 2.2).

A major episode in the earth's climate had been demonstrated by Agassiz in 1840 with his concept of an ice age. Fieldwork quickly discovered ancient glaciations in both the Northern and Southern Hemispheres. Lyell was able to use the evidence of former glacial features elegantly to demonstrate the principle of uniformitarianism in the later editions of his *Principles of geology*. The establishment of human antiquity was also closely linked to the study of the ice ages, in particular the association of stone artifacts and extinct animals, as first shown in the 1830s by Boucher de Perthes on the Somme.[66]

The expansion and contraction of ice sheets in North America and Europe, with corresponding local mountian ice caps and valley glaciation, was regarded by the evolutionists as a key mechanism in the dispersal of plants and animals, especially as it had been shown in 1841 that ice ages produced landbridges. This push and pull model of climatic change depended on geologists accounting for the origins of ice ages. Were they due to changes in the earth's internal temperature or shifts in climatic patterns perhaps induced by sun spot cycles? Several schemes were put forward but by far the most important for natural history came from James Croll in his *Climate and time* published in 1875.

The importance of Croll's work lay in his computation of the origin of ice ages based upon changes in the orbital geometry of the earth. To achieve this he used Leverrier's calculations of changes in the orbits of the planets which had successfully predicted, in 1843, the existence of Neptune. The upshot of Croll's calculations, which are presented in John and Katherine Imbries' *Ice ages*,[67] was the claim that during the most recent glacial epoch, which he thought lasted from 240,000 to 80,000 years ago, no less than eight ice ages occurred, each followed by a warm phase, or interglacial, every 10,500 years.

Wallace devoted a large part of *Island life*, a key work in the development of biogeographical approaches to evolution, to a discussion of Croll's theory and its implications for evolution. He concluded,

> I have further shown that, in the continued mutations of climate, . . . we have a motive power well calculated to produce far more rapid organic changes than have hitherto been thought possible; while in the enormous amount of specific variation we have ample material for that power to act upon, so as to keep the organic world in a state of rapid change and development proportional to the comparatively rapid changes in the earth's surface.[68]

Sadly this demonstration of how a dynamic model of change could be built without an absolute chronology was neglected. Croll's theory of the ice age was abandoned soon after his death in 1890 as revisions of the date at which the last glacial epoch ended caused many to discount the entire thesis and alternatives were actively sought.

By 1909 the evidence of former ice ages in the northern foothills of the Alps had been gathered into a monumental synthesis by Penck and Brückner. They constructed a sequence of four major glaciations based on the evidence of valley terraces whose gravels, they argued, were laid down during the ice age and incised in the following interglacial warm period. They also estimated that the entire Pleistocene epoch, consisting of only four ice ages, lasted some 650,000 years and not the 3 million calculated by Croll. The ice ages were short events accounting for less than 20 percent of this time and separated by longer interglacial episodes which took up almost 50 percent. The remaining 30 percent represented the transition from full interglacial to glacial. Their scheme judged all the glacial periods to be of roughly equal length but the interglacials varied enormously with the middle, or great interglacial, lasting some 240,000 years. Thus Croll's notion of warm and cold periods alternating with a regular frequency was contradicted and instead we find a highly differentiated, rather than repetitive, and a linear instead of cyclical, representation of time. Each succeeding glacial/interglacial stage was differentiated from every other stage by characteristic plants and animals. This provided a system of relative dating which was very useful but which stifled any investigation of how climatic change contributed to the evolutionary process.

The modern synthesis

Both Croll and Wallace got it wrong for the right reasons. Although the estimate of the duration of the climatic cycles has not stood the test of time, this should not detract from the importance of the theory, as used by Wallace, which showed how the repetition of similar climatic conditions can, in the long term, be a selection pressure for speciation. This must now be preferred to the traditional, linear view of progressive change. But when Croll's theories were resurrected and reanalyzed by Milankovitch between 1924 and 1941, the new theory of ice ages using astronomical data was discussed for its calendar information rather than its impact on human evolution.[69]

Milankovitch worked on the principle that glaciation was simultaneous rather than sequential, as Croll thought, in the Northern and Southern Hemispheres. He calculated that eccentricity in earth orbit established the dominant climatic cycle every 100,000 years, during which time the earth's climate goes through a full warm/cold or interglacial/glacial cycle.[70] Within this major cycle another, of 41,000 years duration, is set by obliquity, or the tilt in earth axis, and controls the amount of solar radiation in high latitudes. Finally, a much weaker cycle, but one with greater frequency, is controlled by the precession "wobble" either on a cycle of 23,000 or 19,000 years and which has greatest influence over the low latitudes around the Equator and the extent of differences between the seasons. The interlocking cycles are shown in figure 3.1. The eccentricity curve indicates that in the past 800,000 years there should have been eight glacial/interglacial cycles.

During every 100,000 year cycle forests would grow and then retreat as climate changed. Water levels in lakes and rivers would fluctuate and continental shelves would emerge as dry land later to be inundated. Ice would build up in the high latitudes and on mountain chains. In front of the ice caps would lie broad belts of periglacial and permafrost landscapes. In the long term, all these major changes would occur predictably and frequently. The external forcing of climatic change, as we shall see, reverberated on local adaptation and behavior.

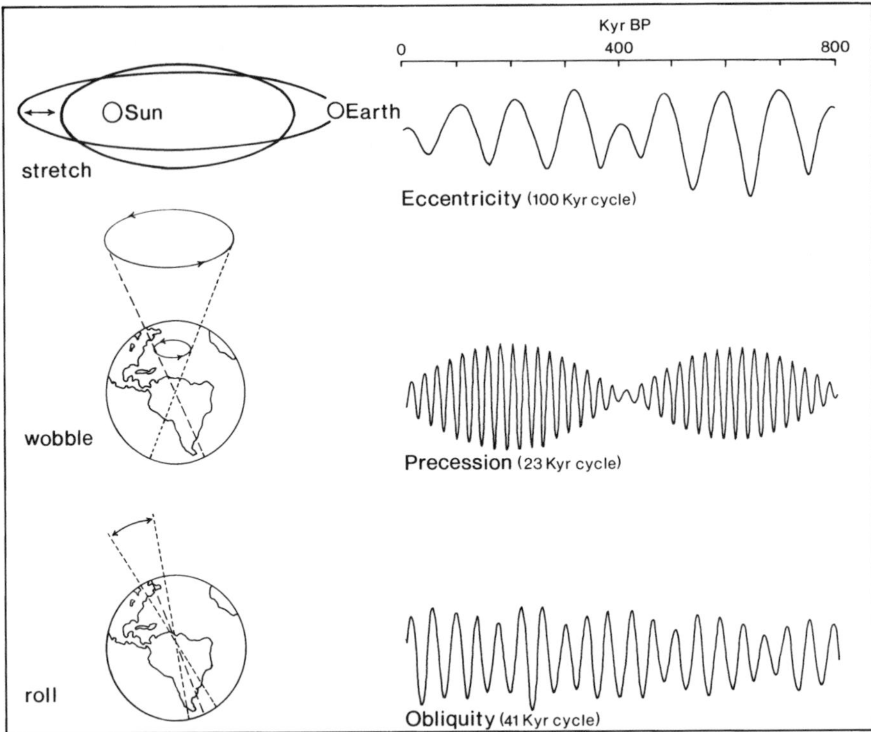

Figure 3.1 Stretch, wobble, and roll—time's cycles, as shown by climatic change, in the last 800,000 years. Eccentricity, precession and obliquity are all changes in aspects of the earth's solar orbit and the way it spins and tilts.

While it was quite certain that Milankovitch had solved the mystery of why ice ages occur and the pattern they took, it was not until confirmation came from the deep seas in the 1970s that the four ice ages postulated by Penck and Brückner were jettisoned.

The data came from the study of the marine sediments recovered from cores drilled into the ocean floors. These sediments are made up of the skeletons of minute creatures, foraminifera, that once lived at the surface of the ocean. Their skeletons are made up of calcium carbonate which when laid down over sufficient lengths of time such as the Cretaceous, would form the White Cliffs of Dover. On the much shorter timescale of the Pleistocene, the significance of these foraminifera is that while alive they absorb through their skeletons the oxygen isotopes in the sea water. On dying they fall to the bottom of the oceans where gradually, and as a continuous record, they build up the muds and oozes.

The two oxygen isotopes absorbed are ^{16}O and ^{18}O. These differ in "weight" and vary according to the size of the oceans. These are small during an ice age when moisture is drawn off to build ice sheets as well as the lighter ^{16}O isotopes (fig. 3.2). The result is an isotopically heavy ocean as measured by a higher

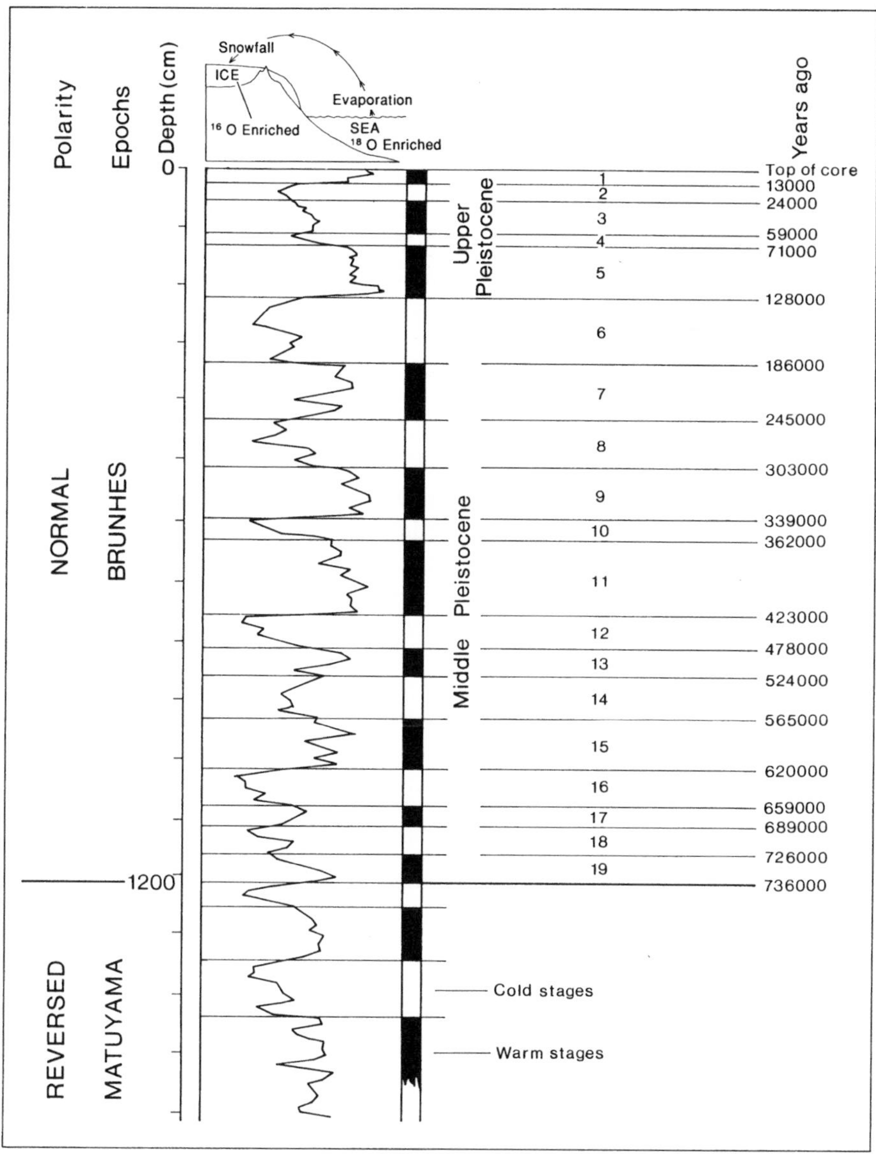

Figure 3.2 When ice sheets are built the oceans become "light" in ^{16}O and smaller in size. The reverse holds for the warm interglacial phases. The repeated effect produces a climatic record for the Pleistocene as shown here in deepsea core V28–238 from the Pacific Ocean. The saw-tooth curve charts the variation in the amount of ^{18}O plotted against depth in centimeters. This in turn reflects the size of the oceans and the ice caps. The "warm" stages are deduced from isotopic evidence for large oceans and the "cold" stages, large ice caps. The change in the earth's magnetic polarity, which helps date the core, occurs at 1,200 cms. There are seven full climatic cycles in the Middle Pleistocene; each cycle has an odd and even stage number. The Upper Pleistocene, stages 5 to 2 consists of a single, but better understood cycle. We are in stage 1, the Holocene. Notice how abrupt the transition is from large ocean and small ice caps to small ocean and large ice sheets. The greenhouse effect is currently believed to be delaying the onset of colder, larger ice cap conditions.

proportion of ^{18}O. The reverse happens during the melting of ice sheets in the interglacials when as the oceans increase in size they become isotopically lighter. These changing conditions are recorded in the isotopic composition of the forams skeletons. To obtain a continuous record of the changes in global climate thus requires finding a location anywhere in the world's oceans where sediments have accumulated, coring them and extracting the forams. The ratio they contain of ^{16}O:^{18}O can then be measured, using a mass spectrometer, and the fluctuating proportions plotted out against depth. The saw-tooth result is given in figure 3.2.[71]

In practice the process is not quite so simple. Several assumptions have to be made about the build up of sediments; whether this occurred at a constant rate and how the configuration of the ocean basin may have affected the record by mixing sediments of different ages. This is particularly important when trying to compare curves from different cores taken in different oceans. Moreover, depth in the core does not provide an estimate of age and this is obtained by measuring the polarity of the sediments. Geomagnetic studies on terrestrial volcanic rocks have established that at certain periods and for unknown reasons the earth has reversed its magnetic field.[72] Quite literally the north pole became the south pole. These fields have remained constant for long periods of time, referred to as epochs, interspersed with much shorter reversal events. The last major reversal occurred in rocks dated by isotope decay methods to 730,000 years ago. At that time the current Brunhes epoch of normal polarity replaced the Matuyama epoch of reversed polarity. In the core shown in figure 3.2 this reversal occurs at 1,200 cm and gives a fixed point on which to build a chronology. It is then possible to do a simple division of depth by age to find out the rate at which the sediments were accumulating and hence provide a date for any point on the core. The peaks and troughs of the ^{16}O:^{18}O ratio are subdivided into those indicating large ocean volume (odd numbered interglacials) and large ice volume (even numbered glaciations). Within the last 730,000 years there have been eight full glacial/interglacial cycles.

The advantages of this record are numerous. In the first place the deep sea data do not reflect local or even regional conditions but in this instance a global event. Secondly, it is a continuous record and so avoids the problems faced by Pleistocene geologists trying to piece together the sequence of events from continental evidence which, in many cases, has been removed by subsequent ice action and at best represents only a fragment of the Pleistocene. Thirdly, different cores can be compared to check the global significance of the record and crosschecked with a similar $^{16}O/^{18}O$ record of cores derived from the continuous accumulation of ice in Greenland and the polar ice caps. This research has shown that Croll's model of cyclical ice ages of roughly equal magnitude while wrong in detail was essentially correct in principle.

The cores indicate the initiation of moderate sized ice sheets in the Northern Hemisphere at 2.5 million years.[73] Extensive ice sheets were not, however, established until some 900,000 years ago. The division between the Pleistocene and Pliocene is fixed by the magnetic reversal known as the Olduvai event and dated between 1.6 and 1.8 million years ago (table 3.2).

Table 3.2 A geological timescale (Myrs = million years).

	Myrs	
Pleistocene	0	Quaternary
Pliocene		
	10	
Miocene		Tertiary
	20	
Oligocene	30	
	40	
Eocene		
	50	
Paleocene	60	
Cretaceous	70	

	Myrs
Holocene	0.01
Upper Pleistocene	0.13
Middle Pleistocene	0.5
Lower Pleistocene	1
	1.5
Pliocene	2

One of the first uses to which this record was put involved testing Milankovitch's theory of orbital forcing on the amount of solar radiation reaching the earth and hence the onset of glaciation.[74] The purpose was to discover whether his predictions of orbitally driven cycles at different amplitudes and frequencies would indeed match the climatic changes as registered in the oceanic record. The results show emphatically that the deepsea cores can be tuned to these major cycles at 100,000, 41,000, and 23,000–19,000 years. However, only the lower amplitude cycles at 41,000 and 23,000 years are directly forced by orbital changes. This is shown by a timelag of several thousand years between orbital changes and their appearance as climatic shifts in the deepsea record. The 100,000 year cycle as predicted by and shown climatically in the cores has no such time lag. Another factor is therefore needed to trigger ice ages through the rhythm of this dominant cycle. In discussing this problem Ruddimann and Raymo suggest that tectonic activity of the magnitude some claim for the uplift of Tibet could be that additional element (see chap. 7).[75] This would have an influence over the jet stream leading to major changes in circulation patterns and the deposition of moisture in northern latitudes where it could be built into ice sheets. A factor of this kind is certainly needed since evidence from much earlier geological periods shows the existence of 100,000 year cycles and yet no ice ages occurred (see chap. 5).[76]

The cores also show that the duration and intensity, or amplitude, of a full interglacial/glacial cycle has changed significantly during the Pleistocene. From 1.6 Myr to 900 Kyr the dominant frequency is for a complete cycle every 40,000 years. These were not severe glaciations. Between 900 and 450 Kyr the duration increases to 70,000 years with the shift to a span of 100,000 per cycle after this date.[77] These last four cycles are marked by severe continental glaciation. It is probable that these trends in the cyclic and recurrent alternation of environments had considerable impact on speciation, as we shall see in chapter 5.

Starting the timewalk

What sort of world have we now created for our timewalkers? As the last section shows, we can produce impressive stories from a column of mud. This chapter has revealed something of how we now create our own remote past with the methods of science.

The study of human origins now starts from a very different set of assumptions than it did when Latham penned his three questions. It is also extremely well informed about process and patterns in the data compared to 150 years ago. The celebration of progress has fallen from the agenda. Living peoples are no longer regarded as living representatives of a past which the Western world once possessed.

But for all these apparently fundamental changes the questions on the agenda remain the same. Why should the study of human evolution be restricted, because of the search for cradles, to some continents at some points in time? Could it be that for all the resources and objectivity which we believe we now bring to the question of human origins we still want to hear the familiar story? "Why were humans everywhere?" is still not the opening line but, as we shall now discover, "Where did we all begin?"

4
The modern cradle for human origins

The triumph of archeological research into the earliest prehistory of Africa was trumpeted by the archeologist Desmond Clark in the Huxley Memorial Lecture of 1974. Titled "Africa in prehistory: peripheral or paramount?" it pointed to the overwhelming evidence from Africa for the origins of hominids, which overthrew the previous view "that the history of Europe is emphatically the prehistory of humanity."[1]

Large brains, proper feet, nimble hands, fire, stone tools, and a range of feeding patterns were all assembled into a lifestyle package prior to any major expansion from Africa almost a million years ago.

The research of the past forty years has indeed been remarkable in yielding up a great many fossil and cultural remains from a broad range of African environments.[2] These discoveries have raised three main issues about the contemporary cradle for the earliest timewalkers. Firstly, it is necessary to examine the habitats from which the evidence comes and the longterm changes these underwent. Secondly, the unsuspected variety of early hominids needs to be described since these discoveries, and the debate over their significance, have fundamentally changed our views of the fossil pathways of human evolution. Finally, we want to know how they made a living.

Fossils in a landscape

The most productive areas for research into human origins have been down the spine of the continent. Starting at the foot of the Ethiopian highlands Pliocene age sediments are exposed in the triangular Afar depression.[3] At Hadar these contain some of the earliest hominid fossils dated to over 3 million years.[4] However, the greatest density of finds occurs between 5° either side of the Equator on the East African plateau at elevations of between 1,000 and 2,500 m (fig. 4.1).[5] In the north are the rich deposits along the Omo river and to the west and east of Lake Turkana. Since 1968 intensive research in the Koobi Fora badlands of East Turkana has produced a rich crop of fossils and archeological remains.[6] Findspots continue along the Gregory Rift and end on the Serengeti Plains of northern Tanzania at Olduvai Gorge and Laetoli. At present there is a gap of almost 3,000 km between these findspots and the limestone uplands of southern Africa where abundant, but very fragmentary remains have been recovered from a number of quarries, caves, and fissures. Most notable among these are the finds from the Sterkfontein Valley.[7]

All these areas have their representatives of a fossil genus first named by Raymond Dart in 1924 as *Australopithecus* or southern ape.[8] No finds of this key fossil genus have so far been made in similar age deposits beyond the plateau region shown in figure 4.1. There have been claims for Australopithecines, or

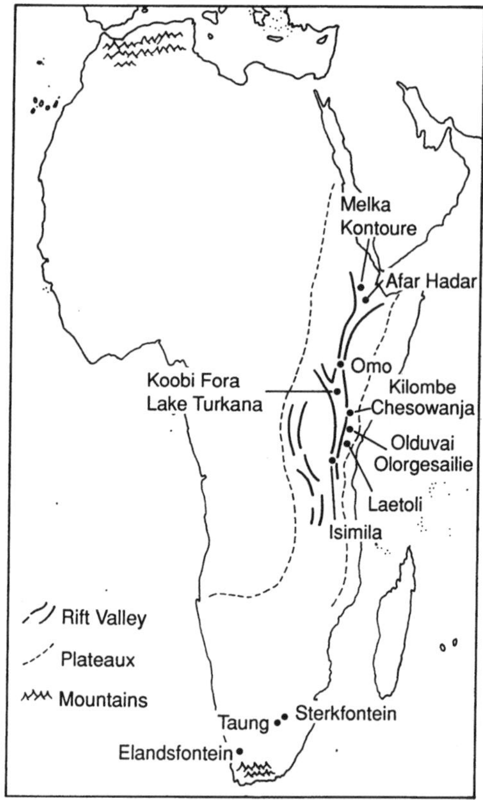

Figure 4.1 Location of early hominid sites in sub-Saharan Africa.

Apiths, from Lake Chad[9] and as far afield as southeast Asia, China and briefly rumored in Spain. None of these is currently accepted as authentic.

Rifts and climate

Two geological features are important. In the first place, Africa is the most stable of all the continental/oceanic plates. It contains large areas of very old rocks which have seen little recent movement.[10] Mountain building associated with plate margins is only found at the extreme edges in the Atlas and southern Cape folded belt.

In contrast to the rock steady continent is the majestic Rift Valley which runs for 3,000 km before continuing through the Red Sea and terminating in the Jordan Valley. Fossil hunting and archeological discoveries have been particularly rewarding along the various sections of the African Rift. The uplift of the plateau through which the Rift cuts took place in the Early Miocene *c.* 20 Myr ago as the African plate reached its present position.[11] This was followed by volcanic activity which produced, among many peaks, Kilimanjaro (5,895 m) and Mt. Kenya (5,200 m). The major patterns of rifting had started by at least 15 Myr.[12] However, the faulting which formed the present topography of the Rift took place between 2 and 1.5 Myr ago during the Pleistocene.[13]

The effect of the faulting, either at a relatively small scale, such as Olduvai Gorge in the Serengeti Plains, or along the line of the Ethiopian, Gregory, and Western Rifts which encircle Lake Victoria,[14] has been to expose sediments containing fossils and stone tools dating to the Pliocene and Lower Pleistocene. The remains are found eroding out of the side of gullies as in the badlands of Koobi Fora and the Afar, or from the exposed geological beds at Olduvai. Patient fieldwalking and acute observation at these exposures can result in surface discoveries. These might lead to the choice of some locations for detailed excavation with the aim of recovering the bones and stones in their depositional context. This is important since much of our ability to reconstruct the remote past from these data rests on a detailed understanding of how such materials became incorporated into either lake and stream deposits or, in the caves of the Transvaal sites, inside limestone caves and fissures.[15]

In the Rift sites the sediments are often interspersed with volcanic tuffs. These are readily distinguishable and form important strategic markers for correlating the position of deposits over some distance. They can also be dated by isotope decay methods[16] and therefore date the fossil-bearing sediments which they under- and overlie. The fossils are never directly dated but rather the geological bed in which they are found. In the caves similar methods of isotope decay dating cannot be applied and age is estimated instead by comparing the animal faunas with those dated elsewhere. More recently, geomagnetic polarity measurements have been used to fix their position within the Pleistocene.[17]

The fossils that occur in great numbers in many of these localities are mostly from animal species. These reveal a long prehistory to the richness and diversity of the contemporary African fauna. The evolution of this community with its many species of primates, antelopes, pigs, bovids, elephants, rhinos, and carnivores including sabre-tooth cats, lions, hyenas, cape hunting dogs, cheetahs, and leopards, is closely bound up with climatic and tectonic changes. The contact at 17 Myr of Arabia, northeast Africa, Iran, and Turkey provided the opportunity for the dispersion of faunas from the previously separated land masses.[18] This contact also led to much drier conditions as the Tethys Sea, which previously had divided the continents, now disappeared.[19] The effect can be traced in the fossil communities of East Africa where between 16 Myr and 14 Myr there is an increase in savannah species such as bovids and giraffes as well as the first appearance of the ostrich, modern forms of rhino, and several Old World monkeys and fossil primates such as *Kenyapithecus*.[20] Other taxa, notably equids and hyenas, arrived from Asia between 12 Myr and 9 Myr ago as the trend toward a drier and colder climate continued.

The uplift of the plateau and the onset of rifting in the Miocene disrupted former vegetational belts of tropical forest, savannah, and desert which once stretched from coast to coast. The trend toward cooler climates that also started in the Miocene found its full impact in the glacial/interglacial cycles of the Pleistocene. These were to have a major influence on community evolution as they regularly altered the vegetation on the otherwise solid face of the African landscape (see chap. 5).

The clearest example of the effect of these climatic cycles is provided by the prehistory of the tropical rain forest that today extends from the West African coast up to the edge of the plateau. This great forest is remarkable for the concentrated diversity of trees, plants, and animals. It has often been argued that tropical forests are diverse and complex because they have been stable for a very long time and so new species have been constantly added.

However, by reconstructing vegetational history from pollen grains extracted from cores driven into lake muds in Kenya and Uganda it is possible to draw different conclusions. At the last glacial maximum some 18,000 years ago, when ice sheets in the Northern Hemisphere covered most of Britain, the corresponding effect at the Equator was the drastic reduction of the tropical rain forests into a few refuge pockets in West and central Africa.[21] At 18,000 B.P. temperatures were lower and aridity very marked in these latitudes. For many thousands of years either side of this climatic marker the forest was severely reduced and replaced by savannah, while the deserts expanded north and south.[22] Within the savannah belt a variety of habitats existed due to rainfall and other factors. These affected the amount of tree cover and the productivity of herbs and grasses capable of supporting herds of grazing animals and predators. The result was a complex mosaic of environments within a local region and considerable variety between regions.

Conditions similar to those 18,000 years ago were repeated during each Pleistocene cycle. Seasonality would have been the main effect for early timewalkers of these recurrent climatic cycles. Today this is a complex matter of elevation, latitude, longitude, and prevailing weather patterns.[23] For example, the equatorial part of the Rift Valley has two rainy seasons neither of which is reliable or predictable. At a local scale the floor of the Rift is generally dry while its sides receive more rain. Lakes and water sources in the Rift are subject to very high evaporation which places such resources at a premium. By contrast, closer to the tropics the rainfall patterns are merged into a single rainy season lasting for about two months. In the long term these seasonal differences were subject to regular changes in the wobble of the earth around its axis,[24] *precession* (see chap. 3), as well as the evolution of the Rift Valley which brought changes to relief and rainfall patterns. As differences between seasons increased in terms of critical limiting factors such as water, temperature, and food, so, it is thought, did the circumstances alter that selected for novel adaptations, either anatomical or behavioral.

Variety of hominids

The starting date for the appearance of hominids is no longer determined by fossil skulls, archeology, or absolute dates. The estimate of 5 Myr as the time when the hominids split from the great apes is based on the immunological distance between ourselves, gorillas, and chimps. These data were first presented over twenty-five years ago by Goodman[25] and later by Sarich and Wilson.[26] They demonstrated the close relationship between the genetic maps of these three species where 99 percent of all DNA is shared.

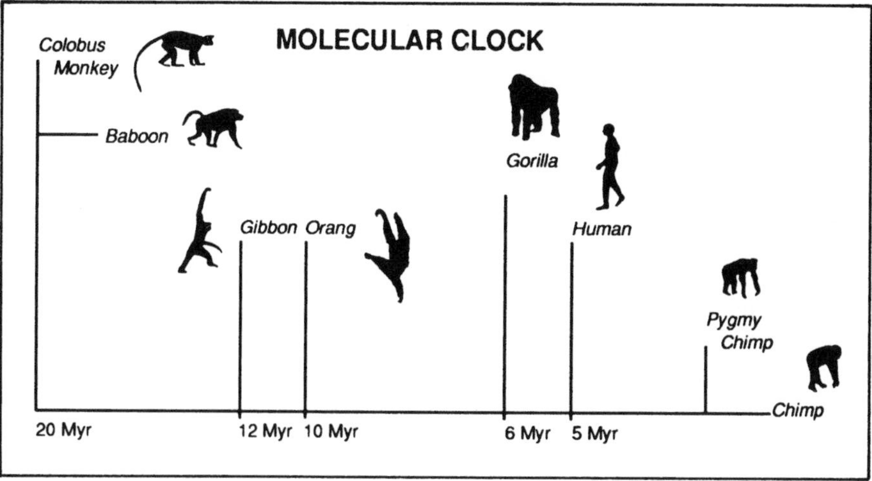

Figure 4.2 Molecular clocks and the great apes. The age estimates are very general, but the relative position of each branch in the diagram is now well understood by comparing the molecular evidence for each species.

Subsequent improvements in genetic mapping of these sister species have shown that chimps and humans are considerably closer to each other than either is to the gorilla.[27] From this molecular evidence the following pattern of branching and sister group relationships can be drawn up (fig. 4.2). Behavioral similarities between humans and chimps—tool using,[28] language capability,[29] and the ability to deceive[30]—reinforce this close genetic similarity to the extent that it has been asked whether apes are descended from man[31] or whether there are three species of chimpanzee—*Pan paniscus* (the pygmy chimp also known as the Bonobo), *Pan troglodytes*, and *Homo sapiens*![32]

The date of 5 Myr was derived from estimates of the rate of change in genetic material. The baseline is set by the division between Old World monkeys and the higher primates. This is believed to have occurred some 30 Myr years ago during the Oligocene. Comparing immunological distance between apes and monkeys against this timescale provides a means for assessing the rate of genetic changes. Hence the age of 5 Myr is calculated. The method is often referred to as the molecular clock which, although it had its critics claiming that it kept poor time, is now widely recognized as a reliable estimate.

Acceptance has come partly with the discovery of fossils from this time period and the rejection of much older ancestors. The genetic maps show that an evolutionary tree that splits the earliest timewalkers from *all* the great apes is wrong. The pattern of branching, called a cladogram,[33] must show instead a succession of branching events (fig. 4.2) for the great apes before anything that is recognizably a hominid enters the fossil record. The final branch, or clade, in figure 4.2 gives us the ancestry of the timewalkers and the taxonomic outgroup of closely related but distinct living species.

The discovery of the link between genetic and fossil data has exposed what some molecular timekeepers call the "*Ramapithecus* fiasco."[34] Until recently it was common to identify a widely distributed Miocene ape as the possible human ancestor. At about 10 Myr there were closely related forms known variously as *Sivapithecus* and *Ramapithecus*, found in Africa, southeast Europe, Turkey, the Middle East, India, and Pakistan.[35] While they are known almost exclusively from teeth and skull fragments their small size and feeding patterns, deduced from dentition, were interpreted as evidence of an arboreal lifestyle in the warm and extensive Old World Miocene forests. Their chronological position nicely fitted the timescale for landscape change and hominid evolution. However, as the genetic data now show, if we stick strictly to the hominids then we only need to go back some 5 Myr to begin the story. In other words the contribution to hominid phylogeny from the much older Siva and Rama apes is irrelevant.[36] Instead, the two Miocene apes have been grouped together with the orangutan and a branching point for this clade fixed at 10 Myr ago.[37] The orangutan therefore has the distinction of being the only great ape currently to possess an agreed molecular and fossil ancestry. The branching points for the other great apes are timed by the molecular clock at 12 Myr for the gibbon while gorillas and chimps branched at about 6 Myr and 5 Myr respectively (fig. 4.2).[38]

Goodbye missing link

A major problem with paleontology is that it has often been a reactive rather than a predictive subject. One of the strongest myths in the study of human origins, as Eldredge and Tattersall point out,[39] has been that evolutionary histories of living taxa are simply a matter of discovering the right fossil. Such a view leads to the constant pruning and felling of ancestral trees as fresh discoveries, particularly those with absolute dates, are brought back from the field. It is also very common to find that problems and uncertainties are put down to the incompleteness of the fossil record with claims of "missing links" to paper over inconsistencies. This appeal to negative evidence is encouraged by Darwin's insistence that new species arose slowly and gradually.[40] The slowness of the process makes it difficult, if not impossible, to spot intermediary forms because all fossils are to some extent changing and therefore transitional.

With hindsight the "*Ramapithecus* fiasco" is understandable since there was a perceived need for a common ancestor for a wide range of fossil hominids. But the result shows the immense difficulties paleontologists have in demonstrating which of their fossils had descendants. We now suspect very few. On the other hand Sarich has been quick to point out that he knows his molecules had ancestors.[41] From his perspective there are no missing links to discover and no reaction needed to new fossil discoveries.

Nowhere is the reactive approach more visible than in the arena of human origins research starting at 4 Myr and ending about 1 Myr ago. In the thirty years since the opening up of the fossil cornucopia of Olduvai Gorge with the discovery by Mary Leakey of a massively robust Australopithecine, *Zinjanthropus boisei*, there has been a wealth of finds from East Africa, Ethiopia, and South Africa.

The flood of data is reflected in the proliferation of newly recognized species among fossil hominids. A survey of theories concerning human evolution in a comparable period of time before 1940 shows the opposite.[42] Here prediction ruled a sparse and slowly accumulating database. For example, the pre-*sapiens* theory predicted very ancient origins for *Homo sapiens*. As a result the discovery of fossils with features judged primitive, such as Neanderthals and the forgery from Piltdown,[43] were never given recent ancestral relationships to modern humans. They were fitted into the scheme of human evolution by giving each one a separate lineage while resuming the quest for the true human ancestor.[44] Small wonder, therefore, that Raymond Dart's southern Apith baby—*Australopithecus africanus*—found in 1924 at the limestone quarry of Taung in the Transvaal, was generally pushed off to one side and left out of the human story for many years.[45]

How many species?

The reaction to the sudden abundance of fossil material and absolute dates was a fierce debate over how many closely related species existed between 4 Myr and 1 Myr. Opinions differed about which ones held the baton for the future human race although the winner was at least agreed. *Homo erectus*, which appeared in Bed II of the Olduvai sequence and elsewhere in East Africa as early as 1.6 Myr,[46] was the direction in which evolution had proceeded. The evidence of a large brain and stone tools combined with close similarities to fossils in Java, found in 1891,[47] and the Chinese cave site of Zhoukoudian,[48] excavated during the 1920s and 1930s, pointed to the winner who broke the tape by leaving Africa. Some recent research, which I shall discuss later, is now trying to question this picture.

The most significant claim for much earlier lineages which were contemporary but unrelated came in 1964 in a paper by Leakey, Tobias, and Napier where they named some of the Olduvai fossils which occurred in the *Zinjanthropus* site as *Homo habilis*.[49] This was in contrast to the very robust Apith discovered in 1959 and with which they shared floor space. It was also in distinction to finds made in South Africa where slender and robust fossils had been termed *Australopithecus africanus*, following Dart, and *Australopithecus robustus*. At Olduvai the volcanic tuff overlying the Zinj floor in Bed I is dated by absolute methods to 1.8 Myr.

The reactive approach to fossil discoveries had scored a notable success. At a date of almost 2 Myr it could be shown that a closely related fossil—the super-robust *Zinjanthropus boisei*—was a sideline to later human ancestry. The contemporaneousness of distinct hominid species was a very significant discovery indeed.

The acceptance of *Homo* and *Australopithecus* as separate genera did not simplify the task of classifying fossil material. Widely separated geographical areas and marked differences in anatomical features—particularly tooth size and facial shape—saw to that. It was clear that some fossils were slender and gracile while others were large, with robust features. But how did these features relate to species within the two lineages? Should all gracile forms, including *Homo habilis*, be called *Australopithecus africanus*, or vice versa?[50] How many species of robust Apiths were there, and to what extent could some of the differences be explained by sexual dimorphism within a species rather than as separate species? How far can size of teeth, brain cases, and faces indicate adaptation either to different environments or the result of isolation among small populations for very long periods of time within the plateaux of East and southern Africa?

The answer to such questions is still far from agreed. It is notoriously difficult to attribute sex to fragmentary material, especially when much of it represents juveniles. The question of demonstrating size differences related to either sexual dimorphism or environmental differences has often been posed without asking what selective contexts would either exaggerate or reduce such potential. Nor has dating always provided the answer.

The difficulties are well illustrated by the discovery at Koobi Fora in 1972 of an almost complete cranium at a date of 2.6 Myr. The gracile 1470 skull, with a large brain (750 cm^3), seemed to clinch the view that *Homo habilis* had great antiquity and hence was unrelated to the Apiths. However, in a joint article the two authors disagreed about what to call it.[51] Richard Leakey claimed it as *Homo habilis* whereas his coauthor Alan Walker regarded it as *Australopithecus africanus*. A few years later the age of the volcanic tuff which dated the site was revised downwards from 2.6 Myr to 1.8 Myr—the same age as material from Bed I at Olduvai where *Homo habilis* was first identified.[52] Subsequent opinion has supported Leakey's view concerning this fossil but further finds at Koobi Fora now suggest that there was more than one species of *Homo* during the period 2 Myr to 1.6 Myr. There were the large brained forms with heavily built faces as represented by 1470. However, another well preserved skull, 1813, had a small brain (only 510 cm^3) and a lightly built face. These, and other Turkana fossils, have been variously classified as either sexually dimorphic examples of *Homo habilis* or as two contemporary species *Homo rudolfensis* and *Homo habilis*. To complete the range of possibilities they are also classed together as a regional population of *Homo* sp. distinct from the Olduvai material which is called *Homo habilis*.[53] Other taxonomists point to differences between these habilines and prefer to treat them as separate species;[54] 1470 becomes *Homo rudolfensis* and 1813 *Homo habilis*.[55]

At Olduvai the plot has been thickened by the discovery in 1986 of OH 62 regarded as a *H. habilis*.[56] OH refers to Olduvai Hominid and 62 is the specimen number. However, the specimen is a very small and lightly built individual in comparison to the more robust *H. habilis* material recovered earlier at Olduvai. However, the weight to stature relationships as well as the estimated lengths and robustness of the forearm bones point to OH 62 having a greater affinity to apes than to modern humans.[57]

This mosaic of fossils and opinion only deals with finds separated by 750 km and roughly of the same age. It does not include the further complications presented by the South African examples of *A. africanus* and *H. habilis* which are dated to 2.3 Myr and later.[58] The current, preferred scheme for all the African material is shown here in figure 4.3.

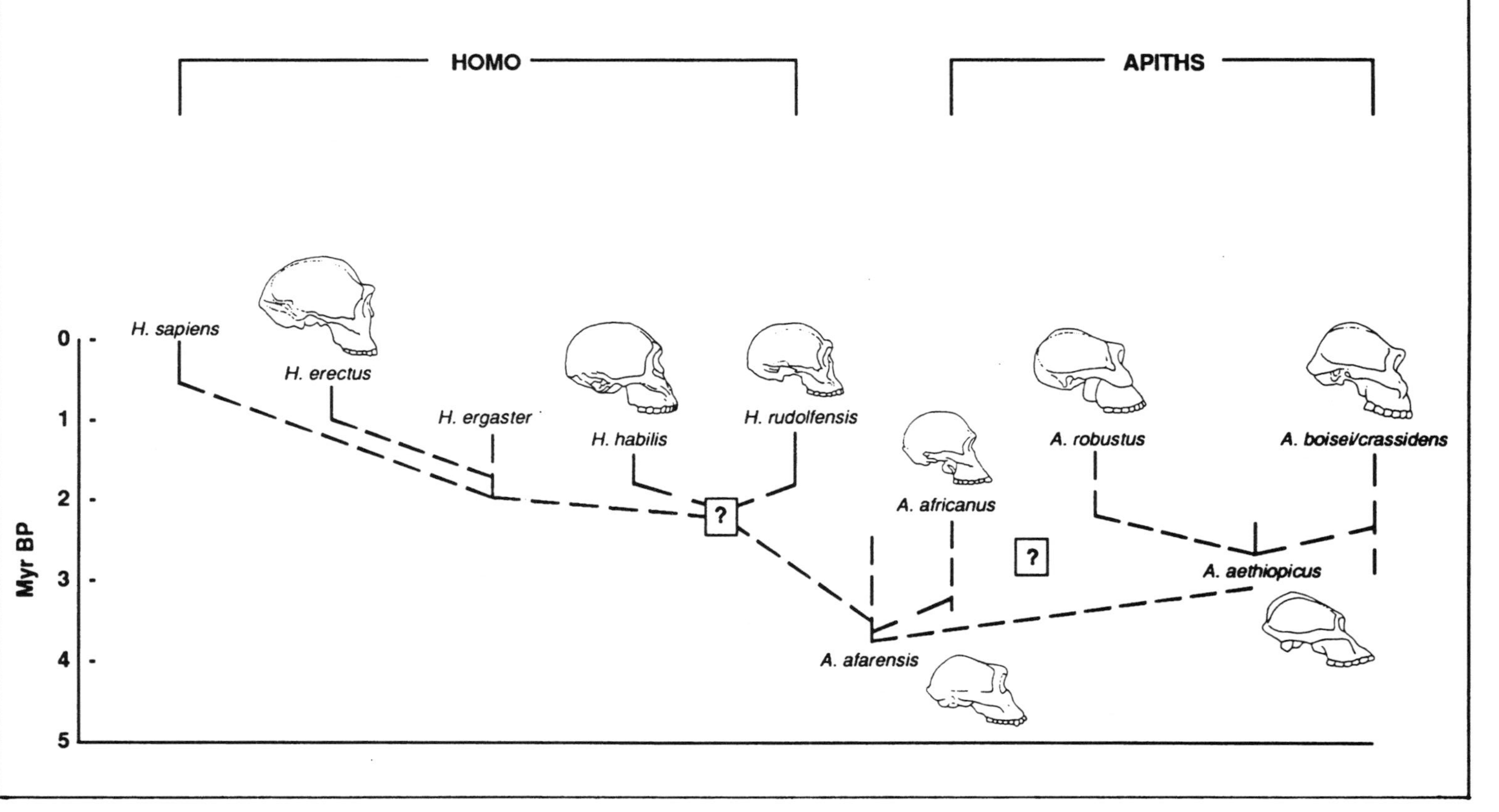

Figure 4.3 Bernard Wood's scheme for the evolution of the early timewalkers (1992 with additions). The scheme emphasizes the radiation of *Homo* after 2 Myr ago. Note the *A. afarensis* skull is a hypothetical reconstruction and Wood prefers to put the robust Apiths in a separate genus—*Paranthropus*. Compare with fig. 4.5.

Branching evolution

Part of the disagreement as to the attribution of remains to species and genera and these in turn to their sequence in hominid evolution stems from the way that family trees are now drawn up according to cladistic analysis.[59] Cladism aims to reveal the pattern of branching within a fossil lineage and as such is not primarily concerned with chronology. Strict cladists, for example, would never draw a family tree. It differs from other forms of classification by stressing the importance of shared derived features. These identify a group of related organisms descended from a single common ancestor. They are known as a *clade*. The term *grade* is used to refer to a group that shares a similar level of adaptation but not necessarily that all-important common ancestor.

Derived, in a cladistic sense, means evolved or new to that group of fossils. Any fossil with a uniquely derived characteristic, be it a feature of the skull or the shape of a tooth, is ruled out of the ancestor game by the simple fact that no other fossils share that feature. Ancestors, as Colin Groves has pointed out, can only be those fossils that possess no such unique features. Few of these have so far been found.

As with all systems of classification the outcome is dependent upon which characteristics are described, the OTUs or operational taxonomic units. Debate will always continue about their selection and evolutionary significance.

For example, alternative patterns of branching can be devised depending on what species and how many species this heterogeneous group of early *Homo*, or habiline, fossils are put in. In a recent study, Andrew Chamberlain and Bernard Wood present three alternative branching patterns.[60] They used ninety measurements on the fossils to generate a matrix of shape characteristics to compare the different fossil species. The control, or outgroup, was provided by the non-human primates. The computer then sorted the fossil species according to the most parsimonious order of branching indicated by these characteristics.[61] Several runs were performed using the different classifications mentioned above for the early *Homo* material. Not surprisingly, different branching patterns emerged depending on which species a fossil was assigned to (fig. 4.4). In the first, all the *H. habilis* fossils are taken as a single group and the cladogram positions them separately from the Apiths. The second plot recognized large and small *H. habilis* in both Koobi Fora and Olduvai. The cladogram reveals a branching pattern that clearly separates the early *Homo* and Apith lineages. The third cladogram examines the model of early *Homo* as two distinct geographical populations and produces an unexpected branching pattern. *Homo* sp. in Koobi Fora becomes a sister group to the robust Apiths while *H. habilis* at Olduvai emerges as a primitive fossil whose branch is situated between *A. afarensis* and *A. africanus*.

Whether, as claimed, such cladistic analyses will reverse the trend from reactive to predictive phylogeny remains to be seen.[62] The simple rule about who can and cannot be an ancestor may well cut down the competitive hype that often surrounds fresh field discoveries. At the moment the most significant, but hardly surprising prediction from this branching sequence of primitive and derived

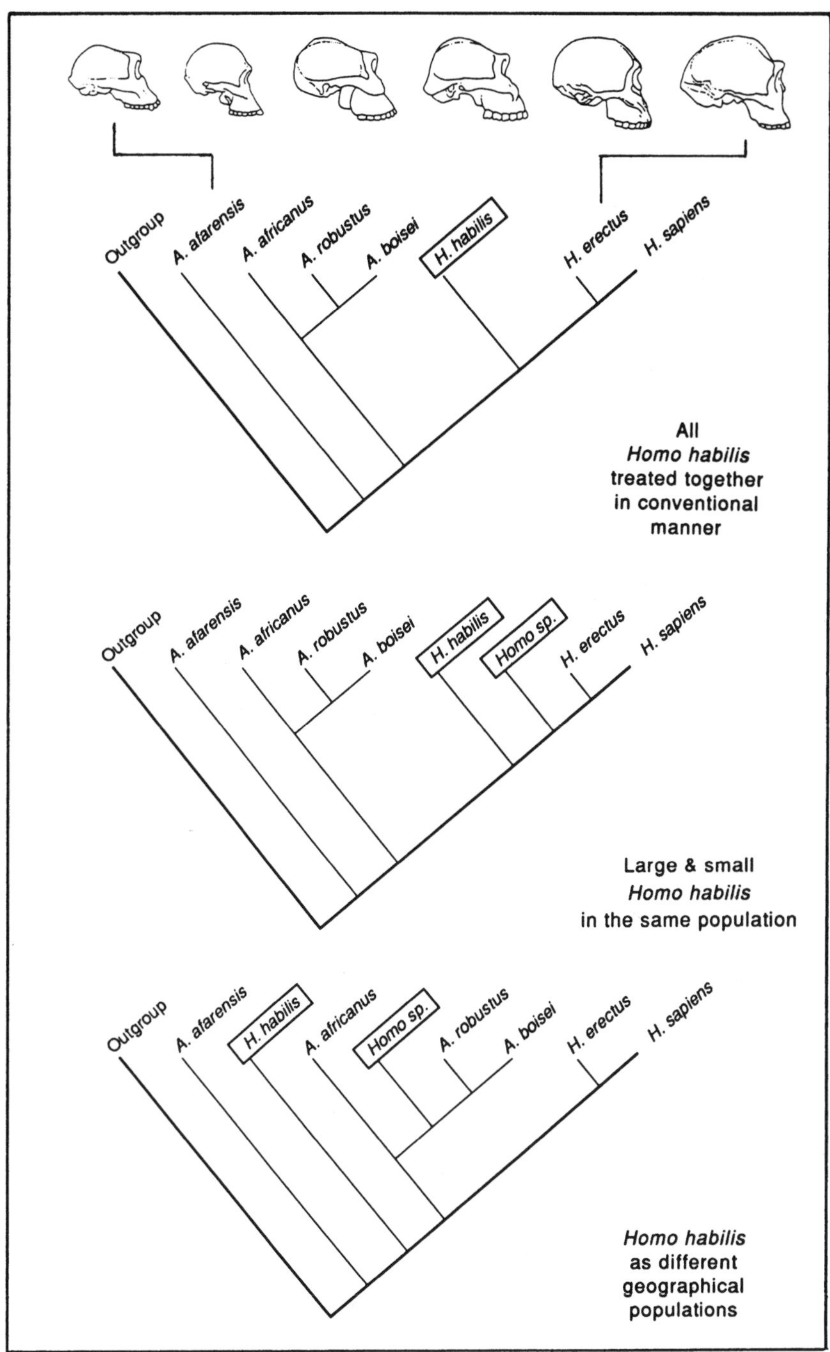

Figure 4.4 Three ways to classify early hominids. Wood and Chamberlain run the options on alternative patterns of branching as revealed by a suite of measurements and anatomical features. The three cladograms depend on the status given to anatomical variation (in particular any derived character states) between fossils classified as *Homo habilis*.

morphological features is to see that in all schemes *A. afarensis* is the sister group to all other timewalkers while *H. erectus* is the sister taxon to *H. sapiens*. There may, however, be some surprises yet in store.[63]

Timewalkers' timetable

How to get out of the maze? Since behavior rather than taxonomy is my main interest I need an account of the fossil record which takes notice of space and time in the distribution of the earliest timewalkers. This will allow the investigation of how behavior developed and the contexts which selected for its change. Tracing this to particular fossil species, however they are recognized taxonomically, is another matter.

The date of 5 Myr provides a starting point while the Apiths and *Homo* lineages are clearly germane to the discussion of where and what hominids were doing.[64]

4 Myr to 3 Myr B.P.

During this period there are as yet only two main findspots—Laetoli, south of Olduvai in northern Tanzania, and Hadar in the Afar triangle of Ethiopia, and one species—*Australopithecus afarensis*.[65] Both areas have produced significant discoveries dated absolutely to 3.7 Myr at Laetoli and 3–3.3 Myr at Hadar. William Kimbel has also classified the skull fragments 2602 from Koobi Fora as *A. afarensis*.[66] This plugs a geographical gap and, at 3.3 Myr old, fits well with the larger collections from Laetoli and Hadar.

At Laetoli Mary Leakey discovered not only a rich collection of teeth and skeletal elements but also a line of footprints preserved in powdery volcanic ash which had subsequently solidified.[67] This track of two adults and a child, treading in one of the adult's footsteps, demonstrates upright walking at this early date and confirms the evidence from long bones and the pelvis found at Hadar. Here a partial skeleton, best known by her nickname of Lucy (AL 288–1), was a small female about 1.07 m tall and weighing only 30 kg. The anatomy shows a fully bipedal gait but with some differences from modern humans. Indeed, these are so distinctive that in his exhaustive study of the limbs Charles Oxnard concluded that the method of locomotion was neither ape-like nor human.[68] Nor was it a midway stage but truly unique. Studies of the hand and feet by Stern and Susmann have shown that these were still adapted for significant arboreal activity.[69]

Lucy was a tiny creature. She represents one extreme of a highly sexually dimorphic population. The average weight of the species has been calculated by Henry McHenry as 51 kg,[70] while William Jungers provides a range between 30–81 kg.[71] This makes them very large primates indeed as a comparison to the great apes confirms (see table 4.1). The height of the large males is more difficult to estimate from fragmentary limb bones but was at least 1.37 m. Assuming that the Laetoli footprints were left by the same species then the length of the paces provides another means to gauge the height of the two adults. Tim White estimates a range of 1.15–1.34 m for one and between 1.34–1.54 m

Table 4.1 Apes, hominids, and humans: sizes.

	Average body weight kg	Range body weight kg	Estimated height m
Pygmy chimpanzee	39	34–43	
Chimpanzee	41	36–45	
Gorilla	118	68–168	
Orangutan	59	40–78	
Australopithecus afarensis	51	30–81	1.15–1.54
Australopithecus africanus	46	33–68	1.36
Australopithecus robustus	48	42–89	1.53
Australopithecus boisei	46	37–89	1.37–1.72
Homo habilis	41		1.17–1.52
Homo erectus	59		1.57–1.72
Homo sapiens neanderthalensis			
male	65		1.59–1.75
female	50		1.52–1.6
Modern Europeans			
male	70	65–75	1.7
female	59	55–63	1.6
Modern Papua New Guinea			
male	60		1.6
female	50		1.5

Sources: McHenry 1988; Jungers 1988; Feldesman and Lundy 1988; Stringer and Gamble 1993.

for the other.[72] If the shorter individual is female, a view some reconstructions favor, then she was considerably taller than little Lucy.

The Hadar and Laetoli fossils were originally combined by Johanson and White into a new species, *Australopithecus afarensis*.[73] The great age and primitive dental features, including a V rather than U shaped mandible, allowed them to place *A. afarensis* as the ancestor to the branches represented by *Homo habilis* and *A. africanus* while later *H. erectus* and *A. robustus* evolved from each branch respectively. The cladistic analyses in figure 4.3 do not, of course, permit such a simple lineage division and, as we have seen in figure 4.3, the tree is now drawn differently.

Not everyone is happy with *A. afarensis* as a separate species. It has been argued that it should be seen instead as a northern form of the gracile *A. africanus* abundant at a later date in southern Africa.[74]

3 Myr to 2 Myr B.P.

The earliest South African fossils date to this period and include the gracile Apiths, *A. africanus* at 3 Myr.[75] Since 1936 the fragmentary remains of at least forty-seven individuals of this species have been recovered from the limestone cave site at Sterkfontein.[76] A third of these were less than 20 years of age. The skulls are gracile with jutting faces. Estimates for body size vary greatly because of the spread of ages but it seems that these gracile Apiths were as sexually

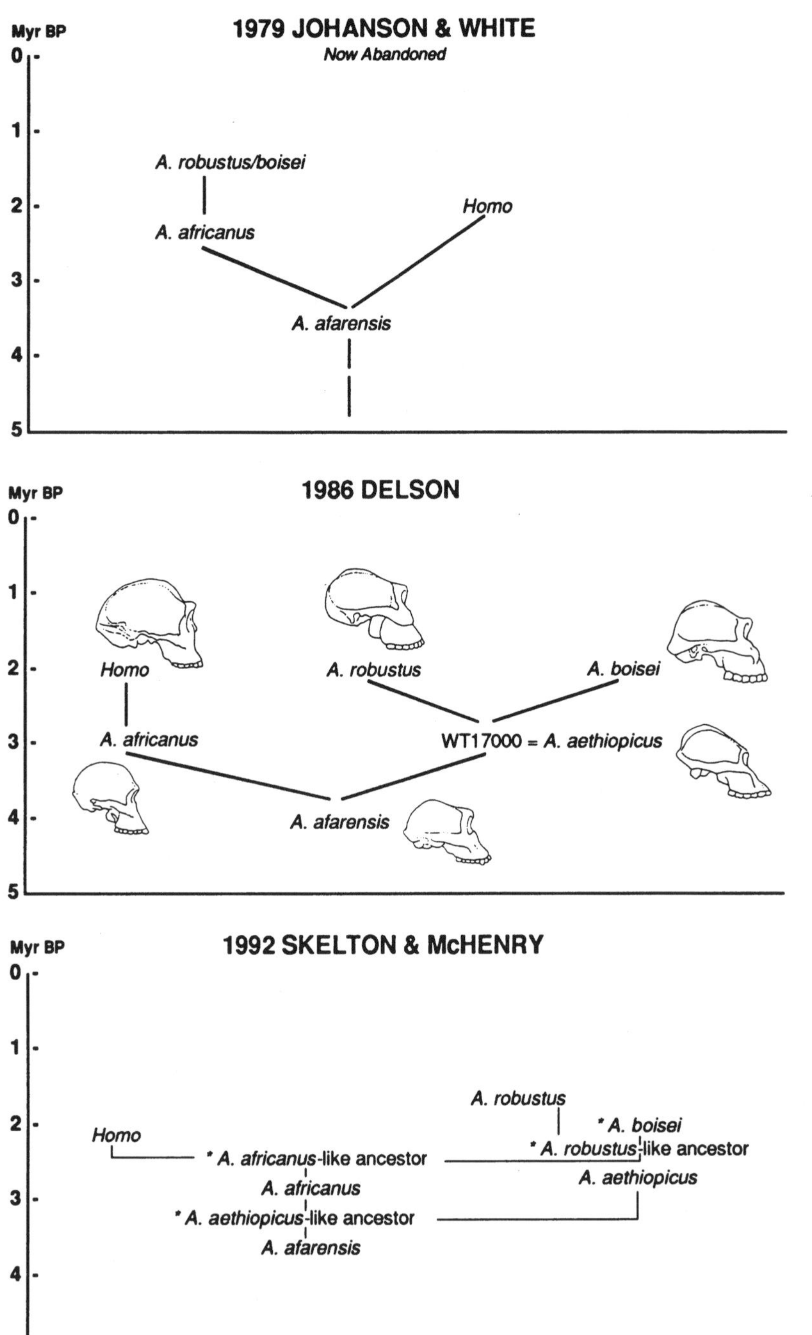

Figure 4.5 Reactions to new discoveries. The scheme put forward in 1979 by Johanson and White followed their naming of *Australopithecus afarensis* from Hadar and Laetoli. Seven years later Delson used more recent discoveries to redraw the diagram as did Skelton and McHenry six years after that. They propose at least two hypothetical ancestors* and a very different role for *A. aethiopicus*. Compare with fig. 4.3.

dimorphic as the much older Afar Apiths. McHenry has calculated an average weight of 46 kg for the species, while the range estimated by Jungers runs from 33 to 68 kg. Brain size was equally variable with a range of between 370–618 cm^3 and an average of 462 cm^3.

It was once thought that *A. africanus* evolved into *A. robustus* (fig. 4.5) since the latter is not found in the same early deposits at Taung, Sterkfontein, and Makapansgat. However, in 1985 at West Turkana a rugged, robust Australopithecine skull, WT 17000, dated to 2.5 Myr was found.[77] This is known as the Black Skull on account of its color. The original description classified it as *A. boisei*, the best known of which is *Zinjanthropus* from Olduvai Bed I at the later date of 1.8 Myr. Eric Delson prefers to call it *A. aethiopicus*[78] and puts it and similar but fragmentary specimens as the ancestor for the later robust and super-robust Apiths (fig. 4.5).[79] Skelton and McHenry disagree.[80] The results of their cladistic analysis suggest a different phylogeny (fig. 4.5) with an *africanus*-like ancestor giving rise to both *Homo* and the robust Apith groups in a very short time. *A. aethiopicus* is sidelined in their scheme.

Finally *H. habilis* is present in deposits older than 2 Myr at Sterkfontein.[81] A small sample of specimens from at least three individuals comes from a level where no other hominid remains have been found.[82] The consensus view is that they are descended from the gracile Apiths, *A. africanus*, as shown in fig. 4.6.[83]

2 Myr to 1.6 Myr B.P.

This is the time of the "big party" in eastern and southern Africa. Graciles and robusts are found living together in widely separated regions. In Sterkfontein Valley in South Africa *H. habilis* occurs in small numbers at Swartkrans at a date of 2 Myr.[84] Robust Apiths, variously called *A. robustus*, *Paranthropus robustus*, and *A. crassidens*, are common at Swartkrans and the nearby Kromdraai cave. The one fossil lacking in these quarry deposits during this time bracket is *Homo erectus*.[85]

In East Africa at Olduvai and Lake Turkana we find small and large examples of early *Homo* at 1.8 Myr. Opinion, as we have seen, is still divided over whether these can be classified further into *Homo habilis*, *ergaster* or *rudolfensis*. Robust and super-robust Apiths are represented by *A. robustus* and *A. boisei*.

Homo erectus is present by at least 1.6 Myr in West Turkana where a nearly complete skeleton of a youth, WT 15000, was found in 1984.[86] The find is older than the *H.erectus* OH 9 from the upper part of Olduvai Bed II.

However, it is this part of the hominid puzzle, which for some time seemed the most straightforward, that is now under attack. R.J. Clarke has recently argued that *Homo erectus* is not in fact found anywhere in Africa but instead evolved from *Homo habilis* populations *after* these had expanded into China and southeast Asia.[87] His scheme recognizes *Homo habilis* but regards a later Olduvai hominid OH 9, always previously classified as *Homo erectus*, as another species to be named after its discoverer, *Homo leakeyei*. This, he argues, is the ancestor for *Homo sapiens* that evolves in Africa and then exits. Groves puts WT 15000 in with the other ancestral contenders that have no uniquely derived features.[88] This is undifferentiated *Homo* sp.

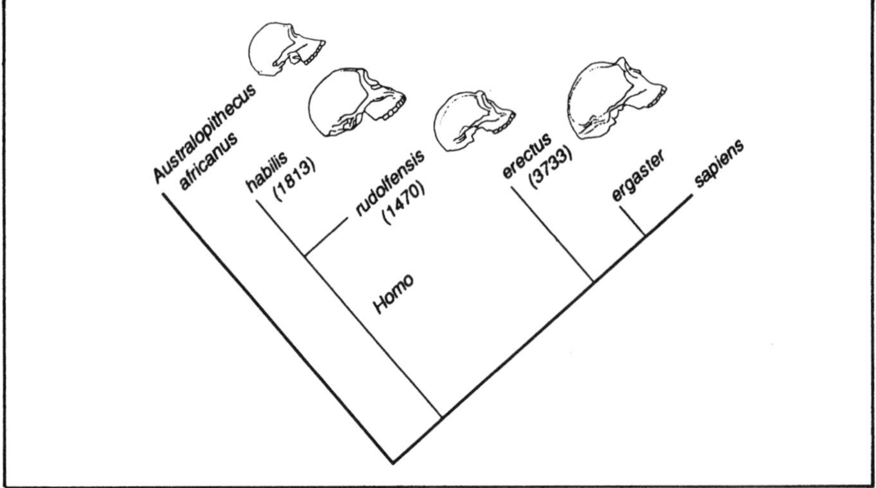

Figure 4.6 A cladogram for the early *Homo* timewalkers and their closest Apith fossil relative (Wood 1992). *Homo ergaster* is currently only known from jaws.

Most workers are unimpressed by this interpretation that *H. habilis* and *H. erectus* were *non*-African species. The majority view, which I favor, is presented in figure 4.6. Alan Turner and Andrew Chamberlain remind us that the important OTUs used in such analyses are those relating to fertilization.[89] Only these will *necessarily* change at speciation. Whether any aspects of these will also be part of the fossil record is not clear. Of critical importance, as I shall examine in chapter 5, is how organisms recognize potential mates and the impact this has on speciation. The consequence, they argue, must be an approach that lumps all these aspects together when it comes to fossil taxonomy. They therefore stick to the traditionally broad definition of *Homo erectus*. Even an avowed splitter such as Groves has remarked that the different species which he recognizes cladistically may not have been reproductively isolated.[90]

So, during a period of 400,000 years and scattered between East and southern Africa there are possibly four species of early *Homo*, if the habilines are subdivided, and three robust Apiths species. Two of these, *A. crassidens* and *A. boisei* are southern and northern species respectively while *A. robustus* appears in both regions. Of course the pattern would be different again if some of the *H. habilis* specimens were classified as gracile *A. africanus* as we have seen, for example, with the 1470 skull from East Turkana.

Size, age, and brains

Size estimates for robust Apiths are considerable. The average weights calculated by McHenry are 48 kg for *A. robustus* and 46 kg for the super-robusts, *A. boisei*.[91] Sexual dimorphism is still very marked. The ranges given by Jungers are between 42 kg and 89 kg for the robusts and 37 kg and 89 kg for the super-robusts.[92] A very small sample of long bones points to heights for both species of between 1.37 m and 1.72 m.[93] The larger body size is reflected in an average brain size of 530 cm^3.

There are the remains of at least eighty-seven individuals in the Swartkrans collection.[94] Juveniles were very common in this sample with two thirds less than 20 years old. The average age at death for this population has been estimated at only 17.2 years.[95]

Estimates of the body size for early *Homo* are difficult because the evidence is very fragmentary. McHenry gives an average of 41 kg while stature estimates on three long bones by Feldesman and Lundy range between 1.17 m and 1.52 m with an average of 1.39 m.[96] However, OH 62 is regarded as one of the smallest hominids, including Lucy, ever to have been found and any size estimate would be very different.[97] The figures for brain size also show a considerable range, from 510 cm^3 to 752 cm^3 for the Koobi Fora specimens. The Olduvai material has a smaller range, between 590 cm^3 and 690 cm^3, although two specimens are badly damaged and the other two are juveniles,[98] which diminishes the measurements' accuracy.

Brain sizes by themselves tell us very little. What we need to know is their size relative to body size. McHenry has calculated the EQ, encephalization quotient, which shows very dramatically (fig. 4.7) the increase in brain to body size for early *Homo*.[99]

More can be learnt when the various estimates for body size are collected together (table 4.1). It is striking that all the hominids are large. There have been almost 4 million years of big Australopithecines as well as early and later *Homo*. The fragmentary fossil evidence also points to considerable differences within species, as is the case with modern humans. The figures given here for Europeans and Papua New Guineans in no way reflect the immense intra-

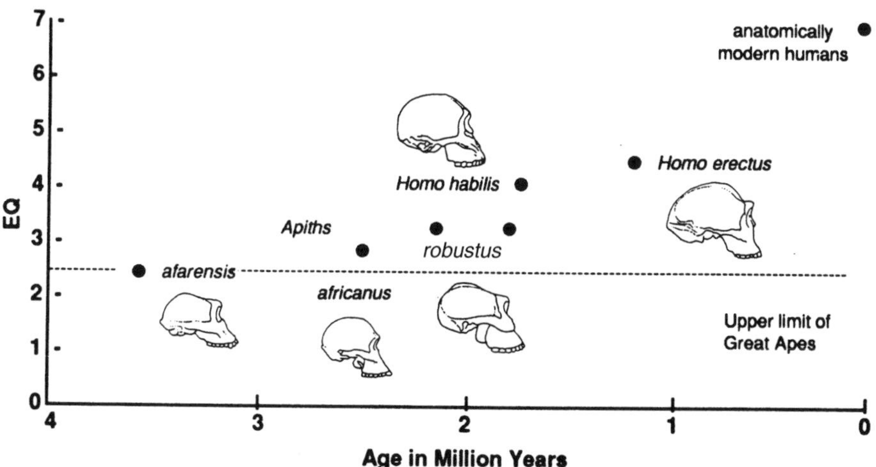

Figure 4.7 Age and brain size measured by an encephalization quotient (EQ). This compares brain size relative to body size rather than comparing them by their absolute values.

population differences we know exist just by bumping into people in the street. This was also the case for the early timewalkers. Small samples will be masking much of the variability. Even so we can see that much of the variation was due to sexual dimorphism. Among the great apes the gorilla is the most sexually dimorphic with the big males over 100 percent larger than the females. A silverback will often weigh as much as 169 kg and the females of 68 kg. The orangutan shows only slightly less sexual dimorphism, as measured by weight. But body weight can vary for all sorts of reasons. Measuring the sizes of the shaft of the femur, as McHenry has done, provides a more reliable measure of sexual dimorphism unrelated to condition.[100] For modern humans the figure is 130:100 (males:females), the gorilla and orangutan are both 154:100. The same measurements on the femurs of the fossil hominids generally exceed these two great apes which are so highly sexually dimorphic.

When the party's over

In the light of this taxonomic cocktail I will make a prediction. As other areas with fossil remains are discovered the regional pattern of hominid diversity, with robusts and graciles living side by side in evolving lineages, will be repeated throughout the period 5 Myr to 1.6 Myr. Why graciles and robusts coexisted is something I shall look at in chapter 5. What we have so far seen is a pattern of hominid and human evolution marked by great variation and many speciation events. While there may be disagreement over recognizing species, this is nothing new, as we saw in chapter 2. We must expect, as Foley has proposed,[101] the classification of many more species than we currently recognize as the biological concept of the species is explored by paleontologists with an interest in understanding evolution rather than taxonomy for its own sake. The timewalkers' path will not be a simple straight line with only a single fossil ancestor walking it at any one time. Whether the clade, derived from a common ancestor, also forms a grade, marked by a shared adaptation, will be the relationship to examine in more detail. It is here that the link between taxonomy and biology becomes relevant.

Nor would I be surprised if similar mixed parties of fossils are found outside the plateau areas. The diversity of species within the two lineages, gracile and robust, will depend to some extent on a combination of tectonic history and climatic impact and how these forces have shaken and stirred the evolutionary process. The mode of speciation involves other processes, as we shall see in chapter 5.

Making a living on limited resources

The cradle contains archeological evidence which bears directly on the behavioral direction of human evolution. There is nothing, however, in the period 4 Myr—3 Myr to add to the fossils and Laetoli footprints.

3 Myr to 1.8 Myr

The earliest stone tools could be as much as 3.1 Myr old, although younger dates of 2.5 Myr seem in order.[102] They come from the Afar region of Ethiopia but are not directly associated with any hominid remains. The tools are simple by any

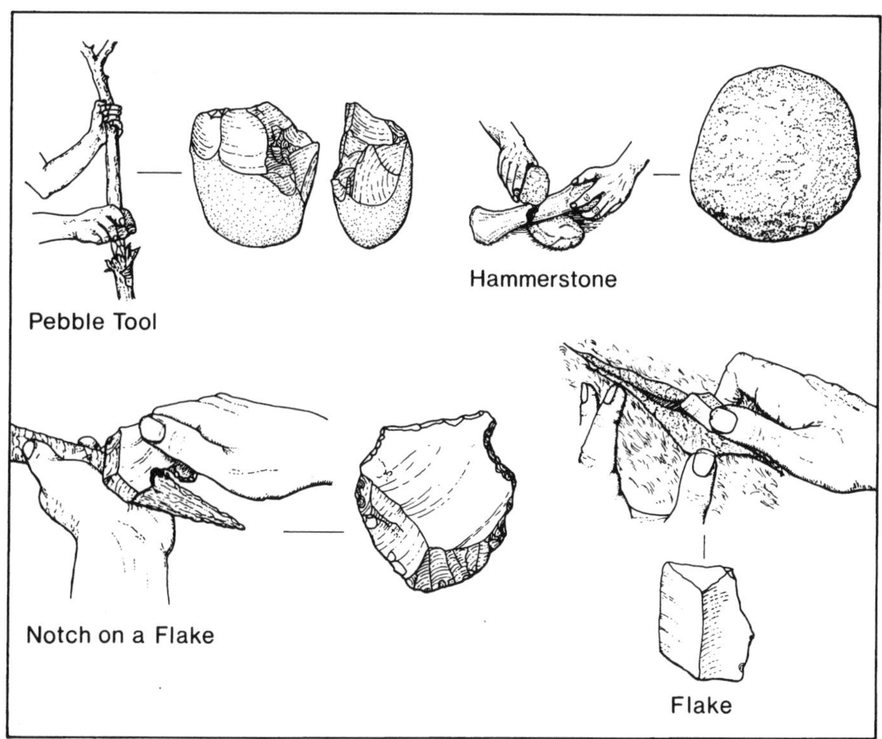

Figure 4.8 A selection of Oldowan tools and suggestions for their use. Similar implements are found throughout the Paleolithic—testament to the need for weight and sharp edges to assist survival.

technological criteria. They consist of river pebbles from which a few flakes have been struck off.[103] These core tools and flakes with sharp cutting edges have been found at a number of excavated sites at Olduvai Gorge. At the Zinj floor Louis Leakey thought *H. habilis* was responsible for their production.[104] The general term Oldowan has been given to similar assemblages of stone tools.[105] These have been found at localities in Koobi Fora, at Sterkfontein and Swartkrans in southern Africa, and at many other sub-Saharan localities.[106] The size of the tools varies from small flakes of a few grams to pebble/chopping tools weighing over a kilogram (fig. 4.8).[107]

The raw materials were always local to the streams and lake margins where the artifacts were dropped. The tools were made from quartzites, basalts, and other metamorphic rocks. While the Oldowan is often referred to as a simple or primitive technology it rather reflects a very low level of effort expended in manufacture. The few blows needed to make a chopping tool or detach flakes are mirrored by the habit of picking up the nearest, rather than the most suitable stone. Raw materials show a very wide range of suitability for flaking ranging from the coarse grained intractable rocks such as those used in the Oldowan to the fine-grained flint and obsidians (volcanic glass), which at a much later time were favored and transported large distances.

The overriding impression from Oldowan stone technology is that it was manufactured to meet immediate requirements.[108] Opinions differ about what these might have been. Finding stone tools and animal bones in excavations has been interpreted as evidence for living floors where large and small animals were killed and butchered.[109] Opposed to this hunting interpretation is the view that early hominids survived as marginal scavengers around carnivore kills concentrated near waterholes and along water courses.[110]

To support this interpretation Lewis Binford has argued that the role of carnivores, such as lion and hyena, is revealed by the parts of the prey carcasses surviving in the archeological sites.[111] Modern studies around African waterholes show a complex but highly repetitive pattern of dismemberment, feeding and transportation of meat on the bone by the neighborhood carnivores.[112] Similar patterns can be traced in the prehistoric bone samples. In this way carnivore behavior is reconstructed. If the hominids were responsible for the bone collections, Binford argues, then we would find very different parts of the skeleton. As a general rule carnivores eat the best parts of the carcass at the site of the kill and only transport poorer, in meat terms, parts of the skeleton to dens for their cubs. Modern hunters generally do the reverse, carrying off the better parts and leaving the low value parts such as feet and heads where the animal was killed and butchered into portable chunks.[113] The various excavated collections of animal remains at Olduvai are, in Binford's opinion, most economically explained as carnivore meals.

The niche that was available for the early hominids was therefore to clean up these carcasses once they had been abandoned. One food resource that would consistently be left even by hyenas with their powerful bone-cracking jaws would be the marrow within the shafts of thick-walled long bones. Binford argues that the selective pressure on behavior to get at this food resource led to the development of stone tools to act as hammers, which combined with simple fulcrums and levers put hominids at a considerable advantage in reaching those parts of the carcass denied to other carnivores.[114]

Evidence for smashing bones open is available. One of Glynn Isaac's excavations at Koobi Fora, FxJj50 dated to 1.5 Myr, produced the splintered sections of a large bovid long bone dispersed over several meters.[115] These could be reassembled and the impact scars observed. Both here and at Olduvai, examination of the surfaces of these animal bones has occasionally revealed the unmistakable traces of cut marks inflicted by stone tools.[116] Carnivore tooth marks are also visible on several of the specimens.

These observations have confirmed Binford's approach to the study of early hominids. He pointed out that although it may seem strange to study hominids by first finding out about hyenas and other carnivores, it is rewarding. He has demonstrated that it is a much more productive route to investigating past behavior than starting with the view that our ancestors dominated the food chain. For example, Su Solomon and Iain Davidson have reanalyzed the early Olduvai evidence and remarked on the large numbers of crocodile teeth in the same, so-called living floors as the hominids.[117] Crocodiles kill and dismember their prey by shaking it in their immensely powerful jaws. This has the effect of spraying parts of their catch over quite some distance. The scattered remains of *Zinjanthropus* and *Homo habilis* in Bed I could reflect how they met their end. Tooth damage visible on the skull fragments of *Homo habilis* OH 7 was most probably inflicted by crocodiles.

The question then becomes whose living floor, since these finds are not primarily the result of hominid behavior. The residues on them come from several sources. Bones from crocodiles and carnivores, stones from hominids and water action. Rather than telling us about early hominids, first and foremost they are a palimpsest of many behaviors which, when unraveled, provide a rich source for understanding the past. Crocodiles in African lakes are no more surprising than hyenas on African savannahs. With such formidable carnivores around there seems little room for either Robert Ardrey's bloodthirsty killer ape or Elaine Morgan's aquatic ape.[118] Either way we will learn more about our ancestors by treating them as part of a diverse animal community, by recognizing their position within and not on top of it.

Research among the modern animal communities of the Serengeti, through which Olduvai Gorge cuts, has also helped to firm up speculations about the possibility of making a living as a scavenger on the well-stocked Pliocene savannahs. By recording the season, location, and the amounts left behind of lion and cheetah kills, Robert Blumenschine has shown that in the dry season and along water courses it would be possible to make a secure existence as a scavenger.[119] In these riparian woodlands the hominids would regularly have come across the remains of heads and untouched marrow bones from medium sized adult herbivores such as wildbeest and zebra. The combination of stone flake for flesh butchery and hammerstones for marrow cracking would have given hominids a considerable advantage in scavenging at this time of the year. Some flesh might also have come from scavenging even larger species such as buffalo, giraffe, and elephant.

Such reconstructions fit in well with the locations of these early sites. Glynn Isaac showed that they are mostly found near bodies of water—lake basins and pans as at Olduvai, Hadar, Olorgesailie, Chesowanja, and Koobi Fora, and river valleys such as the Omo.[120] While this reflects the opportunities for discovery it is nonetheless the case that within lake basins the sites are found along old stream courses where riparian woodlands flourished and scavenging, as Blumenschine showed, would pay.[121]

Isaac's work at Olorgesailie and Koobi Fora was instrumental in showing that this very early archeological evidence should be regarded as scatters and patches on the landscape rather than as sites.[122] The term site is used rather loosely by archeologists who usually have in mind permanent occupation and highly visible remains. This is fine when dealing with Roman villas or Neolithic long houses but does not fit the reality for much of the earlier Paleolithic where artifacts are found sprinkled across landscapes at very low densities.

The patches Isaac referred to represent denser accumulations some of which might be explained by river action collecting stones and bones into one spot. This was not always the case. His detailed excavation at the FxJj50 site shows how little the material in this patch had been moved, requiring a different agency to explain the accumulation. Richard Potts has suggested that the early hominids collected piles of stones together as caches of material for stone tool manufacture.[123] These were visited briefly and usually after food had been obtained. Hence bones were deposited at such locations while repeated visits led to the build up of debris and so to a site. But Potts does not believe that these were campsites comparable to those created by modern hunters and gatherers such as the !Kung San of the Kalahari.

This was not Isaac's view. In a number of influential papers he interpreted these dense patches as home bases from which foraging parties set out and returned with food which was then shared among the group.[124] The critical factor was carrying food and, in his opinion, it was more important as an advance in behavior than what was eaten. Those left behind at the home base could be provisioned with either meat or vegetable foods. It was the commitment to sharing that produced rudimentary social bonds and saw the development of cultural behavior.

At the DK site in Bed I at Olduvai a particularly dense, circular scatter of stones has been interpreted as the platform for a simple shelter that might have been part of one of these home bases.[125] However, others favor water sorting as the natural agency which produced this pattern and have questioned the idea of referring to such patches as either home bases or even living floors.[126] They criticize Isaac for basing his food sharing model too closely on the behavior of modern hunter-gatherers where foraging and settlement are governed by social relationships that we cannot assume existed in the Lower Pleistocene. The denser patches in the landscape are instead the result of the distribution of resources that drew early hominid groups back to some areas more often than to others. They responded to seasonal changes in their local habitat by moving to new food sources as they became available and in this manner would eat their way around the landscape. There were no planned forays from fixed bases to places where plants were known to be growing or where meat could be scavenged and then carried back to provision a central camp.[127]

One aspect still difficult to address concerns the role and amount of plants in the Oldowan diet. No direct evidence exists but the use of seeds, fruits, nuts, and below-ground roots and tubers is likely to have been considerable. Scanning electron microscopy studies of tooth surfaces do show abrasion patterns indicative of chewing such foods.[128] It must be remembered, however, that the different seasons no doubt saw major changes in what was eaten. We have to be certain that wear patterns are telling us more than the abrasive character of the last few meals.

One way to crosscheck is to look at tooth size and structure. The robust Australopithecines have very large cheek teeth, a factor taken to indicate a predominantly seed grinding adaptation.[129] But even so it is not possible to link species narrowly to either restricted environments or diets. Omnivory must be assumed unless proved otherwise.

Many of these issues have been studied in detail by C.K. Brain in his work on the South African limestone fissures.[130] His aim was to unravel the various agencies responsible for the accumulation of materials in the three Sterkfontein valley caves—Swartkrans, Kromdraai, and Sterkfontein itself. These and other cave sites produced many split and apparently shaped bones which Raymond Dart cataloged as bone, tooth, and horn tools used by *Australopithecus*.[131] He argued for killer apes, feeding not only on prey which they had killed and dragged back to their cave but also on each other. These ideas were later elaborated and popularized by Robert Ardrey.[132]

Brain's patient work has helped revise the picture. In the first place he was able to show that large numbers of the Apiths in the cave were the remains of leopard kills. Puncture marks that match the width of a leopard's canines are all too visible in some of the crania.[133] The mouth of the limestone fissures is also a favored, and highly localized, habitat for white stinkwood trees (*Celtis africana*) up which leopards take their prey to avoid competition from hyenas and other social carnivores. As there is every reason to believe they took such evasive action in the Pliocene, then as they fed, so parts of their hominid prey dropped into the cavern below. These very distinctive bone traps answer Brain's question about the Apiths, whether they were "the hunters or the hunted?," in an unequivocal manner. Moreoever, he was able to show that the patterns of bone breakage which led Dart to his claims of an Osteodontokeratic (bone, tooth, and horn) culture were in keeping with the type of damage which carnivores inflict on the bones of their prey during feeding.

However, some loose ends remain after all the natural agencies have been filtered out. There are, for example, a few broken and abraded animal bones at Swartkrans that cannot be matched experimentally by carnivore damage. These Brain interprets as digging implements for roots and tubers.[134] If correct, this would be another example of the environment applying selective pressure to the development of cultural devices. The result was to raise survival by reducing the risk of starvation in seasonal environments.

We have already seen at Swartkrans and Sterkfontein that the hominid samples contain many young individuals less than 20 years of age. Large numbers of baboons are also present. Among the smaller baboons the remains of adults are more common. Converting the age classes into size classes suggests that the preferred prey size for these Pleistocene leopards lay between 13 kg and 45 kg.

It is interesting to see that when the Apiths were the prey this same preference existed. This meant that the leopards preyed on different ages among the robust and gracile Apiths. I have already quoted the age at death figures where the larger *A. robustus* has many more juveniles in the bone collections than the smaller *A. africanus*. Being gracile may well have increased an individual's chances of reaching sexual maturity while for the robusts very few made it to adulthood. The males in each species would have been much more affected by this carnivore preference than the females and it is possible that rather different adult sex ratios existed between the two Apiths' lineages.

No modern carnivore, however, is such a specialist primate hunter. Perhaps, as

Brain suggests, some of the extinct carnivores 2 Myr ago could have filled this role.[135] On the other hand the caves might, as he also argues, have been used by the Apiths as sleeping sites during a seasonal round that led them into the warmer bushveld during the coldest months and back to the highveld of the Sterkfontein Valley in spring.[136] The sites lie at an altitude of 1,400 m and today baboons sleep in a local cave during the cold winter months. Carnivores would be attracted to such roosts and prey on the sleeping primates.

An interesting development in this interpretation comes from recent excavations by Brain and Sillen at Swartkrans.[137] In a deposit dated to between 1 Myr and 1.5 Myr they recovered from among almost 60,000 fossil animal bones a sample of 270 which showed unmistakable signs of burning. Color and surface changes indicate a range of temperatures with the majority heated to >500°C. Antelope bones were the most frequently burnt, but at least one bone from a robust Apith had also fallen in a fire. Their frequency and position point to repeated burning in the cave. The same deposits also contain Oldowan stone tools and bones with cut marks. The only fossil in this part of the site is *A. robustus*, represented by nine individuals. Stone tools, *H. habilis* and robust Apiths are found in older deposits at the site, but without evidence for fire.

In East Africa John Gowlett has excavated an open site at Chesowanja dated to 1.4 Myr. In among animal bones and stone tools made on pieces of lava were some forty pieces of burnt clay. The magnetic anomaly of these lumps is consistent with their interpretation as the result of burning the immediate ground surface by a small controlled fire.[138] It is interesting that the hominid material of the same age from the Chesowanja area is also *A. robustus*.

At Swartkrans the earlier deposit dating to 1.8 Myr with *H. habilis*, *A. robustus* and stone tools raises the question, who made and used the tools? This question was also asked at the Zinj site at Olduvai. Credit was given to *H. habilis*, handy person, on its discovery in 1964. Recent discoveries of hand bones from Swartkrans have allowed Randall Susman to demonstrate that *A. robustus* was fully capable of precision grasping and hence cannot be ruled out as a tool maker on the grounds of lack of manual dexterity.[139] Indeed, the date of 2.5 Myr for the earliest tools at Hadar strongly suggests that species other than early *Homo* were active tool makers and users. The later Swartkrans evidence for fire and stone tools used by robust Apiths now looks very convincing and at the very least warns us against trying to trace the ancestors of later timewalkers through stone technology alone.

Will *Homo erectus* please stand up?

The time between the appearance of *A. afarensis* in the fossil record and the earliest stone tools is almost 1 million years. About the same length of time separates the oldest known tools from the earliest appearance of *Homo erectus* at 1.6 Myr from Olduvai and West Turkana. But as we have already seen (fig. 4.6), there is some concern among taxonomists about this fossil. The key specimens come from Olduvai (OH 9), West Turkana (WT 15000), Koobi Fora (ER 3733), and Swartkrans cave (SK 847) (table 4.2).[140]

Table 4.2 Key fossils from 2 Myr to 1.3 Myr ago, and their designation.

		Existing view	Clarke 1990	Groves 1989	Wood 1992
OH 9	skull	*H. erectus*	*H. leakeyei*	*H. erectus*	
KNM ER 3733	skull	*H. erectus*	*H. leakeyei*	*Homo* sp.	*H. erectus*
SK 847	skull	*H. habilis*	*H. leakeyei*	*Homo* sp.	
WT 15000	skull and skeleton	*H. erectus*		*Homo* sp.	*?H. erectus*
KNM ER 1813	skull	*H. habilis*	*H. habilis*	*H. ergaster*	*H. habilis*
KNM ER 1470	skull	*H. habilis*	*H. habilis*	*H. rudolfensis*	*H. rudolfensis*
KNM ER 1805	skull and mandible	*H. habilis*		*H. ergaster*	*H. habilis*
KNM ER 1803	mandible	*H. habilis*		*H. ergaster*	
KNM ER 992	mandible	*H. habilis*		*H. ergaster*	*H. ergaster*
OH 7	skull fragments, mandible hand bones	*H. habilis*		*H. habilis*	*H. habilis*
OH 12*	skull fragments, maxilla	*H. erectus*		*H. sapiens*	

OH = Olduvai hominid. KNM ER = Kenya National Museum East Rudolf (Lake Turkana) now Koobi Fora. WT = West Turkana. SK = Swartkrans cave fissure. *This fossil is much later in date than the others in the list.

Ignoring what we should call these specimens for a moment, let us see how important they are for the human story. A great deal has been written about the advances this large brained hominid is supposed to represent, and I will look at these in more detail in chapter 7. They are commonly associated with the appearance of new tool types, the Acheulean, as well as being the first timewalkers to leave Africa.

However, set against these achievements are the following three points. Firstly, *H. erectus* is not found with the oldest stone tools. Secondly, for a long time technology did not lead to any appreciable extension of range. And finally, *H. erectus* does not replace the other hominids but shares their environment for several hundreds of thousands of years. The immediate value of such a big brain is not strikingly obvious.

The evidence for this last point comes from sediments at Koobi Fora dated to 1.3 Myr. In these Richard Leakey and his team found a skull of *Homo erectus* (ER 3733) together with a robust Apith.[141] The two fossils could hardly be more different (fig. 4.9). The robust Australopithecine has a small brain case and a massive bony crest down the midline of the skull. This was needed to support the muscles of the jaw with its massive grinding molars. The cheek bones are very prominent while the area of the brain case behind the pronounced brow ridges is pinched like a wasp's waist. The *Homo erectus* skull shows what can be done with this basic architecture and continued down the pathway pioneered by early

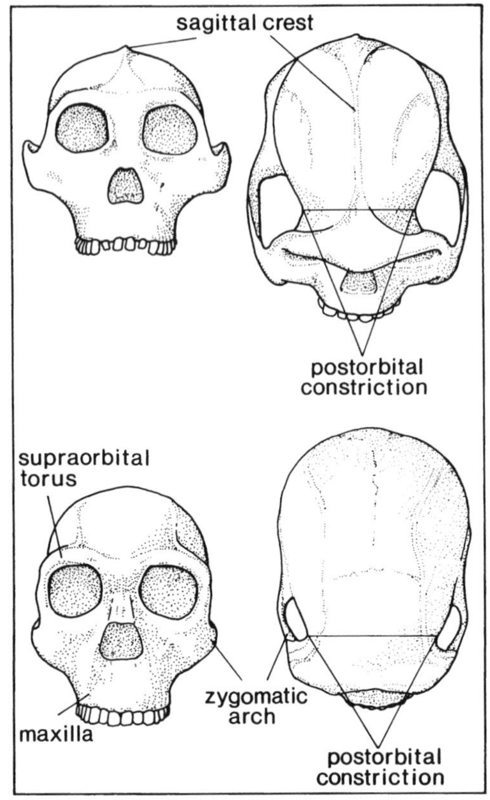

Figure 4.9 The end of the line. A comparison of *H. erectus* (lower) and a contemporary robust Apith from Koobi Fora.

Homo. The skull has a thick vault, flat top, and rugged features. Reduction in tooth size dispensed with the crest and began to pull the face under the brows, which remain massive. The pinched area behind them was now filled out and up, thereby expanding the size of the brain in the frontal areas. Estimates for brain size in *H. erectus* are considerably greater, with a range of between 850 cm^3 and 1,000 cm^3, which puts them at the lower end of the range among modern humans.[142]

The problem the cladists raise with ER 3733 is that it has no uniquely derived characteristic.[143] It is therefore not *Homo erectus* since this fossil does have unique traits. But precisely because it lacks those characters ER 3733 *is* an ancestral candidate for the human lineage. Undifferentiated *Homo* sp. rather than *Homo erectus* may well be a more accurate description. Clarke prefers *Homo leakeyei* for this fossil as well as OH 9 and SK 847. I am not convinced by this scheme since it depends on the undemonstrated departure of *Homo habilis* from Africa. I prefer instead Groves' placement of OH 9 as the African *H. erectus*.

A recent discovery points to the continuing large size of early *Homo*.[144] A nearly complete skeleton from Koobi Fora, WT 15000 discovered in 1984, still earns the nickname "flat head," but even so, and without the help of a pronounced forehead, stood almost 1.68 m tall. What is surprising is his age. Basing the estimate on the pattern of tooth wear and root development, and comparing these with modern humans, the skeleton belongs to a youth aged about 12. However, the length of the long bones from which the height estimates are derived are very comparable to the average found today among white North American *adult* males. This evidence adds further weight to observations that early hominids—both *Homo* and *Australopithecus*—were not only big but had different development rates from modern humans. Among the Apiths root growth in teeth was very rapid with the lower first molar erupting at the extremely early age of 3½ years. As Benyon and Dean point out, among living primates the occlusion of the first molar corresponds to the attainment of 95 percent of adult cranial capacity.[145] Small brain size and faster rates of development are therefore not unexpected. Whether this is the case with a much larger brained hominid such as *Homo* sp. or *H. erectus* remains to be seen.

The size and date of WT 15000 raise a point about phylogeny. At Olduvai *H. habilis* is, at 1.8 Myr old, only 200,000 years older than WT 15000, the earliest *H. erectus/Homo* sp. in East Africa. As we have seen, both large and small habilines existed at this time, and among the later *H. erectus* populations there is good evidence for considerable size variation, no doubt linked to sexual dimorphism.

How should this be interpreted? Is this an opportunity to accept a punctuated, rapid tempo to evolution rather than a gradual evolution towards a new species? The suggestion that *Homo habilis* is not the ancestor of anyone would better fit a gradual view of evolution. Will *H. habilis* eventually prove to be a variety of *H. erectus*—or vice versa? Or might we have to issue more invitations to the party between 2 Myr and 1.6 Myr? Such is the continuing fascination of human origins.

5
Why Africa?

Africa may seem a large cradle, but for some it is still too small. The Russian archeologist Boriskovsky predicts that the cradle can only expand as research in Asia proceeds at a pace to parallel the rate of discoveries in Africa.[1]

But Africa, as we have now seen, has the weight of evidence on its side. In terms of the study of human origins, that is money in the bank. This cradle has weathered the crash of *Ramapithecus* as a hominid ancestor which dealt a blow to claims for a wider crèche in southern Europe, Pakistan, India, and China where such fossils are widespread. The crash was preceded by the dull thud of feet hitting the ground as interest shifted thirty years ago from the Asian apes as a model for human ancestors to those in Africa. Looking back on the move of cradles from Asia to Africa, Campbell and Bernor remark on its coincidence with a major shift in thinking about a key element in human evolution—upright walking or bipedalism.[2] When Asia provided the cradle, then the gibbon and orangutan pointed to bipedalism arising from brachiation. Move the center to Africa and the chimpanzee and gorilla demonstrated another route to the same end by knuckle walking, as elegantly argued by Sherwood Washburn.[3]

Africa apparently has it all—our closest living genetic relatives, the oldest footprints, tools, fire, and a variety of fossil species. But, as I have pointed out in earlier chapters, research into human origins is a fickle field when reputations and careers are based on identifying precisely where it all happened. It can also make monkeys of us all when the cradle is located according to views about the natural order of the world and its peoples. Will future shifts in such opinion make Africa as unacceptable a cradle as Tibet now seems to be?

One way to confront such hypotheses is to answer a very basic but rarely posed question—why Africa? Instead of just pointing to the evidence, the question demands an answer based on our understanding of how species evolve.

Where and how does speciation take place?

According to Endler, speciation can be more rapid in the center of continents than in geographically broken areas.[4] This argument is based on the availability of genetic material for evolution to proceed through natural selection. Centers can be identified, as Darlington argued, by the diversity of life they support.[5] The equatorial and tropical areas would, in his opinion, stand a better chance of producing general adaptations. These are combinations of characteristics that enable a species to colonize many types of habitats and prosper in most. William Brown elaborated these ideas as a theory of centrifugal speciation.[6] Two centers exist for every species, evolutionary and geographic. They are not necessarily coterminous. The evolutionary center has the greatest diversity of genetic material due to larger populations and reliable conditions for breeding.

Fluctuations in population numbers push these genetically more "potent" forms out to the edge. Contraction then isolates them while change continues at the center. The next population push confirms that speciation, shown by reproductive isolation, has now taken place between the center and the periphery.

The size of Africa covers many potential centers. Three of the four subspecies of lion are peripherally located in the Cape and North Africa. These are regarded by Groves as primitive, in the cladistic sense, because of their physical and behavioral features compared to the derived central subspecies.[7] The continuous range of the evolutionary center covers two thirds of the entire continent—ample cradle indeed. However, do we learn anything by identifying it?

The opposite view states that speciation is more likely to occur in peripheral areas, on the edge rather than in the center. Here speciation occurs in another place and is known as allopatry or geographic speciation. Mayr pointed out in 1966 that isolation from the larger parent population can have a bottleneck effect with fewer genes "escaping" from the bottle as part of a founder population.[8] Selection then works on this less varied genetic structure which might also contain some genes in greater numbers than among the parent population remaining in the bottle. Under such conditions the opportunistic process of evolution might lead to rapid divergence, through mutation and selection, and so to new species.[9]

Even critics of allopatric speciation such as M.J.D. White agree that it is the only type of speciation that is uncontroversial.[10] Their criticism is that allopatry is not the only mechanism. The issues are clear when directed against its manifestation as so-called refuge theory, proposed by Jürgen Haffer to explain the diversity of life in the Amazonian rain forests.[11] The repeated contraction of the forests as a result of ice age climate cycles produced refuges into which populations retreated. Once isolated, the rapid speciation of birds, butterflies, beetles, snails, and even, as Meggars once proposed,[12] the cultural differentiation of the Native Amazonians could take place. The similarities with Brown's evolutionary centers are striking.

Critics of refuge theory such as Endler and White point out, quite correctly in my view, that although the forests fragmented and although the tropics are marked by very high numbers of species, this does not mean that without the reduction of the habitat into refuges the accentuated rates of speciation would not have taken place. Even a quick glance at some of the overlapping distributions of modern rain forest fauna, which is the way such ancestral centers are spotted, shows how geographically imprecise these refuges are. These are speciation cradles on wheels, baby carriages without brakes. They are yet another example of origins research which we examined for humans in chapter 3.

Allopatric speciation is not just about refuges but also about the conditions under which isolation occurs. Perhaps, in Ernst Mayr's classic formulation, not enough attention was paid to the patchy nature of large populations.[13] White makes a telling point when he adds in the measurement of *vagility* to assess these demographic units.[14] In ecological time this is the straight-line distance between where an individual lives and dies. In evolutionary time the distance is measured between the point where an individual comes into being through fertilization and where it meets a mate and so creates a new zygote (the fertilized ovum of an animal).[15] Animals may migrate over large distances in search of food and yet find their mates next door, on the same perch or, in the case of bats, shoulder to shoulder on the cave wall. This is also known as philopatry, "love of place," where despite long-distance travel animals return to their birthplace to reproduce.[16] Such low vagility will act to fragment the genetic homogeneity of a continuous population. A small-scale form of founder effect takes place due to the behavior of the organism. On the other hand, high vagility leading to widespread genetic exchange among a large population will maintain existing levels of genetic diversity. In this situation the extreme allopatry favored by refuge theory may well be the route to speciation.

Slow or fast?

Within geological and evolutionary timescales a good deal of discussion focuses on the tempo at which speciation occurred. Darwin described evolution as slow and gradual with imperceptible changes leading to the transformation of animals into new and separate species. Such phyletic gradualism, however, raises the problem of recognizing where one species begins and another ends, not to mention the transitional forms which should link the two.

An alternative, argued originally by Niles Eldredge and Stephen Jay Gould, proposes that the tempo of speciation is rapid and follows long periods of stasis or no change.[17] Such punctuated equilibria can be tested through the branching pattern of species and lineages where attention is concentrated on those points of bifurcation, the moments of cladogenesis. This contrasts with organic tree-of-life models, where everything grades into everything else, and anagenesis—evolutionary change along a single, unbranching lineage—describes the pattern of evolution. With punctuated equilibria there is no need to search for missing links and transitional forms since they never existed.

Critics of punctuation such as Richard Dawkins claim it is nothing more than very rapid gradualism.[18] The evolution of the hominids, which was used as an example of punctuation, could, they claim, be interpreted by phyletic gradualism at varying rates.[19] The supple mathematical propositions of catastrophe theory enable Zeeman to argue that abrupt change can always be more parsimoniously explained as a continuous, steady process.[20] While this latter hypothesis contributes to our understanding of the nature of change, it suggests that both viewpoints concerning the tempo and mode of evolution can be accommodated in a single model. Since the propositions are so fundamentally different this in turn would suggest that the significant differences between them have yet to be adequately modeled by mathematicians. As Elisabeth Vrba has noted, supporters of punctuated equilibria suggest that faster phenotypic evolution *is* associated with the branching of the tree of life. In contrast, those who favor phyletic gradualism, while acknowledging that there are phases of fast evolution, do not propose this specific association with lineage splitting.[21] Here is a crucial distinction that should not be muddled.

The challenge of vicariance

A central tenet in the Darwin/Wallace tradition of biogeography, discussed in chapter 3, is that species arise and then disperse. The means of dispersal are many and various. Landbridges were often invoked to explain separate distributions of closely related forms. On the other hand, Darwin presented evidence that icebergs could move plants around the Arctic regions thereby accounting for similarities in the flora of South America, Australia, and New Zealand.[22] A century later Simpson elegantly demonstrated that landbridges were not necessary to account for the unimaginable.[23] He calculated that if the chances of a group of animals crossing a barrier, such as an ocean, were calculated at one in a million in any year then some 700,000 years must elapse before the chances of crossing become equal to the chances of not doing so.

An alternative biogeographical tradition to that of Darwin and Wallace has been assembled from the elements of continental drift, the model of allopatric speciation and cladistic taxonomy. Under such a synthesis vicariance (or replacement) biogeography has attempted to set a radical agenda to what several of its supporters see as a discredited imperial tradition based on the idea of centers of dispersal.[24]

A key figure in the vicariance synthesis was Léon Croizat, an independent thinker who in several massive, glutinous volumes set out the framework for what he called a panbiogeographical approach, even though by his own admission in his 881-page "summary", *Space, time, form*,[25] "fully 99% of the pagination of my works is wind, trifle, piffle, tripe, rot, stuff, in sum, entirely unworthy of the attention of a serious scientist."[26] Distilling these tomes down to manageable proportions allows a few simple and useful ideas to emerge.

The first is the observation that the distribution of modern and fossil taxa is frequently disjunct—separated by oceans, mountains, and other barriers. In his panbiogeography Croizat joined up such fragmented patterns by what he called a track. This simple device reveals that for widely different species and life forms there are a number of repeated tracks. He concluded that common processes in the history of the earth have produced similar tracks for widely different taxa such as plants, insects, birds, and mammals. Others have pointed out that his tracks follow the rims of ocean basins where separation and fragmentation through tectonic activity is to be expected. As a result Croizat concluded that "Earth and life evolve together."[27] The geological processes that shape and mold the earth are responsible for the pattern, diversity, and creation of new life forms upon its surface. This view would have been familiar to James Hutton in the eighteenth century as he proposed his cycles of geological time.[28] It is a view evident in today's Gaia theory where the earth is seen as a single organism by James Lovelock.[29]

The second concept to emerge from Croizat's thesis involves a simple formula,

$$\text{Evolution} = \text{space} + \text{time} + \text{form}.$$

This reflects his view that biogeographers are "students of the effects of space and time on the course of organic evolution."[30] They can set about this task because geographic distribution contains the record and dispersal interprets it. This may seem little different from Wallace or Darwin but Croizat had his own definition of dispersal as "translation in space and form making." Translation in space does not mean migration and may not even involve such a process. Instead, it refers to the distribution of the taxon or taxa under investigation. While the terms are unnecessarily confusing, the concept behind them is again simple. By tracing his tracks for all living taxa Croizat determines that "Nature forever repeats."

The cause of the repetition reflected in the geographical distribution of plants and animals is speciation by what Croizat called immobilism. This is the heart of vicariance biogeography with its notion that species evolve *in situ*, that is by replacement, and that this occurs through separation.

As an example, faced with explaining the variety of species and subspecies of birds and tortoises in the Galapagos Islands, Croizat argued that it is a classic case of how a widespread parent group has evolved into highly localized varieties due to the fragmentation of an originally homogeneous population by earth history,[31] in this case volcanic eruption forming new islands and changing the shape of others. Furthermore, this has happened many times in the past with unrelated taxa, such as the iguanas, which did not migrate to the islands but were always present. The earth history of this part of the Pacific has repeatedly produced the conditions allowing form making, his phrase for speciation, to occur *in situ*. Alternatively, Darwin explained the varieties of finches and tortoises as new adaptations which emerged after their pioneering ancestors arrived as they subsequently radiated into the new and available niches. Reproductive isolation could then develop leading gradually to new species.

The dogma of vicariance, as its detractors have dubbed it, therefore argues for a passive response by organisms to changes in physical geography. Vagility plays no part in strict vicariant texts. It is a non-selectionist creed since there is no adaptation but rather the play of earth history upon the evolution of new life forms. Speciation is akin to an exaptation of geological processes, the unintended consequence of deeper rhythms unaffected by potentials for migration, dispersal, or colonization (table 1.2).

Those loyal to the older dispersal tradition stress the active movements of organisms to account for changes in biotic distributions.[32] The disagreement, as Erwin points out, centers on the major question of whether separation necessary for allopatric speciation to take place occurs either through dispersal, leading to adaptation and subsequent reproductive isolation, or by geological interruption.[33] According to dispersal theory the appearance of new animal taxa only comes some time after the creation of a barrier, while the vicariant thesis would expect them to be the same age as the barrier.[34]

The coevolution between an organism and its environment, which John Odling-Smee favors, provides another approach to this question of how reproductive isolation, which could lead to speciation, occurs.[35] By pointing out that animals construct niches and pass on information about them to the next generation, he argues that the physical environment is not the only force for selection. In other words, animals, to some extent, change themselves. They are not just passive lumps of genetic putty molded into new forms solely by the environment through the mechanism of natural selection. Instead of simply adapting, organisms are also building. In this sense niche construction by animals can be seen as exaptive or, put another way, as a form of active vicariance. Fragmentation of a once continuous population might require changes in vagility from high to low values. This might or might not need changes in migration and dispersal patterns. In this case separation of once common breeding populations stems from behavioral as much as physical isolation. Longterm survival is not just a game played by animals against the environment but also one played against themselves.[36]

Ways of looking at the question

Center or edge, here or there, slow or fast, active or passive—the list of differing views is long. I will now return to the question, why Africa?, and examine it by looking at the environmental evidence for separation and climatic forcing before turning to the behavioral issues of dispersal and colonization.

Separation and rift: an adaptive story

The manufacture of the African cradle is generally put down to a salinity crisis at the end of the Miocene (table 3.2).[37] This followed the closure of the Straits of Gibraltar which resulted in massive evaporation in the Mediterranean so that it became like the Dead Sea. Huge deposits of salt were rapidly laid down and during several dry land phases animals moved between Europe and Africa.[38]

This Messinian salinity crisis did not necessarily put a big cork in a very large bottle, even though the addition of deserts blocking the northeastern corner must have helped. It did, however, coincide with the appearance, between 5 Myr and 6 Myr ago, of the East African savannahs toward the end of the longterm deterioration in Miocene climate which, according to Desmond Clark,[39] confined hominids to the tropics. Many feel that such a catastrophic event had sufficient knock-on effects to produce in sub-Saharan Africa the environmental conditions for local speciation.

Within this part of the African bottle the general Miocene cooling led to the fragmentation of the forests, when applied to the plateaux, and more generally to the disruption of the zonal belts of vegetation across the continent. The critical balance between forest and savannah tilted in favor of the latter so that today it comprises some 65 percent of all African vegetation. The savannahs are highly varied and extend in a massive arc for 13,000 km from Senegal to the Cape.[40] According to Kingdon,[41] the suitability of East Africa for all animal speciation depends on the proportions of this environmental mosaic rather than its composition, which has varied little since the Miocene.[42] The factors controlling this balance between forest and savannah and the isolation of animals in vegetational refuges would be:

1. the physical barriers stemming from uplift and rifting;

2. the geography of the deep rifts, lakes, and mountain massifs that act to isolate populations within a regime of great ecological diversity determined by altitude and climate;
3. the climatic cycles of the Pliocene and Pleistocene that have repeatedly isolated animal populations adapted to either wet or dry conditions.[43]

These controlling factors are obviously not unique to Africa. Only in the Rift Valley is the local combination of tectonic and climatic activity believed significant for the high rates of mammalian speciation.

The northern savannahs provide the cork in the bottle. They are relatively simple and homogeneous in contrast to the southern savannahs which are divided up into a complex pattern of mosaics. The vegetational differences are reflected in the mammals endemic to the two zones. In the north there are only ten species, in the south twenty-three. On this evidence R.C. Bigalke concludes that the northern exit from tropical East Africa "provided fewer opportunities for speciation and may have been unattractive to potential colonizing species from the south."[44]

Not everyone agrees. Karl Butzer has argued that the diverse group of hominids, described above in chapter 4, was found in eastern and southern Africa precisely because it was "an area least affected by the large-scale environmental dislocations of the end-Tertiary [late Miocene to Pliocene]."[45] An adaptive radiation into the varied opportunities of the vegetational mosaic led to hominids evolving side by side, rather than in isolation by allopatry.[46]

Walking out of forests

At these large scales it is difficult to decide on the mechanisms involved in speciation. With a finer focus the disruption of local habitats at least provides the classic conditions for allopatric speciation through isolation, either by dispersal or vicariance. Climatic cooling was not the only mechanism that led to the fragmentation of habitats into a regional mosaic of highly varied types. It was assisted in East Africa by rifting and faulting and the impact these had on altitude and topography. Intense fragmentation was the result.

The break up of the Miocene forests has been seen as the selection pressure on hominids to evolve novel mechanisms for dispersal outside those areas which still supported continuous tree cover.[47] During the Miocene there were fewer arboreal habitats and larger gaps between them, while in the Pleistocene Kingdon has shown how the climatic cycles repeatedly produced three main areas of forest refuge in East Africa. In the Miocene, cooler climates led to greater seasonality on the plateaux while the rifting produced many microhabitats. The result was a landscape of island-like habitats; moist, forested areas capable of supporting hominids set in a sea of open, drier conditions. Whether populations now became marooned on their islands (ideal settings for allopatric speciation), depended on their response to selection from these environmental conditions. They could opt to stay and feed within the patches of forest, concentrating either on shoots and leaves at ground level, a pattern followed by most gorillas, or feeding on fruit in the trees like chimpanzees.[48]

These two species have, of course, been subject to the constant fragmentation and regrowth of their habitat. Neither, as David Pilbeam points out, furnishes an exact model for a prehominid ancestor.[49] This, he suggests, would have been as or even more arboreal than the chimpanzee, although less well adapted to knuckle walking, and not as bipedal as the Apiths.

Another option was to move between the patches and use the open spaces in addition to the islands of trees. The distance between food patches and the dangers involved are thought by several authors to have provided the selective pressure needed to produce a novel form of dispersal strategy, namely bipedalism.[50] However, dispersal is not just about locomotion and travel as we shall see later.

Other selection pressures need to be found. Peter Wheeler has recently suggested that bipedalism emerged as an adaptation to searching for food in equatorial latitudes.[51] Quadrupeds, such as the African predators, have difficulty keeping cool during the hottest part of the day since their four-footed gait exposes a large surface area to the vertical sun. This is why many of them rest in the shade and hunt at other times, often at night. Furthermore, the brain is difficult to keep cool, increasingly so as this organ becomes larger. Many of the grazers have solved this problem by evolving a nose that acts as a "radiator" by evaporating water from the moist linings of the nasal chambers. Panting increases air flow and hence evaporation. The blood supply to the brain is kept cool through the pattern of carotid arteries beneath the brain which act as a "heat exchanger" by transferring heat to the sinus where it can be cooled.

In the absence of a muzzle, bipedalism helps solve some of the problems. An upright stance under the noonday sun reduces the exposed surface area, loses heat faster and takes advantage of breezes. The highly evolved human sweat glands make a further contribution, as does head and shoulder hair which has the effect of coping with heat stress by acting as a shield. All of these help to keep the brain and body cool.

The niche that such adaptations to thermoregulation open up is the capacity to forage at midday, when competitors are taking their siesta. This is the time when the early hominids could reach waterholes, scavenge carcasses, find plant foods, and kill other small animals with some degree of success. But Wheeler points out that bipedalism allowed hominids to forage farther faster. They could explore more patches for food and water. They were able to extend their ranges.

Nonetheless, the ranges they foraged across were not their exclusive niche and, as James Steele has shown, carnivore pressure must have had an important selection role on group size for protection, social competition, and, as we shall see in chapter 6, on the evolution of intelligence.[52]

Climatic forcing and speciation: exaptive partners

The hominid/chimpanzee split 5 Myr ago coincided, as Malone points out, with the height of the fragmentation of the environment in the Miocene.[53] This is also the timing of the Messinian salinity crisis. To what extent are these climatic and environmental events related to the appearance of hominids?

I have already discussed in chapter 3 Wallace's idea that climatic cycles forced the pace of evolutionary change.[54] The longterm cooling of global climate has often been cited as a forcing mechanism in the production of new species. Northern animals were driven south and southern ones had to compete more keenly and more frequently. The increasing northern cold was diagnosed by Lull in 1917 as an impelling cause for migration of the primates and hominids from the roof of the world in Tibet. For Lull the pulse of life, the expression points of evolution, had a fundamental principle, "that changing environmental conditions stimulate the sluggish evolutionary stream to quickened movement."[55] Life, in the form of speciation, was forced along.

Faced with such conditions animals either migrated with their habitat or adapted through behavioral means to such dramatic changes. This would lead either to the ebb and flow of a species within an area or to continual occupation. According to Elisabeth Vrba, the latter would be the generalists, able to persist in an area through the climatic oscillations.[56] They differed from the specialists whose ranges were determined by the contraction and expansion of preferred habitats and food resources. For example, as the tropical rain forests shrank to their glacial refuges so, presumably, the ancestors of chimpanzees and gorillas kept to and tracked the dwindling habitat rather than adjusting to new, open conditions. But primates were not the only animals coping with these major climatic shifts. By the end of the Miocene the sub-Saharan fauna included many new grazing species and carnivores. The opening of the forest led to larger herbivores with novel mating patterns linked to these more open conditions.

Among grazers the prevalence of either monogamous or polygynous mating systems depends on such factors as body size and the abundance and location of food resources. In open environments today group size among herbivores tends to be larger as a defense against predators while sexual dimorphism, as measured by body size, is much greater among the polygynous groups. It is likely that some grazers developed large horns as armaments in habitats where food and mates were sufficiently concentrated so that a male could defend a territory and control females.[57] Other grazers did not. Unicorns aside, the lack of horns among horses can be explained by their single stomachs. Unlike the multistomached ruminant they meet their grazing needs by covering much larger ranges which makes such territorial defense impossible.

Elisabeth Vrba has examined in detail the timing of speciation among the African antelopes, gazelles, and buffalos, or bovids as they are collectively known.[58] Today, Africa has some seventy species of bovids which account for 40 percent of the world's hoofed mammals.[59] The trademark of the bovids are their horns which are a striking visual component and assist in the recognition of species not only by zoologists and paleontologists, but more importantly by the animals themselves.[60] This *recognition concept*, developed by Hugh Patterson,[61] is one way to explain how reproductively isolating mechanisms can emerge in a breeding population. Closely related species can often continue to produce fertile offspring. The important point is that they choose not to. Recognizing mates through either visual and olfactory cues, or via greeting and courtship behavior sets up effective barriers. Eventually this can lead to changes in the pattern of gene flow and the reproductive isolation of populations that should still be homogeneous. This pathway to speciation is therefore exaptive rather than adaptive in character.

How does this work? Vrba points to a marked evolutionary pulse for the bovids between 2.4 and 2 million years ago which coincides with the onset of major glacial and interglacial climates (chap. 3).[62] As an example, the antelope tribe Alcelaphini, which includes such species as wildebeest and hartebeest, has twenty-eight living and fossil species the majority of which appeared during the past 3 million years.[63]

The accelerated pace of bovid speciation is matched by the adaptive radiation of the hominids, and Vrba suggests that these major speciation events were forced by the onset of climatic rhythms which accelerated the interval between arid and wet conditions.[64] The recurrence of alternative environments, each with their different distribution of patches of food, was all important. It was not so much the regular switch between forest and savannah that led to an increase in speciation among some animal lineages, but rather how the resources were distributed in those alternative environments. If they were clumped in the forests and clumped on the plains then little selection for change occurred and as a result those lineages show slow rates of speciation. For example, the aardvark that feeds exclusively on termites and ants finds its food in many different types of environment. Not surprisingly, aardvarks are found throughout Africa, but only in Africa, and the Pleistocene climates never bothered them.[65]

How can we explain the very high speciation rates among the bovids of East Africa? Vrba makes the crucial point that speciation is an effect of evolution at the macro scale of combinations of evolutionary processes such as natural selection at the micro scale of organisms and their genes.[66] It is therefore exaptive not adaptive. Specialist grazers can be very sensitive to climatic change replacing their favored environment with an alternative. Unlike the aardvark their food supplies are not found in a wide range of different vegetational types. Vrba argues that macroevolution selects those characters in an animal that increase the range of diet so that any shifts in environment can in future be met by generalists who adjust their feeding strategies accordingly.

Consequently, the elaboration of horns was neither an isolated nor a primary element under selection from 2.5 to 2 Myr ago when a host of new adaptive niches became available. They may have been required for defense but then became exapted as part of the recognition process. The effect of selecting generalists at the macroevolutionary scale produced the conditions where the potential for diversity in headgear was unleashed. As the horns became differentiated the number of species increased rapidly since it was through such visual means, the recognition effect, that they were defined.[67] As a result the diversity of life's patterns, represented here by the horns which bovid species have evolved, are incidental rather than adaptive to new conditions.

Occupancy is the name of the game in either adaptive or exaptive radiation. An unforeseen and hence exaptive result is the selection for those elements of the phenotype, such as horns, which lead inevitably, but coincidentally, to a high species diversity. Since the occurrence of gazelles and antelopes is predominantly, but not exclusively, African, these high rates of speciation have to be placed in the context of macroevolutionary patterns in the sub-Saharan part of that continent. Christine Janis has pointed to the rarity of fancy horns in the North American ruminants and argues that in these more northerly latitudes the fragmented mosaic environments, so characteristic of East Africa, never existed.[68] In North America more arid and seasonal habitats led to wider spacing between patches of food so that territorial defense by alarmingly armed males was rarely possible. Here the development of more general diets did not result in the incidental effect of a major pulse in speciation because there was no comparable exaptation for those all-important visual clues, such as horns.

Dispersal and healthy behavior

Valerius Geist has also studied large northern mammals, principally mountain bighorn sheep, whose horns are impressive armaments.[69] He adds an adaptive twist to the exaptive tale by linking speciation and dispersal to a general theory about health.[70]

The foundation of his theory rests on the observation that body tissues have different growth priorities. Those that give phenotypic distinctiveness such as horns, antlers, and coat coloring are generally a low growth priority. Faced with food shortages animals will use bodily reserves to support critical life functions rather than improve their appearance. However, it is through these phenotypic differences, as we saw with the recognition concept, that species recognize themselves and are recognized by others.

The investment in and elaboration of such "ornaments" are therefore closely tied to the dietary resources in the animals' habitat. Geist argues that individuals are constantly seeking to disperse under the selection pressure that impels them to encounter new environments in the hope that these will reveal rich food supplies. Such favorable environments will allow low growth tissues to be supported, reproduction potential among individuals to be enhanced, and so, finally, speciation to occur. Hence his theory that "health is maximized when diagnostic features of a species are maximized phenotypically."[71] Speciation is good for you because it occurs in environments that allow the full potential of the phenotype to be realized. As Peter Grubb puts the case for evolution among the African mammals, what matters is the phenetics, not the genetics, of speciation.[72]

Two types of phenotype are described by Geist.[73] The *maintenance phenotype* represents the holding phase. Population numbers are generally balanced to the available resources. The dispersion, often achieved through spacing mechanisms involving territories and their defense, leads to a balance where food supply is optimally regulated. There are few shocks to the system since the number of young produced is strictly limited to what the habitat can support. Elsewhere, V.C. Wynne-Edwards has argued,[74] this represents a case of group selection where individuals cooperate to regulate their food supply. According to him, this increases their reproductive fitness against that of non-cooperators, who exploit "similar habitats for personal gain without regard to the consequences."[75] Cooperation results in greater survival of offspring. This in turn justifies high levels of parental investment. According to this hypothesis social behavior is crucial as the reference framework by which animals evaluate the environment which includes food, competitors, predators, and how members of their own species are packed in. For Geist these will be saturated environments, densely packed and as a result additional resources are scarce.[76] Such environments place animals under "K" selection (chap. 3) where in ecological time environmental pressure is buffered through behavior. Niches are constructed and relations regulated. The diversity of species and saturation by population numbers lead, in Geist's view, to the maintenance phenotype producing communities of specialists. Similarity breeds content, as Vrba argues.[77] The genetic architecture of the maintenance phenotype will be controlled, as Peter Parsons has demonstrated, through regulatory genes.[78] Small populations existing in or invading a habitat similar to the parental one are unlikely to speciate since there will be scant selective pressure on the system of how a species recognizes its mating partners. Exaptive radiation, and evolutionary pulse of new species, does not take place.

Geist's alternative is the *dispersal phenotype*.[79] The key concept remains the same: individuals seek every opportunity to disperse and, when successful, encounter environments which for them are super-abundant in food because there are few competitors and predators. These are not like the parental homelands. Under the dispersal phenotype reproductive output is increased as well as exploratory behavior and mobility. Changes in a few structural genes with large additive effects support the status of a colonizing species. They need to be resilient rather than fine-tuned as is the case with the regulatory genes buffering the maintenance phenotype. As Parsons argues, the effects of additive genetic variability is maximized under conditions of extremely *hard* selection.[80] Hard selection sees many instances of local extinction. The survival threshold is a grim reaper. In contrast, *soft* selection weeds out a certain proportion from each local population, thereby maintaining a more varied genetic matrix within the region. Occupancy continues in each local area so that recolonization is not on offer.

Phenotypic development is maximized not only by the rich resources which can support low priority tissue but also through the sparcity of population. Genetically distant mates are encountered and founder effect occurs. As a result speciation will be rapid. These are described by Geist as generalists. Individuals can be large—animal giants are a feature of the dispersal phenotype. Neoteny, offspring born at increasingly early stages of development, is a pronounced feature.[81] The changes in reproductive output point to an organism that is under "r" selection in ecological time (tables 3.1 and 5.1).

Table 5.1 Two phenotypes from the perspective of two different but related concepts of time.

	Ecological time	
Maintenance	*Phenotype*	*Dispersal*
"K"	Ecological selection	"r"
Less	Environmental variability	More
Stable	Ecological character	Resilient
Behavioral	Route to fitness	Genetic
Specialists	Behavior	Generalists
Density dependent	Population	Fluctuating, expanding
Calculated	Migration	Non-calculated
	Evolutionary time	
Maintenance	*Phenotype*	*Dispersal*
Central	Population position	Peripheral
Low	Population vagility	High
Regulatory	Control genes	Additive
Soft	Population selection	Hard
Specialists leave	Local response to environmental fluctuation	Generalists stay
Less	Seasonality	More
	Colonization	
Adaptive	Radiation	Exaptive
Climatic/external	Forcing/selection	Behavioral/internal
Low	Evolutionary rate	High

With the two phenotypes illustrated in table 5.1 we have another answer from Geist to the question, why sub-Saharan Africa? Specialists are found in rich productive environments while generalists are in the less productive. Hence, "Zoogeographical dispersal is likely to be a one way street from the tropics to the arctic, from the humid to the dry, from the productive landscapes to less productive ones."[82]

Macroevolutionary effects: stasis and bipedalism

Few lineages make the whole or even a significant part of Geist's global journey. They also change along the track as the increasing body size and armaments of the large, horned animals of the north—ibex, wild goats, and sheep—display.

For the moment let us just contrast Geist's model with the macroevolutionary effect of speciation put forward by Vrba.[83] Both have used the descriptions generalist and specialist with selection favoring the former either to broaden its dietary range or maximize potential for dispersal. The effect of such selection is, in both models, a high evolutionary rate within a lineage (table 5.1).

The trick, of course, is explaining how, in geological/evolutionary time, a lineage switches from being an upwardly mobile, go-getting generalist to a security-conscious, pension-watching specialist, or vice versa. Mayr referred to this switch as the "genetic revolution," where the genetic balance in a population was disrupted by the loss of systems that had previously acted to buffer it against change.[84] He argued for the inertia of large numbers, where gene flow between the local populations of a species would provide such a balance. High vagility must be assumed for it to work. The inertia control would be considerably relaxed in those peripheral, isolated populations where dispersal leading to speciation is thought to occur. Most importantly, any new species would emerge with their own balancing systems, both behavioral and genetic. The implication is, therefore, that the dispersal phase is short lived and that maintenance phenotypes regain control in the absence of any further selection. Put another way, the behavior involved in niche construction with its investment in learning can be swamped by the genetic route to fitness but, in the long term, stability is restored.

Vrba and Geist both have a similar solution to the removal of these checks and balances. The climatic pulses at the onset of the Pleistocene ice ages 2.5 million years ago put the finishing touches to the fragmentation of the East African forests that began in the Miocene. Selection for generalists on the open savannah environments now proceeded apace.

Geist also predicts that waves of new species should coincide with climatic and geological revolutions. At these times the magnitude of the changes are such that the behavior patterns linked to the maintenance phenotypes can no longer buffer the genome—the total genetic material inherited from parents—against selection. Prior to this any minor alteration in an animal's environment were dealt with by corresponding changes in behavior since this represents the cheapest means of adjustment. Rather than alter the genome under such minor selection to produce, say, upright walking, the response would now be to modify behavior as far as the obvious constraints of anatomy and function allowed. But massive environmental change swamped such behavioral defenses leaving the genome exposed to selection for change rather than stability. At such times the potential for dispersal is finally realized and adaptive radiation results. The alternative is much higher rates of extinction among local populations. Selection, according to this adaptive argument, favors the generalist. The exaptive alternative would be that selective pressure was temporarily suspended thereby encouraging a great deal of genetic plasticity that only starts being selected when the environment has returned to some stable equilibrium.

This does not mean, however, that the mechanics of hominid migration, upright walking, made such changes possible. Instead bipedalism, as the instrument of colonization rather than migration (table 1.1), may be another macroevolutionary effect; incidental rather than intentionally adaptive. Our mechanism for migration does not explain our history of dispersal. The issue is whether we have been designed, or had a hand in that designing by building through learning our own environments which thereby made natural selection a less direct force for change.

Specialists and generalists

How do the early African hominids fit in to this bigger picture of speciation? Is their radiation into two genera and several species the result of incidental effects forced by the climate? Or did the fragmentation of the forests select for rapid speciation as adaptive opportunities arose?

A good deal hinges on accurately assessing the types of environment in which the fossil species have been found. Despite the difficulties Vrba has made a division, not agreed by everyone, where the smaller Apiths, *A. africanus*, belonged to wetter, more closed woodland vegetation.[85] In support of this model there is some skeletal evidence, discussed by Susman and Stern.[86] This shows that just as *A. afarensis* before them, they were still adapted for a significant amount of climbing.[87] By contrast the larger forms *A. robustus*, *A. boisei* and *H. habilis* were adapted to drier conditions and more open vegetation. This second group would be subject to greater seasonality in food shortage and the distribution of critical resources such as water. This is not a crisp distinction between forest and plains species but rather took place along the sliding scale between moderately open to open savannah environments.[88]

The Sterkfontein Valley in South Africa provides the evidence. The deposits where *A. robustus* and *Homo* have been found are linked with bovids from dry open habitats while the gracile Apiths lived under wetter, more wooded conditions. The leopards that fed on both groups are a good example of a specialist carnivore able to cope with alternative environments because they do not mind what they eat so long as the parcel of food is the preferred size (see chap. 4). Such specialist feeders show little brand loyalty. They compare well with the aardvark; as long as it is termite for dinner they are not bothered by which type of termite.

When the early hominids were feeding they could either stick to a particular package size, subject to the amount of time and energy it took to search, find, and consume it, or to some reliable brand names such as grasses and fruits which come packaged in distinctive colors, smells, tastes, and textures. They also had the advantage of omnivory for solving problems of seasonality and inevitable shortages. As mentioned above, the options they faced ranged from being within feeding-patch specialists to between-patch generalists where the open areas are as important for foraging as the food resources in the clumps of trees. The latter option means that the risk of food failure within a patch is spread more widely. What is needed, as we shall see below, is a migration mechanism to implement such a strategy.

Macroevolution (evolution above the species level), as Vrba suggested, will select for those species that expand their breadth of diet:[89] the generalists who can move between the bush and the dry, open plains as well as cope with longterm climatic change.[90] This is the role that Vrba assigns to early *Homo* while the robust Apiths (*A. robustus* and *A. boisei*) are cast as specialist vegetarians whose massive molars were well suited to masticating the tough plants of the open grassland.

If this scenario is correct then it follows that one effect would be a rapid rate of evolutionary change among *Homo* while the fossil Apiths would exhibit much less change. In chapter 4 we encountered robusts and super-robusts (*A. boisei*) from a number of findspots and as early as 2.5 million years. Largely unchanged forms are also found side by side with *Homo* at Koobi Fora KNM ER 3733, as late as 1.3 million years, providing more than a million years of living in the same areas. This spans the onset of glacial climates and greater seasonality, and yet does not produce any appreciable speciation within the robust fossil lineage. On the other hand, irrespective of how *Homo* is derived, there has in the same period of time been either accelerated evolution from *A. africanus* through to *Homo* sp. or greater frequency of speciation events within that lineage.[91]

This may sound like a rather circular argument to prove an evolutionary point from a fossil record that is often bitterly contested. A second test of the model offers much more hope and is central to the interests of this book. This follows Vrba's prediction that the regional histories of these generalists and specialists should differ quite markedly.[92] With the recurrence of those alternative environments the narrow specialists should ebb and flow out of and into an area. As their preferred foods became rationed due to climated change then population became locally extinct and recolonization only occurred when in due course conditions changed for the better. On the contrary, the regional record for the generalist early *Homo* would show that they hung on through the climatic changes and left behind evidence for the continuous occupation of an area.

This will be a difficult hypothesis to test. Sampling these ancient landscapes is never easy. Grant proposals to search for negative evidence are usually unsuccessful. Moreover, negative evidence may result from the destruction of material rather than accurately reflect the distribution of hominids. However, I shall show with later timewalkers that it is possible to come to such conclusions in areas with longer histories of research and shorter time spans.

Can you force hominids to change their ways?

The link between a major shift in the pattern of climate and speciation was explored by C.K. Brain in 1981.[93] His observation is one of the first uses of the new deepsea chronology, not only as a timescale but also as a model for longterm environmental change initiating the conditions for allopatric speciation.[94] The timing is certainly suggestive (fig. 5.1). Climate appears to be acting once again (chap. 3) as the forcing agent in evolution. Elsewhere Neil Roberts has made a similar correlation between a shift in the tempo of Middle Pleistocene climatic pulses and the appearance of hominids outside sub-Saharan African between 800,000 and 900,000 years ago.[95] The first appearance of timewalkers outside of Africa coincides with the radiation of the spotted hyena and lion out of Africa and wolves from the Middle East. As Alan Turner points out, this faunal wave which swept them north took place at the beginning of the Middle Pleistocene some 730,000–780,000 years ago.

But do such correlations reflect climatic forcing and support Vrba's view that "evolutionary events are a direct function of environmental change?"[96]

The answer is that forcing can only be demonstrated by a timelag. Climate takes time to respond. Ice ages are not instantaneous. The only immediate effect of such climatic forcing on speciation events or range extension would be extinction. In chapter 3 I discussed the Milankovitch theory of the ice ages. Here we saw that the lower amplitude cycles are indeed forced by changes in the earth's orbital geometry since there is a clear timelag between the astronomical event and the pulse in the cores which signal it took place. But this is not the case

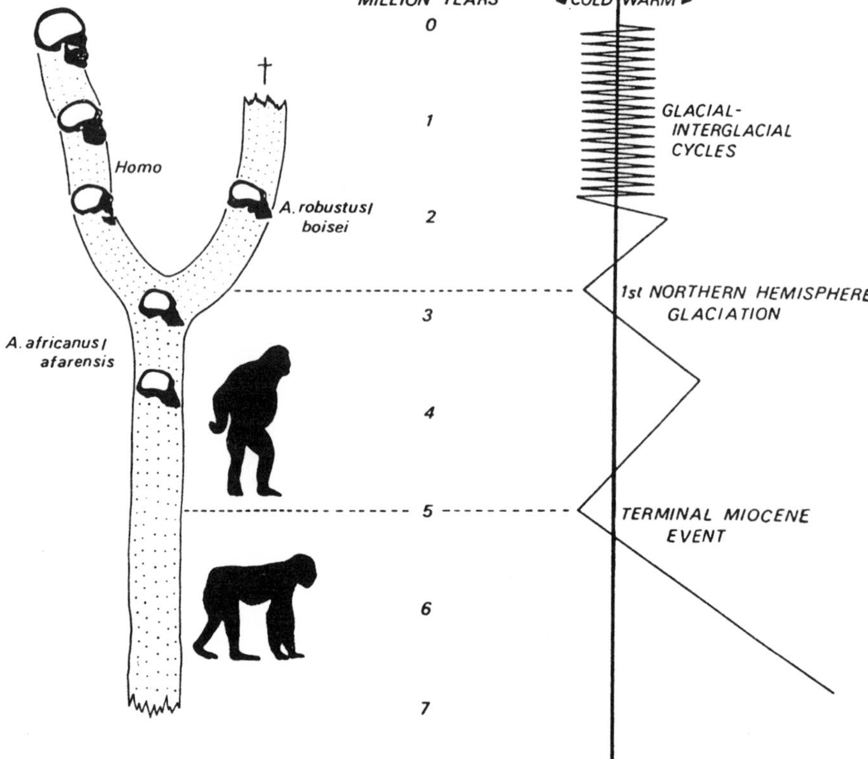

Figure 5.1 C.K. Brain's suggested link between speciation and ice age rhythms. The end of the Miocene is linked to bipedalism while the first northern hemisphere glaciations force the split between *Homo* and the robust Apiths.

with the 100,000 year cycle. This is completely in step with the deepsea climatic data. The geometry of the earth's orbit cannot be a forcing agent if it is so well tuned. Consequently, W.F. Ruddiman, M.E. Raymo,[97] and others[98] have looked to tectonic history as the reason for the onset of glacial climates at 2.5 Myr.

This example of correlation and causation serves as a reminder about how we should use the environmental data from the deepsea cores. The timescales may well be blurring the picture. We may be confusing different processes and explanations in our quest for precision. Scientific accuracy, however accurate, does not mean that the age old link between climate and life (chap. 3) is a sufficient system of cause and effect to explain variation. The same criticism applies to the out of Africa correlation suggested by Roberts and supported by Turner who shows that hominids were participants in a major biogeographical event.

Let me turn the tables for a moment. If the 2 Myr to 1.6 Myr old "party" was a major speciation event then there was almost half a million years of at least twelve full climatic cycles, each one 40,000 years long, *before* speciation occurred. Can this extended timelag be used to claim a climatic cause for the pulse of hominid and bovid speciation? Why not just say that enough time had elapsed to make it probable? Opinions will differ according to preference for either time as an arrow or cycle and any causal effect each concept might have.

One way to judge the various claims would be to answer Vrba's prediction about the settlement histories of different hominids. If cyclical changes in climate account for the selection of generalist strategies, where some hominids persist rather than leave a locality, we would have to sample for archeological remains at a resolution of less than 40,000 years and in small geographic areas—for example the Sterkfontein Valley. At the moment this is beyond the accuracy of our dating and fieldwork techniques at such remote periods. But what a spur to research!

Tracks outside Africa

Why not Asia?

I have now given some reasons for the question, why sub-Saharan Africa? Evolution is a restless, opportunistic process. Accordingly a combination of tectonic and geographical features together with the effects of cyclical patterns of climate resulted in the rapid speciation of the mammal communities of eastern Africa. It appears that the early hominids were probably, in Vrba's phrase, some of the "founder members" of the extensive African savannah populations that developed in the late Miocene.[99] To account for this radiation the pattern of speciation appears to be exapted. Selection for new adaptations undoubtedly took place but it is not necessary to invoke it for every novelty noted in either the fossil or archeological records.

When Africa is compared with the Old World tropics of southeast Asia the most striking difference lies in the equation between savannah at altitude and tropical rain forest. Today the southeast Asian rain forests are not associated with extensive savannahs, which are only found in limited parts of Sri Lanka, Papua New Guinea, and the Philippines. These would only have increased during times of forest retreat during glacial periods as the Sunda Shelf was exposed by falling sea levels.[100]

The area is tectonically unstable. Almost 40 percent of the tropical rain forest lies in the highly active seismic zone. In Africa the figure is only 1 percent. Southeast Asia has the highest diversity of plant species even among tropical rain forests. Here 10 percent of the world's plant species, 25 percent of all genera, and 50 percent of all families are found. In northern Sarawak 780 species of tree were recorded from a single 10 hectare plot. The United Kingdom has thirty-five native tree species.[101] It is probable that this astonishing diversity in plant life is a result of repeated fragmentation either by tectonic activity separating communities or by the flooding of the Sunda Shelf and the repeated marooning of plant species on the archipelagoes and peninsulas of the region. Consequently, Africa specialized in the production of large mammal species. In Southeast Asia the same climatic rhythms led to speciation in plant life. Very many rare species evolved as these forces of isolation and fragmentation were repeated throughout the Quaternary. The radiation of the lineages of grazers in East Africa has no counterpart in this region as animals predominantly followed specialized feeding strategies within closed, forested habitats.

Vicariance revisited

Can vicariance biogeography as championed by Croizat do any better in accounting for Africa as the hominid cradle? What is his alternative to the imperial tradition of Darwin and Wallace?

In drawing up his track for early hominids, Croizat reveals the weaknesses in his case. His hominid track (fig. 5.2) links human evolution with the general pattern of mammalian radiation. It could be tied in with the configuration of plate tectonics where geological vicariance would be expected to be greatest. I have already done as much in discussing East African rifting and volcanic history.

The flaws in his method are obvious. Firstly, all he has done is to join up the findspots of known fossils. This seems little advance on spotting cradles by the same technique and drawing arrows as though direction explains the pattern. We would have to change "why sub-Saharan Africa?" to "why this Old World track?" As a result Croizat's track is open to other correlations than his claim that "Nature forever repeats" will allow. Perhaps it represents nothing more than the immobilism of the British Empire as it became the Commonwealth and the dispersal of research effort within a new political framework?

Secondly, by ridiculing any form of movement except local movement, he obviously regards taxonomic and chronological relations as of little significance: "No more did *Australopithecus* trot out to Java—there to alter into

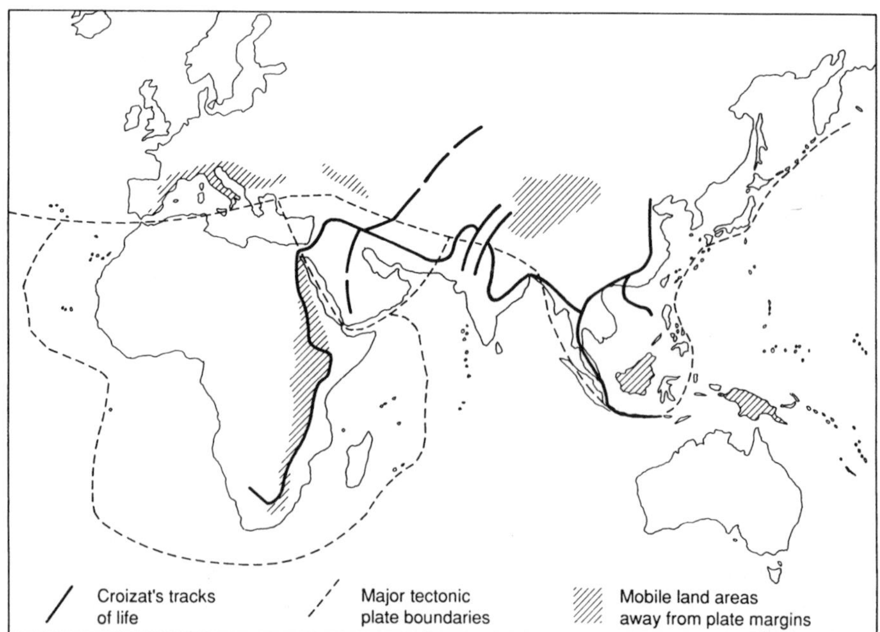

Figure 5.2 Croizat's track of life along which in-situ, vicariant, development could take place. Notice how this follows tectonically active areas, including the plate margins and earthquake regions. The track links up the main areas of early timewalker finds.

Pithecanthropus—than did *Pithecanthropus* voyage to South Africa there to sire *Australopithecus*."[102] The early center in East Africa is the point on the track where vicariant events led to speciation. He implies that this could have occurred anywhere along the track, and later did on the Java/China segment. As Peter Bowler has pointed out, this marks Croizat as a latter-day polygenist with a belief in parallel evolution in separate geographical centers.[103]

Croizat's conclusions are, therefore, not a fruitful avenue here. I have only returned to them because his unorthodox challenge at least makes us pause to consider how ingrained in human origins research is the biogeographical notion of centers of creation and routes of dispersal. I am also aware that Croizat's track (fig. 5.2) does not seem as absurd as many of the changing claims and evidence for cradles around the globe which I reviewed in chapter 3. There we saw how the cradle has been moved backwards and forwards along his track as opinions about the world, as much as the discovery of data, have changed the location. So why not follow his lead and put all these points together into one continuous track to head off any future moves of the cradle as new evidence is turned up? This would fit the spirit of origins research urged by Boriskovsky.[104]

Furthermore, even though no hominid remains earlier than *Homo erectus* or *Homo* sp. have yet been found outside of sub-Saharan Africa, there are hints of a more complicated picture. Robin Dennell has recently found a core tool with a possible date of 2 million years from Riwat in Pakistan.[105] A team of Georgian and German archeologists has found at Dmanisi in the Caucasus a very robust mandible, probably *Homo erectus*, with an estimated age of 1 Myr to 1.6 Myr.[106] R.J. Clarke presents a taxonomic case (chap. 4) that *Homo habilis* not *Homo erectus* left Africa.[107] This could almost double the timespan of timewalkers outside of Africa. Obviously, none of this evidence at the moment seriously challenges the African cradle. But that is not the point. Croizat's advice, quirky as it is, that we should abandon origins research with a narrow regional focus may be timely, if unwelcome, because of its radical implications for the hold of the imperial tradition.

Adaptive or exaptive colonization? The direction of evolution

I can now summarize the colonization process which led to timewalkers leaving Africa. Was this a result of adaptation or the consequence of exaptation? What was the role of climate? Is Neil Roberts correct when he describes the Sahara as a pump, sucking population in during wet phases and spitting them out when the climate returned to its familiar arid conditions?[108]

I call this the watermelon model. Eventually some of the pips, timewalkers, are spat far enough that they encounter the temperate grasslands and so thrive (fig. 1.1). Here they realize their preadaptation for the resources of the northern plains. All that has kept them back is a hostile barrier, the Sahara. All that they do when they arrive is continue their previous sub-Saharan adaptation for almost a million years (chap. 7). Enough time, and a bit of climatic help in watering the Sahara, is all that is required. Outside forces open the door and the timewalkers scamper through.

Under such an adaptive scenario, expansion was always a thwarted goal until modern humans evolved the solution. The reasoning is shown in table 5.2.

Table 5.2 Adaptation and colonization.

Hominids/humans designed by their history (adapted) for their current role of unlimited geographical expansion: bipedality, brains, omnivory, large size, culture

but
checked by the environments through which they must disperse, given that we were originally a sub-Saharan clade: e.g. deserts and ocean barriers

So

either	*or*
Selection to overcome these limitations: social structures, larger brains, technology	wait for a change in conditions: wetter phases, landbridges

Results:
1. fulfill adaptation to expand
2. continue to expand as selection produces new adaptive outcomes
3. the timing and direction of expansion depend on outside forces such as climate or the animal communities of which hominids were a part

It is assumed that all timewalkers, indeed all animals, are adapted for unlimited geographical expansion. This, as Geist told us, is apparently a law of life, a tenet of natural selection.[109] We explain the special qualities shared by the various hominid clades, for example big brains, upright walking, large size, and omnivory, as adapted to longterm local survival which involves a constant pressure for geographical expansion. Selection is directed to an eventual goal—global colonization. Is then, as Peter Parsons asks, the limit to range expansion set by a lack of genetic variability for further adaptation to novel habitats?[110] Or is behavior the limiting factor, not directly controlled by genes?

All sorts of animals, including timewalkers, were on the move during the ice age. Alan Turner makes the important point that some fell out of the race at different stages.[111] Spotted hyenas never reached the New World although there was obviously food for them. It was not their inability to migrate or, presumably, their genetic variability but rather, as he claims, that they were unpressured to do so. If carnivores seem to be on the move at roughly the same time as timewalkers that might seem to favor the watermelon model. They were adapted to expand and the conditions happened to allow it. This may seem more parsimonious than arguing that every carnivore and hominid evolved similar behavior independently and at the same time. Convergence of this nature would clearly be unlikely. However, it is equally parsimonious to argue that every species coopts the behavioral opportunities offered by virtue of their membership of a diverse animal community. The combinations provide their own checks and balances to anything which the climate and habitat, within reasonable limits, might be forcing them to do. Otherwise, why are there so few major dispersal and speciation events compared to the very many repeated changes in climate and geology?

When treated as exaptation the teleology of global colonization recedes and the process reemerges. Rather than pre-adapted, inevitable, directed or selected the process is more subtle, contingent and stochastic (table 5.3).

Table 5.3 Exaptation and colonization.

Hominids/humans are fit for their current role (apted) which is to expand geographically if and where possible. This is due to a suite of adapted elements rather than any one element: bipedality, brains, omnivory, large size, culture

but
extent of possible expansion limited by social behavior which does not require expansion to take place

So
cooption, under selection from social life, of behavior designed for an alternative role: e.g. alliance networking, memory enhancement

Results:

1. secondarily adapted to current role of geographical expansion and to assist in overcoming environmental checks
2. expansion less predictable in terms of timing and location as it is not primarily dependent on outside forces such as climate that we can identify independently of the fossil and archeological evidence

Due to a set of evolutionary opportunities—accidents is too strong a term but conveys the flavor—all timewalkers have the necessary elements, such as big feet, which make them fit for, rather than fitted to, expansion. None of them was designed for geographical expansion at the scale of colonization (table 1.1). Being bottled up in sub-Saharan Africa for over 3 million years cannot be explained as waiting for the desert cork to pop. In the same way, we did not wait another million years for the invention of the boat to reach Australia.

What makes expansion impossible? What produces the staccato rhythm in world colonization (table 1.2)? The answer is a lack of integration of the various elements that literally put the world before our feet. This integration only comes from social behavior. Such behavior provides purpose. Without this the production of offspring and their expansion under favorable climatic conditions, however rhythmical, leads only to ephemeral geographical gains.

Wildebeest could walk from the Serengeti to Stockholm. New York as well as Nairobi could have packs of hyenas. Chimpanzees have the brains and the omnivorous tastes to accomplish similar journeys. Like the early timewalkers, they are not equipped for such travel by the possession of either a single or even several appropriate gifts or elements. They are not limited by the environment but by themselves. At some point around a million years ago some timewalkers opened the desert door and moved north. The key they used was the evolution of behavior that coopted the elements adapted to migrate and turned them into the trail of colonization. Thus exapted they left Africa. I shall now look at the evolution of these behavioral changes.

6

Social climbers and migrant workers on the early African savannahs

Understanding why timewalkers left Africa is like asking why the chicken crossed the road. Were they escaping or just indulging in colonization on the Mount Everest principle that "since it's there, let's climb it?"

There are many similar anecdotal reasons why bipedal animals such as chickens and humans make journeys. None of them provides a satisfactory answer from the longterm, evolutionary point of view. Accept them, and we are back to Jack London's human drift, Lubbock's creeping weeds, and Milton's "wand'ring steps and slow." Driven by hunger, exiled through fear, and pushed by the relentless increase of population, human evolution and global colonization simply become a matter of enough time working its inevitable purpose to remove every uncivilized facet from our species.

But as we have repeatedly seen, these pictures of the past are contradicted by the facts. The earliest timewalkers did not cross the road for several million years. Even if we just focus on *Homo* sp./*Homo erectus*, the evidence shows them firmly rooted to sub-Saharan Africa for at least 500,000 years before they made up their mind to venture out. What we need to understand is why humans and animals migrate and colonize.

Migration and minds

The answer given by Robin Baker when a chicken is found on the opposite pavement is framed in evolutionary terms, "It seems likely that a chicken crosses a road because the other side is sufficiently more suitable than the side it is on for the risk involved in crossing the road to be worth taking."[1]

Risk in evolutionary ecology is a precise term. It refers to the failure of an organism to meet dietary requirements and as such acts as a powerful selective force on behavior. Very often the outcomes from managing such risk are described as optimal solutions.[2] Typically, they take the form of an animal locating itself to food resources so as also to maximize its reproductive success. An optimal location in a habitat takes into account the costs of searching for, pursuing, and consuming food. These costs can be measured in terms of time. The animal is under selection to optimize between the expenditure of effort and the food returns it will get. Time is not money in such situations. Rather it reflects a different currency, energy, as represented by the calories that have to be expended in order to obtain those required to maintain bodily functions and a healthy individual. Since adult animals are responsible for feeding themselves a clear link emerges on an ecological timescale between foraging behavior and potential reproductive success.

An optimum solution to these basic survival problems should provide an individual with more time than its competitors to devote to acquiring mates, defending territories, raising young, and so generally increasing its reproductive fitness. Alternatively, an animal may opt for a feeding strategy that provides secure rather than optimum returns. This might be achieved by severely limiting the number of offspring so taking the pressure off food resources.

Baker put these conditions into a simple equation. Migration will occur when

$$h_1 < h_2M$$

where h represents habitat suitability and M the migration factor. Habitat suitability is any location that can be occupied by an individual.[3] Its value is measured by potential reproductive success. The migration factor M takes into account the expense of migration by comparing those behavioral costs which relate to differences in potential reproductive success between two habitats H_1 and H_2. The distance involved and the predators along the way will also influence the migration factor, M.

The migration equation helps the chicken make hard choices—to cross or not to cross. The difficult part, however, comes in deciding whether animals calculate such risks or whether migration of this kind is entirely non-calculated. Baker maintains that both forms of migration are common among animals but that calculated migration will be more advantageous, and hence selected for, because it represents a substantial reduction in risk. As he points out in his exhaustive study *The evolutionary ecology of animal migration*, all animals have the potential for migration, the act of moving from one spatial unit to another.[4] It is no longer sufficient just to ask how species "A" migrates. What is important is asking why species "A" migrates and not species "B".

Looking at this problem in terms of evolutionary risk provides a solution. Selection will favor individuals in a population who can measure H_1 and H_2 in terms of their habitat suitability and the migration factor. To do this they will need to assess h_1, h_2, and M through those immediate factors in their environment with which they correlate. One way this might be done is by auditing environments via density measures of population numbers. In practice, as argued by V.C. Wynne-Edwards, this means that space, in the form of territories, is substituted for the food it contains.[5] Animals do not count the food available. They do, however, respond to territorial displays since these indicate how many competitors there are for the supplies available in a particular area.

Animals construct niches partly by anticipating changes in the local environment and by investing in the acquisition and communication of such knowledge (chap. 3). Other, more passive means will be through the changing seasons where the length of day acts, according to Andrewartha and Birch, as a resource "token" that triggers a positive physiological or behavioral response from an animal.[6]

Migrating animals will also require the capacity accurately to compare h_1 to h_2M and so determine whether it is less, equal to, or greater than the latter. Here is strong selection for increased intelligence as a way for individuals to reduce migration risk.

Finally, Baker argues, individuals will be selected who use methods for assessing the habitat as well as any other strategies that help to increase the ratio of calculated to non-calculated migrations they undertake.[7]

Intelligence has now entered the story. The immediate context for its selection deals with the acquisition, processing, and storage/retrieval of information about habitats which will be advantageous to an individual's fitness in ecological time. The critical element for timewalkers (and where we are not just dealing with the seasonal return migrations of birds or the annual round of a gorilla troop), rests on the additional contexts arising in evolutionary time when colonization results in speciation (table 1.1). This is the nub of the matter as we were previously reminded by Simpson (chap. 3), since temporary expansion (dispersal) and longterm survival (colonization) in those new ranges are two different and unconnected abilities.[8] They may, however, become connected if calculation occurring in ecological time has an effect in evolutionary/geological time (fig. 6.1).

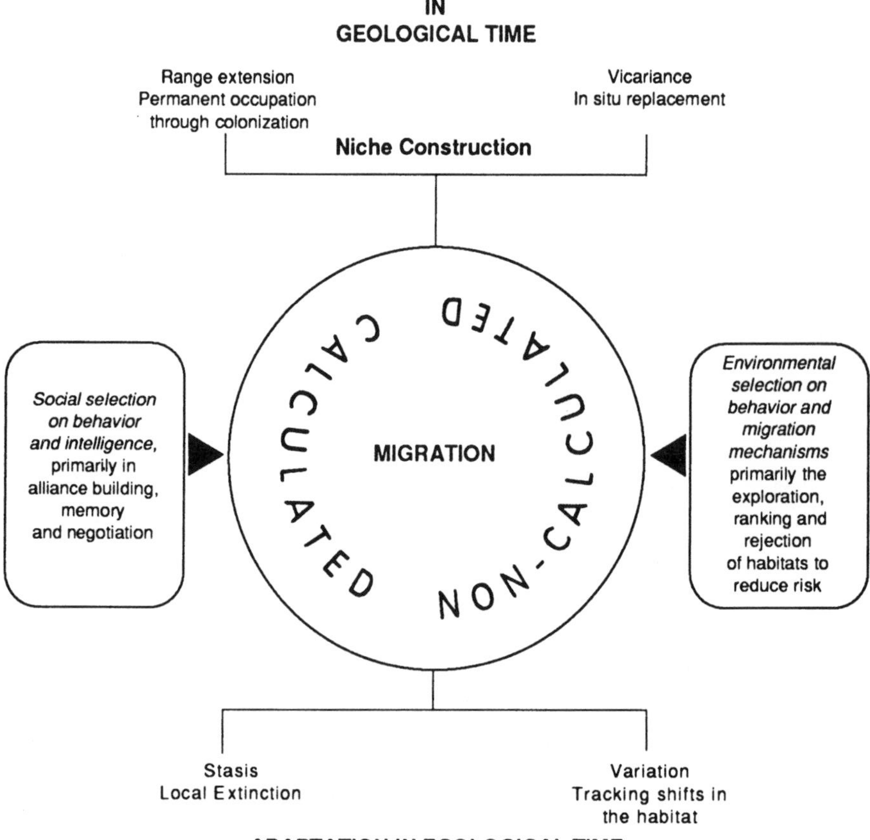

Figure 6.1 Two forms of migration and their social and environmental selection.

Because of the language of natural selection the acquisition of the intelligence to handle survival information has to be adaptive. This is undoubtedly so in shortterm ecological time. However, the current utility of such intelligence behavior does not automatically suggest its historical process. Just as our knowledge about global colonization does not mean we were adapted for this singular mammalian distribution, so to with the development of intelligence. Should the "law" of unlimited geographical expansion (chap. 5) be matched by one for intelligence? Is intelligence an unattained goal for all other species, and earlier timewalkers, because of some practical check or limiting factor as potent as the Saharan cork in the African bottle?

Once again an exaptive reason is more likely. Existing behavior was coopted for alternative purposes. Fit by reason of its form and not the history of its origins. The repertoire of hominid behavior was not directly selected to allow calculated migration to proceed so that eventually colonization could occur. Timewalking behavior emerged instead in the only context that confers meaning on such activity—social life.

The brains to do the job?

Humans are generally considered to be more intelligent than chickens when it comes to migration. Even though we both walk bipedally, our brains are so large and complex they seem to guide the actions of our feet.

Comparisons to chickens are, of course, completely unfair. The only reasonable comparison is to other large mammals and particularly to our sister species among the primates. Our brains are big but not the biggest in the world—that distinction goes to whales and elephants where they are scaled up in accordance with the animal's total size. As Passingham has neatly shown, we have a large brain based on the primate pattern.[9] This is the purpose of calculating the encephalization quotient (EQ) as a standard for inter-species comparisons (fig. 4.7). Our EQ is big for our size of primate—three times as large as might be expected. When compared to the chimpanzee some significant proportional differences emerge. In the hindbrain the cerebellum is greatly enlarged while the (literally) gray matter of the neocortex, which covers the cerebrum in the forebrain, is more extensive when compared to total brain volume.

Passingham points out that size alone does not tell us much about a brain's processing power.[10] The human brain is, however, more efficiently organized than that of other primates, with different association areas assigning different, though related functions, to the asymmetrically structured hemispheres. It is this arrangement that marks the cognitive gulf between ourselves and the chimpanzees. This conclusion is underscored by the fact that brain size varies significantly among humans. Male brains, which are slightly larger on the whale and elephant principle, have 95 percent of their range between 1180 and 1704 grams while for the same proportion of women the range is 1033 to 1533 grams. Other mammals and especially fossil hominids show similarly wide variation within their clades (chap. 4). The claim that Byron was a great poet because he had a brain of 2000 grams is a nonsense, especially when there is some uncertainty as to whether his huge brain was weighed in English, Neapolitan, or Venetian pounds![11]

But the most striking feature of the human brain is its postnatal growth. A newborn macaque has a brain that is 60 percent of the weight of its final adult size. For the chimpanzee the figure is 46 percent. The brain of a human baby is only 25 percent the size of an adult's. Growth at the foetal rate continues for the first two years then slows down over the next three. What is interesting is that humans, chimps, and macaques, as Passingham shows, have the same rate of increase before birth in the size of their brains relative to their bodies.[12]

This proportional growth in brain size at different stages is a most significant aspect of our phenotype. When compared either to other primates or to early hominids we retain in adult form many juvenile characters. The classic feature is our round head, with the face tucked under the brows which makes us look like a young chimp. The trend in human evolution has been toward ever increasing early stages of development, known as neoteny.[13] Some biologists, most notably Groves, argue that brain enlargement was just a consequence rather than a cause of this process.[14] If correct, this would relegate natural selection to nothing more than an external role of fine-tuning internal genetic factors to do with growth. The latter would direct evolution by molecular laws alone rather than through the more traditional avenues of adaptation under selection from the outside environment.

When put this way it is indeed difficult to suggest what timewalkers did that required and selected for such a large brain and a long period of infant dependency. What were the advantages of all that neocortex and a brain with more CCs than a Harley-Davidson motorbike? Kurt Vonnegut, for one, is unimpressed in his novel *Galápagos*:

> When I was alive, I often received advice from my own big brain which, in terms of my own survival, or the survival of the human race, for that matter, can be charitably described as questionable. Example: It had me join the United States Marines and go fight in Vietnam.
>
> Thanks a lot, big brain.[15]

Nor were the big brains of *Homo* sp. telling them to colonize the world one and a half million years ago.

But what Groves ignores entirely is the interface between an organism's genetic structure and the environment it has to make a living in. This buffer is, of course, provided by behavior. The interesting point is not that evolution is driven by genetic laws but rather that it is infinitely varied as a result of dealing with survival through the fuzzy medium of behavior. Certainly, animals are engaged in reproductive strategies where Darwinian selection demonstrably takes place. But they are not just machines to replicate genes. They are social as well as reproductive units for the simple reason that social life is where such strategies are played out.

Tools for thought

It is, however, commonly argued that the evolution of large brains, and presumably the intelligence to use them, occurred in the context of gradually controlling the environment and developing technology as a means of adaptation. This adaptive idea has a very long history.[16]

It has also been supported from less traditional sources. The pioneering fieldwork among chimpanzees, initiated by Louis Leakey and so brilliantly executed at the Gombe stream by Jane Goodall,[17] might have questioned assumptions about human uniqueness when it came to tool using but also promoted the conclusion that chimp intelligence was stimulated by the use and manufacture of objects. Since her first sightings of chimps modifying twigs as probes to tease termites out of their mounds a wide range of tool using activities has been noted. In West Africa Kortlandt observed chimps using anvils and hammerstones to crack oil palm nuts.[18] In their study of Ivory Coast chimps the Boeschs discovered that the adult females were both more efficient and frequent nut crackers, selecting and sometimes transporting stones to do the job.[19] These fascinating insights do not, however, make a convincing case that chimps and, by inference, the earliest hominids developed large brains in order to make termite probes and Oldowan chopping tools. Birds make far more elaborate nests without any corresponding increase in brain size and intelligence. As Nicholas Humphrey pointed out in an important paper on the social function of intellect, making and using objects such as simple tools involves learning by imitation and emulation by association.[20] Such aping behavior—at which primates excel—may act to restrict individual creativity and needs no planning ahead. In his opinion the subsistence technology of chimps and the earliest hominids is a substitute for, rather than a product of, intelligence. Owen Lovejoy has reached the same conclusion, that "tools are used to manipulate the environment and are thus a vehicle of intelligence, not necessarily a cause."[21]

The implications of this argument are very considerable. Graham Richards has cogently remarked that bipedalism does not so much free the hands as enslave the feet.[22] The dexterity of all four "hands" of a chimpanzee seems a lot to lose for the ability to stand upright in order to hold a stone tool. Technology is not all it is cracked up to be in accounting for change. For an archeologist the fact that stone tools may not be telling us very much is potentially devastating. If selection for larger brains and increased primate intelligence does not come from organizing objects in the food quest then what is the fitness value of intelligence? What is the advantage in being as clever as a chimp or, as Vonnegut might say, as stupid as a human?

Social intellect and Machiavellian intelligence

Humphrey developed his critique of the technological explanation for the development of hominid intelligence to propose that the advantage of intellect lies in the social life of animals.

> The life of the great apes and man may not require much in the way of practical *invention*, but it does depend critically on the possession of wide

> factual knowledge of practical *technique* and the nature of the habitat. Such knowledge can only be acquired in the context of a social community – a community which provides both a medium for the cultural transmission of information and a protective environment in which individual learning can occur. I propose that the chief role of creative intellect is to hold society together.[23]

Chimpanzees illustrate this social function all too clearly. They exhibit technical proficiency alongside what many fieldworkers regard as extravagant socializing. This, however, is not necessarily evidence for either limited social foresight or scant economic planning; fiddling before Rome burns. Instead, it can be argued that during good times chimps optimize their food gathering so as to free as much time as possible to engage in these social games and contests. This is not pursuit of leisure time for its own sake but rather provides the space for gaining social knowledge based on transactions between individuals as well as the information of complex webs of relationships that define the group. These are conveniently regarded as *alliances*, founded on negotiation and expensive in terms of time to maintain and service. The precious resource of time will, however, be budgeted to accommodate them since it is through their manipulation that an individual increases its inclusive fitness.[24] This concept recognizes that cooperation enhances the reproductive value of the individual *and* its relatives. The concept takes into account the shared genes between relatives. It proposes that individuals have a genetic future even though they may have no offspring. For example, they may assist in parenting or group defense and in this way contribute to the differential survival of a cohort of genetic material, some fraction of which they share through relatedness. Behavior may be based around such kin relationships rather than individual selection.

The appeal of such sociobiological explanations for apparently aberrant behavior such as altruism is obvious. The genetic sums add up. Better to die for a sister than a second cousin because the degree of genetic relatedness is much greater. But the evidence that these genetic calculations do underpin or even explain the variability of human social behavior remains tenuous. Why intelligence should emerge remains unexplained. Why its social uses are so varied remains unknowable through the arguments of inclusive fitness alone. It will remain so as long as the trend continues to reduce everything to an adaptive genetic goal.

Manipulating people rather than their genes or the objects they wield is what requires high intelligence. People are structural, behavioral, and genetic packages. Unpredictable and maddening, but also malleable and open to control through a host of devices. They construct and are constructed by society.

Being clever with objects, as Alison Jolly pointed out over twenty years ago, almost certainly evolved in the context of primate social life.[25] If this was the case then constructing society provided the evolutionary context for primate intelligence. Richard Byrne and Andrew Whiten have taken her argument further and nicely encapsulated it as Machiavellian intelligence.[26] The difficult part of alliance building is to keep track of your alliance partners. Successful negotiation depends on experience and memory. Even in small groups the number of alliance partners increases geometrically and with them the amount of social data that need to be processed, retained, and used at the telling moment. As Byrne and Whiten remark, Machiavellian intelligence involves a larger working memory from very recent social data as well as more elaborate longterm memory for social happenings, interactions, and relationships.[27]

An example will help to clarify what occurs. It is taken from Frans de Waal's study of *Chimpanzee politics* based on observations of a captive group.[28] In this group there are three males—Yeroen, Nikkie, and Luit—competing for the dominant position of alpha male.

In a shifting power struggle Luit allies with Nikkie to depose Yeroen as dominant male. Although he is allied with the females their support in the face of a long campaign of attrition gradually fades leaving Luit as the alpha male. Once in this position de Waal notices that Luit adopts a very different policy to stabilize his newly acquired position. He switches from supporting winners in fights to helping losers become winners. His intervention on the losing side jumps from 35 percent before to 87 percent a year after he becomes alpha male. Through this controlling role he gains respect and support and tries to prevent a similar two-male power bid by cultivating Yeroen and leaving Nikkie out. However, Yeroen is not finished. A year later he allies with Nikkie and helps establish him as alpha male in Luit's place. But once in this position Nikkie fails to make that transition from assisting winners to supporting losers. As shown by the greetings from the females Yeroen commands all their respect and support while his partner, though feared, is isolated.

> Sometimes it seemed that Nikkie was being used as a figurehead, and that Yeroen—experienced as he was and extremely cunning—had him in the palm of his hand. The broad basis for leadership rested not under Nikkie, but under Yeroen. The older male had a coalition with the females to pressurize Nikkie and a coalition with Nikkie to keep Luit in check. . . . Nikkie stood on the shoulders of someone who was himself very ambitious.[29]

The role of the king-maker, as de Waal remarks, is nicely described by Machiavelli's advice to his prince that

> He who attains the principality with the aid of the nobility maintains it with more difficulty than he who becomes prince with the assistance of the common people, for he finds himself a prince amidst many who feel themselves to be his equals, and because of this he can neither govern nor manage them as he might wish.[30]

The selection for the evolution of such Machiavellian intelligence is, as Alexander Harcourt suggests, the efficient use of alliances in contests of this nature.[31] The trick, as Whiten and Byrne point out, is to make and keep alliances with the right individuals.[32] Combine together and a powerful context for the selection of intelligence to acquire, process, compare, store, select, and retrieve social information is created. Memory is required to keep track of these alliances forged in the past while some planning is also needed to pursue longer term social goals. Compared to this, the use of stones as hammers to crack nuts or marrow bones seems a very poor explanation for the evolution of intelligence.

Intrigue rather than creativity is also the function that Donald Symons assigns to the human brain.[33] The selection for intelligence in his scheme lies in the calculation of reproductive success. Estrus, the period of sexual receptivity, is concealed in humans. Its hidden nature has contributed to our highly complex sexual lives. Ideology has of course intervened to deny the reality of the situation.[34] For example, Western society has as its role model an ideal family based on monogamous pair bonding. It is common to find claims that such marriages are "natural." Authorities as diverse as Desmond Morris[35] and Owen Lovejoy[36] have both reconstructed early hominid society based on monogamy. On examination such first families seem little more than modern Western origin myths. Reality, either Western or non-Western, reveals otherwise. The plethora of marriage systems—not to mention sexual practice—lends no support to their models. Human sexuality cannot be compared to the monogamy practiced by gibbons where infidelity is unknown.

Nor is our complexity matched by the chimpanzee where females are often solitary while the males form multimale groups.[37] The females reach sexual maturity by 12 years but are usually sterile until 14.[38] Intervals between births vary from five to seven years. Their brief periods of sexual receptivity are advertised by extreme swelling of the vaginal lips to which groups of males are attracted. The rarity of reproductive opportunity is emphasized by the latters' large testes, relative to other primates, and correspondingly high sperm counts. Outside such brief liaisons the females and males lead separate lives and build their own alliances.

The reproductive cycle of chimps suggests much simpler sexual lives than humans, even amidst their intricate socializing and alliance construction. In this context, Symons reminds us that natural selection does not promote human happiness.[39] Marriage is based on economic alliances to link individuals, families, and households rather than on an erotic transaction founded on genetically prescribed behavior. Moreover, intrigue need only be celebrated as a virtue if we stick to the equation that "natural" must be good. The games people play are the context for the evolution of intelligence and where Machiavellian rather than Queensberry rules apply.

What this insight confirms is that stylistic creativity, expressed so massively in material culture and the thousand and one ways to design a toothbrush or a stone projectile point, is an effect rather than the cause of such intelligence. The case for exaptation rather than adaptation, seems strong.

However, many believe that the social/sexual environment is not the only area selecting for intelligence and larger brains. Katharine Milton, for example, favors sharing rather than cheating as the strategy for early *Homo*. She argues that bigger brains need more energy so that higher quality diets will be selected and indeed are required by humans because of our specialized small intestine.[40] In this case selection came from dietary requirements favoring more efficient methods for searching and locating patches of high quality food. Greater niche breadth would be an advantage and so favored the evolution of more intelligent generalists.

It seems unlikely, however, that social games stepped in just to utilize an overexpanded brain. Nor is this an instance where a process such as neoteny produced an unintentional enabling device such as a big brain. Non-selection is not sufficient cause.

Instead it is the interplay between elements in the total package of what was a hominid or is a chimpanzee that matters. Brains did not lead the way any more than feet or stomachs. The elements are obviously connected but not necessarily historically dependent for their present expression. Exaptive behavior may result from their respective developments within the package. Adaptation of behavior and physical structure may follow.

For example, in chapter 5 I presented Peter Wheeler's model that bipedalism may have developed as one way to keep a larger brain cool. In turn this meant that a midday foraging niche could be exploited. The amount of time spent searching for widely spaced patches of food was therefore increased. This strongly suggests that bipedalism postdated the evolution of large brains in early hominid societies. Two legs did not carry would-be generalists into new, open environments where bigger, more intelligent brains were then selected for, as Milton has argued, under the pressure of more efficient searching for higher quality foods. But an unforeseen consequence of living in these open conditions would be the social requirements for intelligence.[41] These needs, as we shall see shortly, would be intensified as individuals and subgroups of the breeding population had to fragment on daily and seasonal foraging rounds.

In other words, the entire package was not developed in the trees and then launched like a first edition onto an open market, to be followed at regular intervals by later editions. Neither the direction nor leading elements of evolution—the gifts Landau identified (chap. 1)[42]—can ever be specified with accuracy. To do so involves either falling into the trap of claiming that everything is and must be adaptive or arguing with hindsight that developments are due to the function they will eventually fulfill, the teleological doctrine of pre-adaptation. Colin Groves puts it nicely when he caricatures the twists that now surround the explanations for the Apiths' funny walk:

> the australopithecines were not-yet-perfected intermediate stages, well out of the trees and waddling across the savannah with their eyes fixed on that distant day when their inefficient present would become a fully adapted future and it would have been worth it.[43]

The alternative is even less acceptable. The outside force of the ice ages is summoned to explain the social cleverness of modern chimpanzees in their forest habitats. In a vein reminiscent of Montesquieu (chap. 3),[44] Kortlandt and van

Zorn argue that chimps have become dehumanized.[45] They once lived in more demanding, open environments that made full use of their intelligence. Their social games mark some form of degeneration from what could only have been a chimp golden age. "Glorious it was to eat bananas in that dawn. . . ."

Society and colonization

Spotting teleology is easy. So too is identifying a historical tradition to plead the strength of a modern case. But how to illustrate my package of elements and webs of behavior formalized in alliances?

Let me return to my opening question, "why were humans everywhere?" My solution is to examine the consequences of migration for colonization (table 1.1). As I concluded in chapter 5, social behavior is the integrating force, the locus for understanding. The temptation now is to single out an area of behavior and give it special treatment as a prime mover behind human evolution. The obvious candidate would be migration itself. But behavior is not readily partitioned into separate elements and putting the explanatory money on a prime mover only works in a one horse, adaptive race. An integrating concept is needed and society, for all its faults, is the best candidate.

In order to use it we need a working definition. Wynne-Edwards provides a useful description of animal society as "a group of individuals competing for conventional prizes by conventional methods. . . . It is a brotherhood tempered by rivalry."[46] The "prizes" are also important factors for selection of behavior be they mates, food, or predators which an animal must successfully avoid. Cooperation establishes obligations as well as benefits. While the social ties thus established are often competitive they are also, as we have seen with alliances, negotiated.

Within this framework we can see that *one* of the ways to gain the "prize" either by competition or negotiation is through migration. As we have seen, selection will favor those individuals who calculate the risks involved rather than moving haphazardly because they have to. In ecological time this can be seen as another difference between "K" and "r" selection strategies (chap. 3 and tables 3.1 and 5.1). As Jared Diamond discusses, there are two ways to avoid the ever-present threat of local extinction.[47] Resources can either be concentrated to prevent it from happening or else turn the tables on the inevitable by expanding. The "K" strategy with few young, long periods of dependency, and high information processing generally describes the former path while the rapid and constant increase of population under "r" selection, matched by high levels of mortality, fits the latter.

Calculated migration proceeds to known destinations. These are decided upon through prior knowledge, perception, and what Baker calls *social communication*.[48] The problem which any animal faces is the fact that the area it knows and can assess is finite. In other words the information it can use to solve the equation $h_1 < h_2M$ is patchy and limited. Social life provides the framework for the dissemination and evaluation of information from a wider area. Indeed, society requires it. Cooperation whether through competition or negotiation gathers the data and provides a frame of reference that establishes its meaning. Society, even in the rudimentary definition supplied by Wynne-Edwards, is simultaneously defined by and responsible for defining its much needed life blood of information.

In many other instances non-calculated migration as described above is the answer. Put simply, the costs saved by not running a complex information system based on social behavior are channeled instead into producing offspring who find out the hard way, as population numbers force them to expand and take their chances. Enough survive to make the strategy workable but the wastage rates are severe. The result is comparable to the genetic route to fitness described in chapter 3. The melon pip model for getting early *Homo* through the Sahara would look no further for its biological justification.

This explanation for migration would ignore the perspective supplied by evolutionary time. As Andrewartha and Birch comment, "Evolution involves a compromise between doing the best in today's world and keeping some reserves for an uncertain long-term future."[49]

For large mammals like the Apiths or early *Homo* the genetic route would have been a disaster. If we assume small population numbers, then long gestation periods followed by considerable parental investment made them an expensive "K" selected species. Life was not cheap even though it may, by our standards, have been short. Consequently, time devoted to gathering and disseminating information would be critical to justify that evolutionary investment. In the absence of major physical and technological armaments, information would be a major tactical resource in winning the "prizes."

Social communication and a model society

But how was this information acquired? What exactly is social communication? Bees, for example, are adept at indicating through their dance at the hive entrance where pollen can be found in the surrounding countryside. While marveling at the bee, our maps and language divert attention away from communication in highly social primate groups which involves all the senses and is formalized into behavior such as grooming, greeting, mating, and dominance rituals. Graham Richards calls these an animal's life routines that are embedded in their physical and social environments.[50] One function of such behavior is to deal with the information gained by individuals as the group fissions to cope with the state of local food resources. Learning by imitation and emulation of behavior through association opens up the opportunity for dance, gesture, and other mimicry to narrate the habitat and its resources in an intelligible way by even the earliest timewalkers. Such stories would also cover meetings with strangers and so provide those all-important measures of population density in the local area.

Emulation is in fact the earliest recorded hominid behavior, as shown by the right-hand track of the Laetoli footprints where a child trod in an adult's footsteps. The repetitive and limited shapes of the earliest stone tools are further evidence of aping behavior.

Group fission may be slight in some habitats such as rain forests where plant foods are patchy but predictable. The more cohesive nature of gorilla groups probably reflects this structure. However, in more broken and open environments the pressures increase not only for individual foraging but also for the extension of group knowledge about the habitat. An individual would increase his/her fitness by belonging to a group where such information was shared rather than hoarded. This is in contrast to Owen Lovejoy's model where food provisioning by males leads to home bases in order to capitalize upon monogamy as the original sexual phenotype.[51] But we have seen above that no such rigid marital phenotype exists today, so the claims for it once existing are irreparably weakened. If provisioning took place at all it was to feast on information.

Two weeks in the life of an early hominid

How was more and better information acquired by early African hominids to solve the equation? The answer was through social interaction. I will now illustrate my argument with a simple reconstruction in ecological time (fig. 6.2).

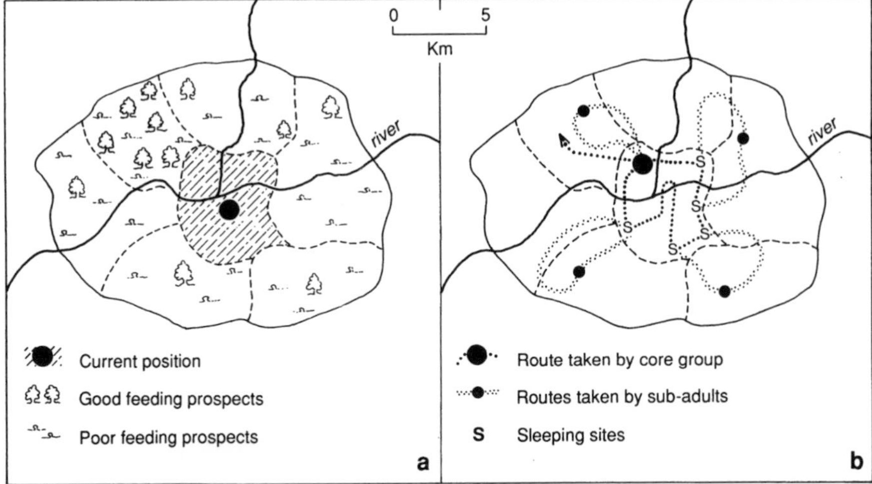

Figure 6.2 Two weeks in the life of a group of early timewalkers on the African savannahs. Their lifetime range (a) is divided into seven habitats. These vary with the seasons. Important factors are water, plant and animal foods, trees, sleeping sites, predators and other hominid species. During our two-week "visit" the problem that faces the core group is how to gather the information to make the move to the best resources as food starts to run out in their current habitat. A solution is offered in (b) where two migration routes are shown. The core group moves between safe sleeping sites and feeding grounds. Several subadults take different routes and are temporarily out of contact with the core group. These might be individuals or small parties of subadults. In this solution the decision to move to the habitat with the best feeding prospects is taken on the "look" of the subadults who have just returned from that area after an absence of a few days. This would be an example of calculated migration.

Let us begin with a group living on the open plains. For my purposes it is both the reproductive and social group although contacts with other hominids, some recognized as potential mates others not, broadens the social scope. The group might reasonably vary in size from twenty to fifty Apiths or *Homo*. The core is females of reproductive age and dependent offspring. They set the social pattern since raising young depends on food resources which are not evenly distributed either in the area or during the year. The longterm success of either the individual or the group depends on the female feeding strategy since they bear the reproductive costs of gestation, nursing, and, to a large extent, parenting. Their strategy is to get to the best food patches and to move round the area from one quality patch to another as these become available. If they can achieve this then they raise their reproductive potential.

Their problem is their size. They are small relative to the males. Sexual dimorphism is very marked.[52] They can be controlled in a group of perhaps fifteen to twenty by a single alpha male, either through violence on his part or because he provides the best protection against predators or other males.[53] This of course is the nub of the male strategy—how to gain access to the fertile females.

The sexes therefore organize their behavior on different principles.[54] In order to fulfill either strategy they need to cooperate to reduce risk. The females want to get the best food resources with the minimum threat to their offspring and themselves. This interest also serves the alpha male since his offspring have a better chance of surviving. He has to facilitate this while preventing access to the fertile females by other males and protecting them from hazards such as the local wild dogs and hyenas.

All sorts of combinations are, and probably were, possible. Males might well cooperate to defend and so share sexual access. Females in less open conditions might opt for a more solitary existence where they could protect themselves by climbing trees.[55] Establishing which is right is not the point of my reconstruction. My purpose is to illustrate how information could be gathered and communicated to calculate the movements between those quality food patches. This is a female led model of timewalking behavior.

One condition of living in a seasonal habitat without food storage is that the group splits up, fissions. One consequence is that the members cover a much wider area. They can therefore collect in their lifetime an immense amount of information about conditions and resources, all of which can potentially be marshaled to cope with risk. If memory can be extended beyond the lifetime of an individual, or even if individuals live longer to add another generation to the group, then the amount of information available to solve the equation $h_1 < h_2M$ increases exponentially.

The problem facing the group is that not all individuals have the same opportunities for exploration. The alpha males are restricted in their movements by the females which in turn are limited by their relationship to the males as well as by the constraints on mobility imposed by predators and the necessity of staying near food supplies. How, therefore, does this core group acquire that information upon which it will base its calculated move around the landscape?

The answer is through alliances formed with one section of the group, the subadults. These are most likely to be males who may be sexually mature but have not yet achieved the necessary social position where they can gain access to the females. There is strong social pressure for exploratory behavior among these subadult males. They are old enough to forage and feed for themselves. They are also better off out of the "house" and so away from the alpha males since this reduces the violence attendant on sexual rivalry. By migrating either singly or in small groups they learn the local area and who and what is in it.

Exploration in the context of calculated migration now takes on an important role. When exploring, individuals are constantly comparing, evaluating, rejecting, and ranking the habitats they pass through. On the basis of this information they can return with the core group to occupy a site explored on an earlier occasion.[56]

But one thing must be made clear. The impetus for exploration comes not from some sort of adaptive curiosity.[57] Instead it stems from the nature of the cooperative alliances, negotiated and contested between the core and peripheral members of the social group. The alpha males and females require the information in order to calculate their next move. They are aided in this by tolerating the subadults who can roam and so acquire the data. The young males in their turn need to keep a line open to the females since there lies their best genetic future. Alternatively, their inclusive fitness has been raised as a result of sharing information gained from exploring.

Making the point

How could this information be distributed and used? Putting to one side the contentious question of language among early hominids (chap. 8), I can see three ways, in addition to dance and body language, of literally making the point.

Firstly, there is the life history, ontogeny, of the individual as it moves from dependency in infancy through the prereproductive phase of subadulthood to adulthood. Among early hominids infancy, as shown by tooth eruption, was shorter because brains were smaller (chap. 4). Moreover, judging by the few estimates of life expectancy for Apiths of between seventeen and twenty years, their effective reproductive phase, marked by their achieving social rather than sexual maturity, was often short relative to their average lifespan. The combined result was to extend the subadult phase to a significant proportion of an individual's lifetime.

Ontogeny provides powerful visual clues to be used in social communication. The message decoded socially would be that these are the individuals who know about the suitability of habitats, their experience is to be trusted, their example followed. If they failed to come back because a leopard had got them, well even that was information of a kind. The alpha males had, of course, once been exploratory foragers themselves and this experience would allow them to evaluate information, calculate the migration equation, and husband their females across the landscape.

A second means of social communication is through the individuals' phenotype, how healthy they appear. An individual, at whatever stage of ontogeny, will, in comparison to his or her peers, show subtle phenotypic differences. For example, de Waal describes the change in Luit when he became the alpha male "Like Yeroen before him, Luit constantly had his hair slightly on end, so that he appeared big and powerful. He looked magnificent."[58]

The body is an excellent medium for communication. It commands respect through size and condition and relates to the optimum use of resources. The foraging success of individuals as fusion of the group's members occurred will be all too apparent. In this way the current phenotype of the subadult males—fat or thin, sleek or mangy—provides a map of the suitability of the habitats over which they had recently been foraging and so suggests the next move for the core group. As John Speth concisely argues, nutritional status also varies in a systematic fashion with season.[59] The toughest season is at the height of the dry, a time when, as Robert Blumenschine has shown, scavenged resources would be plentiful.[60] Speth points out, however, that unless fat sources were guaranteed the hominids would soon run into severe nutritional problems if all that was available was protein in the form of lean meat. Oil rich plants might have formed a more important resource at this time even though scavenging yielded good returns. During the wet season both plant and animal foods are widely distributed and nutritional stress less of a factor.

Applying these findings to my model suggests that the information from exploration is more critical at some times of the year than at others, with the condition of subadults most variable during the hard dry season. Making the right decision about where to move the core group to find resources is vital at this, the most limiting time of the year, and coincided with the clearest message from the look of the youths.

Finally, social communication among hominids involves artifacts. These can potentially code a great deal of social information as I briefly discussed in chapter 2. Actions are truly governed by objects whose power comes from the social contexts which define them.

It is a giant leap from our use of objects as signs and symbols to Oldowan pebble tools. Nor would I suggest the latter contained such subtle layers of information. Indeed, it seems improbable to me that this was even an original reason for the use of stone tools. It is more likely that an incidental effect of chipping sharp edges and selecting hammerstones during foraging was to build in to migration strategies a latent means of social communication. Style and design are very social values. The materials are there for exaptation as and when social contexts require these pieces of stone to have meaning. The selection for such communication comes from the advantages for the individual to relate their exploratory findings to the group so that the migration equation can be satisfactorily solved. This is not altruistic behavior on their part but rather the best strategy for their own chances of future reproductive success.

So, my spatial model for the early sub-Saharan hominids consists of a local area, core group with females of all ages and dominant males. They did not necessarily go around in a permanent group since adults were responsible for their own daily foraging, which is perhaps what the Laetoli trio also records.

Instead, the members of the core group occupied a smaller range, met each other frequently during foraging, could come together more quickly, and, as suggested for the Sterkfontein Valley (chap. 4), met regularly at sleeping sites and other "safe houses." The subadult males would fission for longer, maybe days at a time.

Individuals fed themselves from a range of vegetables, nuts, seeds, fruits, and animal tissues. Sharing of food, especially meat, is not impossible but would only appear in contexts where partnerships and alliances were being negotiated or reconfirmed. The impetus to such behavior therefore stemmed from the social life of the early hominids rather than from dietary necessity or ecological opportunity. Using the Machiavellian metaphor, this core group witnessed power struggles and the movement of females between males as their fortunes fluctuated. Intrigues abounded and contests were keen. The local area had no home bases only some sleeping sites. The range of the core group consisted of several such local areas and migration between them took place because of seasonal factors. Information about these other areas and where to move was provided by the exploratory behavior of subadult males who were bound by other alliances into the core group.

This sketch is a starting point to see how these generalists, as defined by their broader diets, started to disperse and speciate rather than just migrate on the more open environments of the sub-Saharan savannahs. A suggestive pathway for change opens up. Selection for information from exploratory behavior led to larger ranges, longer periods of absence by the subadult males, and so to increased investment in socializing to provide coherence, and hence positive communication when it occurred, through the intensity of interaction.

The advantage of complexity

I have now presented my reconstruction for the migration behavior of early timewalkers. I make no great claims with it. I am well aware that much of it rests on plausibility since most of the behaviors I have suggested leave no evidence. Even those which might be inferred from the fossil remains leave areas of doubt. They remain no more than plausible since I am trying to describe very short snatches of ecological time from a vantage point 2 million years in the future.

I can, however, elaborate the model in evolutionary time. Here I can do better than appeal to plausibility as I bring a different perspective to bear. I am no longer interested in a snapshot of what went on but rather the changing, general themes.

One such theme raises the issue of social complexity, not at any one point in the timewalkers' story but rather as a general principle of organization that underwent change. Complexity is as difficult to measure or discuss as intelligence. The reference point tends to be our own experience. The remote past will, by default, compare unfavorably. My version of social complexity sees it as enabling colonization. It is, if you like, my equivalent of the Saharan cork in the sub-Saharan bottle. The difference of course is that a threshold in social complexity would be internal rather than external to the timewalkers.

As an example, contrast for a moment my reconstruction of these mobile Apiths and *Homo* with Irwin DeVore and Sherwood Washburn's pioneering study thirty years ago of another social primate: "Many baboons live a lifetime without being more than three miles from where they were born and without ever being out of sight of another baboon."[61]

Major repercussions follow. These include low requirements for information to guide migration between nearby food patches and extremely low vagility. Is it any wonder that Clifford Jolly has found two different species of baboon producing fertile offspring that phenotypically form a third species?[62] Nor should we be surprised that baboons are poor colonizers.

None of this means that baboons are socially uncomplicated. Shirley Strum and William Mitchell have intensively studied one group known as the Pumphouse Gang.[63] They discovered that these baboons depend more for their evolutionary success on manipulating social contexts than on aggression.[64] This is, of course, achieved with only a modest sized brain and a lower EQ than the great apes. Friendships and partnerships between the sexes are far reaching since the investment may only pay off far in the future. From these observations they propose that two routes exist to such manipulative, negotiated social complexity.

The first path is a result of social fusion, *living together*. The constant presence of many potential competitors in the Pumphouse Gang produces strong selection for social manipulation. These groups contain both relatives and strangers. It is a question of living at high densities and in locations where for reasons of food and defense there is little or no opportunity of escape.[65]

The alternative to such complicated lives would of course be to move away, fragment the group, and reduce tensions. Their second route cites the chimpanzee case where fission is forced on them by the distribution of resources. The social group consists of several solitary females whose foraging ranges are overlapped by a roaming multimale band. The group as a whole has to deal with the problem of *temporary strangers*. Strum and Mitchell see the answer in the elaborate appeasement and reassurance gestures used by chimps as well as in their greater intelligence and skill in manipulating members in groups whose membership is constantly changing.

Another theme is the pursuit of different strategies by females and males. There are sound reasons why this should occur but in most accounts of human evolution the balance shifts towards one sex rather than the other as the leading force for change. Hunting favors males as does a model of aggression. Foraging and the role of parenting promotes the importance of females. None of this is very helpful since all it betrays is our current view of the sexes. It is more helpful to recognize that society is based on the negotiated outcome of two very different strategies.

Sexual dimorphism provides an example. Willner and Martin have examined the relationship between brain and body size in order to question the view that marked sexual dimorphism in gorillas and orangutans (table 4.1), as well as early

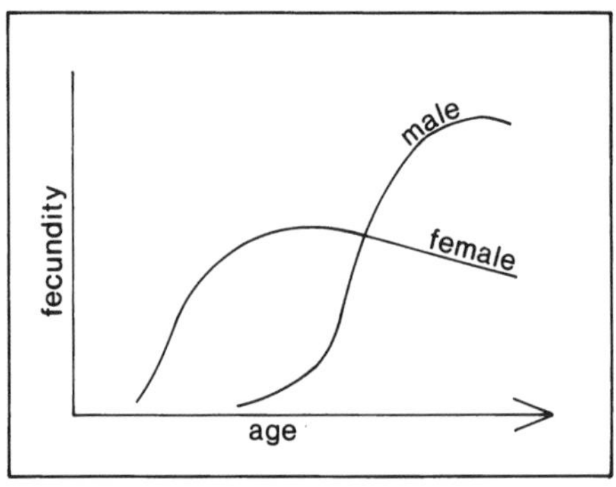

Figure 6.3 Different patterns of male and female breeding. This applies to many mammals and included early timewalkers. The earliest onset of female breeding usually corresponds with smaller adult size. Males have to wait until they have reached social maturity before their full breeding potential is realized. Such different life histories would have enormous repercussions for species size, sexual dimorphism, population dynamics and social behavior.

timewalkers, is due either to males getting bigger to protect females or to compete more effectively among their peers for access to females.[66] As they point out, both sexes are probably under selection for optimal size.[67] The sexes also have different life histories with sexual and social maturity occurring, as we have seen, at different ages.[68] This can be shown in a simple graph (fig. 6.3). Males carry on growing long after they have reached fertility. Females, however, rapidly achieve their full adult size which means that they can begin breeding at much younger ages. In some sense the girls have outsmarted the boys by stealing a march on them in the reproductive stakes. Eventually on reaching maturity some males will have the opportunity in polygynous groupings to sire more offspring than any single female can produce. But this does not happen for every male.

A divergence in male and female strategies seems most likely. Females become smaller and males, on occasion, larger as they respond to selection pressure. Moreover, part of this pressure comes from the response of the other sex rather than external factors in the environment such as plentiful food or more open conditions. The result must be that behaviors, such as enhanced intermale competition or defense against predators, are instances of exaptation not adaptation.

The measurement of brain size reveals what happened. Willner and Martin found that among the great apes, female brains are larger than expected when compared with species where males and females are the same size. This is not the case for the males. It seems that for these large primates marked sexual dimorphism stems from body size reduction in females rather than correspondingly large increases in male size. Here is an instance of divergent strategies producing change. The interaction between them produces a rich area for exaptive outcomes.

A further evolutionary theme concerns *sex ratios* among those species where the sexes start breeding at different ages (fig. 6.3). For modern humans the sex ratio at birth is 105 males to every 100 females. This figure varies and is usually different from the adult ratio. Why this should be has attracted a great deal of comment.[69] What interests us here is recognizing another tactic that exists for each sex to pursue divergent strategies. The outcome has interesting implications for colonization.

The heart of the issue comes down to reproductive cost. Female mammals in good condition produce more sons because only then can they afford to meet the higher preweaning costs of this sex. Sons can also outreproduce daughters. When mothers are in poor condition then the lower costs of producing daughters swings the ratio. So, in a genetic futures model the female strategy is to relate condition to sex ratio.

The other aspect about sex ratios relates to which sex disperses. Males are usually favored for this role, particularly those which have not yet reached social maturity. The selection for these males recognizes that they may only find suboptimal habitats as they disperse. Since young males are not yet breeding this does not matter. They can live on "junk" food. If the breeding females dispersed into the same areas they would be put at a disadvantage. Their best strategy is to stay where they were born, in the optimum habitats occupied by their mothers. Hence vagility would be low as the young males return "home" having now reached social maturity and begin to contest for the females. Any loss of dispersing males evens up the sex ratio between birth and adulthood.

Hamadryas baboons, chimps, and humans are, according to C.N. Johnson, exceptions to this pattern.[70] With these three species males stay where they were born and females disperse. His answer to this unexpected pattern is that by forging alliances males can begin mating before they reach social maturity. Multimale gangs are effective against lone high-ranking males although the likely result, as we saw earlier with Nikkie, Yeroen, and Luit, is for shifting patterns of alliance that also involve the females.

The three species differ importantly in terms of scale. Baboons cover small territories while the ranges of chimpanzees are generally less than 4 km^2.[71] These two primates cover the social spectrum from living together to coping with temporary strangers. Nonetheless, both are highly limited in the migration distances they cover, the degree of vagility, and the spatial scale of their social universes.

None of these aspects can be accurately estimated for earlier timewalkers. I can only suggest that they covered most of the social spectrum. Moreover, I would expect the various species of Apiths and *Homo* to differ socially as much if not more than between the two genera.

At any one time neighboring populations of the same, or different species, would have shown the range of strategic options as conditions varied in terms of food and as the sexes pursued their different goals. This in turn would have set up different conditions for migration and information. The result differentiated between essentially very similar animals in overlapping and adjacent groups. At any one time they might recognize each other as potential mates within a negotiated social world; at another—and with no significant change in phenotype—the recognition failed and meetings led to aggression or avoidance. One situation that never existed was a stable set of species each with a neat once-and-for-all-time parcel of behavioral traits. For instance, just as monogamy is not and never has been a universal, "natural" pattern for humans, so the assumption of male philopatry as a behavioral norm is equally unfounded. Flexibility is and always has been the timewalkers' pattern. There should be little surprise about alternate gracile and robust forms. The longterm success of the graciles is more apparent than significant.

Social life on the savannah

Can we reduce the outcomes from this spectrum of social opportunities to Vrba's opposition of a generalist *Homo* with a specialist Apith, 2 to 1.6 Myr ago?[72] Viewed in this way it now seems that the major difference between them was not technology or the size of their brains, nor even their diets or gracility, but rather the scale of their societies. Scale in this instance is measured by the area individuals covered and monitored through the organization of social behavior, as well as by the numbers of interacting individuals. The evolutionary advantage is very apparent given the cyclical shifts in vegetation during the Pleistocene. Under such changing conditions the Apiths ebbed and flowed from parts of the landscape while *Homo* maintained occupancy by calling on a wider array of information to direct migration. The need, as we saw earlier, came from society itself.

But while this would distinguish the robust savannah Apiths from early *Homo* it is still the case that the latter operated within finite limits. Otherwise we should find *Homo* outside sub-Saharan Africa at much earlier dates than we do. What we see is that the appearance of *Homo* sp. occurred within sub-Saharan Africa and was followed much later by *colonization* out of the continent.

We are therefore dealing with two unrelated processes. Selection for generalists produced the savannah lineage of *Homo*. However, within this lineage we see several species such as *H. habilis* and *H. rudolfensis* but no major colonization, no escape from the sub-Saharan bottle. The evidence, I believe, is conclusive that the extension of range, which occurred after a million years ago, was the product of selection on the social organization of these savannah groups not on their skulls and brains. An expanded web of relations, here termed alliances, incidentally overcame some of those barriers to migration and the potential for colonization was released. The construction and constitution of these alliances using the elements of omnivory, mobility, dexterity, social intellect and its effect on memory and manipulation produced unintentional results.

Large, upright walking hominids with big brains were all necessary but not sufficient elements for the extension of early hominid ranges.[73] The explanation lies instead in the constant selection pressure for calculated migration since this reduced risk and benefited the fitness of individuals who belonged to a more complex social unit. Dispersal leading to colonization may well be important for speciation since it is at such moments that phenotypic differences are emphasized. But the African evidence is quite clear on one point—major changes in social behavior guiding migration were rarely correlated with the appearance of new faces in the fossil record.

7
800,000 years down the Old World track

Any footsteps out of Africa involved survival in ever more seasonal environments. The growing season shrinks as latitude grows, while daylight hours for foraging wax and wane in the course of the year. Plant foods are often available only during short seasons and herd animals facing the same problems migrate large distances in search of alternative supplies.

None of this means that food was scarce. Far from it. The low and mid latitudes outside sub-Saharan Africa had more than enough plant and animal stocks to meet the dietary needs of early hominids living at very low population densities (fig. 1.1). The problem was not one of useable energy but rather of how it was packaged and when it was available. The limits to colonizing such new territories were not set by inadequate technologies but rather by behavior that could exploit only a limited range of seasonal variations. The scale of expansion was set by the scale of social behavior. The dispersal of the first Ancients, *Homo erectus/Homo* sp., from Africa about a million years ago is an example of behavior changing as a result of continuous selection pressure to extend range as one route to longterm survival. It was achieved through a series of exaptive solutions leading to adaptive success. Throughout this chapter I will be exploring alternatives to my model for early hominid society and subsistence on the savannah and woodland margins. In the absence of paleontological agreement I shall refer to the fossils by their more familiar tag, *Homo erectus*. The time range we are considering lasts from 1 million to 200,000 years ago—the time of the Ancients (table 1.2).

What is hunting?

A key element in this expansion has often been seen as the development of hunting, in particular of big game hunting, where naked Ancients are depicted in popular books slaying elephants in marshes and impaling cave bears on the end of wooden spears. The mastery over protein and the desire for red meat are regarded as crucial in the expansion of a super-predator into an unsuspecting world.[1]

In one respect only the big game hunting model has a valid point. For Ancients in mid latitudes it made good sense to move up the food chain and live off animals. This solved the problem of winter dearth by using animals as living storehouses. At the extremes of the Ancients' world—in southern Africa, northern China, Central Asia, and Europe—a move to greater reliance on animals as food seemed inevitable as useable plants, nuts, and fruits declined in abundance. But an omnivorous diet does not inevitably lead to a highly programed predator. The traditional view of hunting as killing animals of all shapes and sizes *at will* needs rethinking, as well as the idea that it permitted colonization by the Ancients from whom fully modern humans eventually evolved. The alternative is radical, as Lewis Binford asks, "Is it possible that hunting and all that it implies in terms of planning may well be a part of the emergence of our humanness in a modern sense?"[2]

A good deal depends on how one defines such a slippery term as hunting and on the tests that can be applied to the archeological evidence. Here the timewalkers' evidence sets right immediately a basic misconception. If the Ancients had been equivalent to modern hunters when they left Africa then the punctuated, staccato tempo to global colonization would not exist (table 1.2 and fig. 1.1). It would not have taken the best part of a million years. Instead, world colonization should have been rapid and on these timescales instantaneous at about 1 Myr ago.

But what is a modern hunter? The quotation from Binford stresses the all-important word "planning." This feature is not of course restricted to people in the modern world who live by a hunting and gathering lifestyle. It is a feature all people possess. Hunters and gatherers have their own history and are not prehistoric survivals. They are a reminder to a high-tech, urban section of the world that alternative social, economic, and survival realities do exist. They show how modern humans at low population densities make full use of mobility and planning to cope with survival problems, as the following example will help to clarify.

In ice age Europe the major seasonal factor for the Ancients would be the winter, when plants stopped growing, some animals hibernated, and the all-important herd species such as red deer, horse, and reindeer migrated in search of pasture. Unless these hominids could keep up with such herds over hundreds of kilometers (which seems unlikely), they were faced with the problem of surviving for five to six months in areas which, while abundant with resources in spring and summer, were seasonally reduced to a few resident species of megafauna such as musk oxen and possibly woolly rhino and small groups of giant deer.

Hunters in today's Arctic environments might respond to the problem in the following ways. They could kill a surplus of reindeer while these were in the area during the summer and rapidly process them in order to set up stores for the winter. Alternatively, they could move to more predictable winter resources such as seal fishing on the pack ice, only to return to land hunting when the snows melt and the animals reappear.[3] But as we shall shortly see, capturing these animals is a costly business.

Arctic peoples live in high risk habitats. The penalties for mistakes are such that there are no second chances. This intense selection is apparent in the technological sophistication of transport systems such as sleds and snowshoes as well as the hunting equipment. It might be argued that there is no choice in the matter. Living in the Arctic or ice age Europe needs a sophisticated technology to combat the cold, the ice and snow. At one level this is true, although the lack of advanced technology for the European Ancients suggests, as we shall see below, that they relied on other solutions. Even among recent Arctic technologies there is great variation in the sophistication of tools and hunting gear even though cold, ice, and snow are everywhere.[4]

Robin Torrence has shown that the most complex hunting gear is associated with overcoming problems to do with time.[5] When reliability is at a premium, as at a seal breathing hole when only the animal's nose may appear for a brief moment, the response is to overdesign, put in back-ups, invest effort in making well crafted tools and have special tools for each task. The selection pressure to produce such costly systems lies in reducing risk.

Hence the most complex technologies in the Arctic are those that are used to hunt seals and other marine mammals where the difference between success and failure can often be measured in seconds. When time is less of a constraint then investment in materials and design is adjusted accordingly. Peter Bleed refers to these as maintainable systems.[6] Tools and toolkits are simpler, more easily transportable, with components based on modular design so that users can quickly and easily replace broken parts. Furthermore, such systems can be used for unanticipated and even inappropriate tasks such as using a bow to spear a rabbit when all the arrows are spent.

Both reliable and maintainable systems stress the element of planning, where nothing is being left to chance. The object is to reduce risk at every possible opportunity. The optimum conditions that favor one technological system over another will obviously vary due to a host of factors such as time, distance to be traveled, season, availability of alternative foods, possibility of transport by river or across snow, and the size of hunting parties. It is not surprising that animals, being mobile, are the object of more complex food getting technologies than plants.

When planning ahead, decisions are weighed about how to budget scarce resources, such as time and labor, against the pressures of meeting dietary needs today and in the future, when local resources will change with the seasons. Moving people, changing group size and its membership, setting up and using stores are all elements in any modern strategy. The mobility of hunters and gatherers and the fluid character of their local groups are instructive in seeing how risk is buffered.

The tactical use of storage and technology depends on acquiring and processing information in order to make sure that people are in the right place at the right time and stay there for the appropriate duration. This is decided by striking an optimum balance for the rate at which resources are harvested to meet immediate and future needs. Foraging costs generally rise the longer a single resource is exploited and it is the information about the next move that provides the cut off point for diminishing returns. It is not a case of eating everything before moving on. Nor is the optimum balance always determined by a very favorable ratio of calories expended for calories extracted. In many cases security will be the goal and when stores are being planned individuals will be prepared to work longer at harvesting and processing in order to guarantee the future.

At the same time the alliances between individuals and groups provide the struts in a social framework along which individuals can move in search of alternative resources should theirs fail. Among modern fisher-gatherer-hunters such partnerships based on exchange, marriage, and family extend over huge areas. They involve time being set aside for visiting, feasting, and competitions such as song contests and poetry recitals. As Leah Minc has shown, these story tellings not only confer prestige on an individual but also serve as a repository of knowledge about how to cope with periodic subsistence crises.[7] Where longterm survival information is coded in such oral traditions it is common to find it sanctified and linked to ritual performances. Unlike ordinary storytelling this leads to accurate repetition so that the vital information is not lost or embellished by the present generation. Periodic crises in the Arctic may not always happen in an individual's lifetime, but on a longer timescale they certainly will. Social communication of this nature represents an extreme example of how information is stored and survival enhanced through a group memory. John Pfeiffer has called this the tribal encyclopedia.[8] Obtaining, updating, and preserving such knowledge thus involves complex and timeconsuming social practices understood to be indispensable for longterm survival.

Lean meat and seasonal diets

Coping with problems of time and the seasons by planning through the use of storage and technology was not a feature of the early sub-Saharan hominids. Artifacts were expedient—made, used, and then thrown away. They were not part of either maintainable or reliable planned systems. Storage was entirely natural. It lasted the few days a carcass could be used and the longer shelflife for nuts and below ground roots and tubers. As we saw in chapter 6, omnivory allowed them to overcome nutritional stress during the hard times of the year. This was done by living in areas that contained appropriate resources which stored themselves as well as being found within a small range.

Any increase in latitude tends to disrupt such convenient local concentrations of basic resources. The low latitude dry seasons were tough periods of the year but would have been a very different proposition to the long cold seasons of mid latitudes. For example, John Speth points out that East African grazers have very low proportions of carcass fat compared to those in northern latitudes.[9] This represents the different "in-house" responses by animals to the severity of seasonality on their nutrition. The human response to lower calorie intake in both severe dry and cold season habitats is to concentrate on obtaining either fats or carbohydrates to compensate. At such times of food stress and weight loss a protein diet of lean meat, so beloved of the big game hunter scenario, can intensify a downward spiral. One example will suffice from the experiences of a trapper in Utah in 1830:

> We killed here a great many buffalo, which were all in good condition, and feasted, as may be supposed, luxuriously upon the delicate tongues, rich humps, fat roasts, and savoury steaks of this noble and excellent species of game. Heretofore we had found the meat of the poor buffalo the worst diet imagineable, *and in fact grew meagre and gaunt in the midst of plenty and profusion*. But in proportion as they became fat, we grew strong and hearty.[10]

This example of what, in the sub-Arctic is known as "rabbit starvation," highlights the fact that eating large quantities of lean meat by itself will exacerbate dietary stress because the body inefficiently metabolizes such an energy source. The pursuit of a balanced diet means that animals must be selected for their fat condition if some of the dietary consequences of seasonality are to be overcome. Bone marrow and bone grease, which can be obtained by boiling, fluctuate over the year just as the animals' body fats in quantity and quality.[11] Once again, recognizing and pinpointing such animal resources is a task calling for planning where risk reduction provides the hone of selection. Hunting is not a catch-as-catch-can existence where either the nearest animal is clubbed to the ground or driven over a cliff or even the closest plant food picked and eaten.

The problems of weight loss and dietary deficiencies increase as the major limiting season changes from dry to cold, and plant and animal resources are scattered over larger areas and at lower local densities. Storage is one way out of the problem. Planning is the key. But even then the ethnographic literature of mobile foragers in northern and continental lands contains many grim records of failure.

Another form of storage is of course based on the body itself. The largest individuals of a species are expected to be found in cold, northern environments. This is known as Bergmann's rule and is normally explained by the fact that the cost of thermoregulation decreases with increasing body size. The bigger you are in a cold environment the less expensive in terms of energy it is to keep warm and maintain basic bodily functions. But large size is also selected for in hot environments since, as with elephants, it produces a smaller relative surface area for absorbing heat both by radiation and convection.

Stan Lindstedt and Mark Boyce argue instead that large body size will be selected for in seasonal environments because it increases fasting endurance.[12] When faced with a seasonal dearth animals respond by using stored fat. As an animal increases its size this becomes a greater fraction of body mass. They show that at low temperatures a doubling of body size among small mammals leads to a 60 percent increase in such fasting endurance. Body size therefore correlates with seasonality and is largely independent of temperature.

Applying this principle to hominids we should expect them to become larger as they pioneered ever more seasonal environments. In this case they coped with at least some of the limiting factors through biological rather than social means. *Homo erectus* is indeed larger than earlier sub-Saharan hominids (table 4.1). Among the Ancients it is noticeable that those from Java are smaller than their northern counterparts in China. It is most clearly seen in head size. For brain size the Java sample has a mean of 896 cm^3 and a range of between 815–943 cm^3. The Beijing figures are 1043 cm^3 and 915–1225 cm^3 respectively. The differences are usually put down either to the greater age of the Java specimens or to variation between isolated regional populations. I will suggest that it has as much to do with differences in seasonality between the habitats they occupied. As animals and timewalkers moved along Geist's one-way street from the tropics to the Arctic they entered increasingly seasonal environments which resulted in some phenotypic effects on body size and cranial shape (see chap. 8).[13]

Finally there is a corresponding reduction among the Ancients in sexual dimorphism which represents a substantial increase in the size of females. It will be interesting to see as the number of specimens increases when and where these larger females appear. I would predict it will only be apparent as colonization proceeds outside Africa and after 1 Myr ago.[14]

Limiting factors

Seasonality is difficult to quantify on the timescales with which I am dealing. While relative values might have stayed the same, the absolute values must have fluctuated along with the climatic cycles of the Pleistocene (chap. 3). Latitude provides a rough guide to temperature and the growing season while factors such as rainfall will further differentiate between areas in the Old World. Figure 7.1 provides a simple division into five types of environment characterized by climatic and thermal stress. In the 800,000 years covered in the remainder of this chapter it is the extremes of dry cold and dry heat that appear to have been avoided by the Ancients. These of course correspond to habitats where the cost of making a living is high (fig. 1.1).

While the physical environments pioneered after 1 Myr, as well as those that were not, posed problems such as what to eat in winter or how to cope with unpredictable rainfall patterns and water supply. These are not sufficient reasons

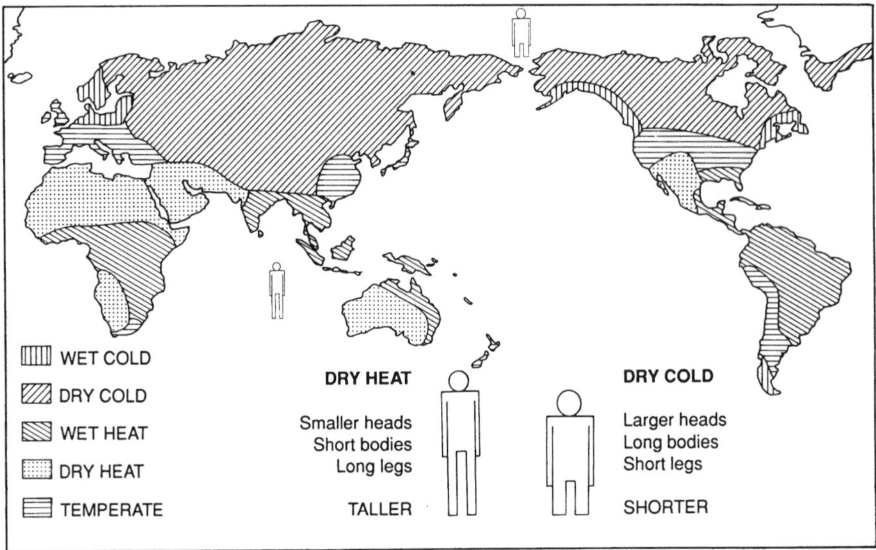

Figure 7.1 A division of the world's habitats into different climatic stress zones. These are believed to influence adaptation in body and head size, skin color and other phenotypic responses to the environment. Until 50,000 years ago timewalkers avoided the continental dry cold conditions. The divisions drawn up by Beals et al. (1984) correspond to our present, warm, interglacial conditions. Areas of dry cold would have been even more extensive under the colder parts of each climatic cycle.

to explain the course of colonization. There was plenty of food in such environments for small populations. During each climatic cycle lack of water was not a constant feature. The limits were set instead by the realities of social life in such habitats. On the one hand, there may have been the problem of stasis posed by living in large groups for some months of the year tethered around waterholes. Alternatively, the group had to fission for long periods of time in search of widely dispersed food supplies during the limiting season. Problems of conflict and recognition would have existed if the Ancients were still organized around a core group with subadult information gatherers.

The solution would be for individuals to combine the roles of information and subsistence foragers. But this could only be achieved if the social networks were somehow intensified to cope with greatly increased separation, which could lead to a failure to recognize group members, as a penalty of extended fission. Any reduction, however minimal, in the separation of roles by age and sex during local migration marked a significant shift toward colonization that could cope with any limiting factors.

Put another way, we can see that social relationships such as alliances, partnerships, and kinship are a form of storage. They are an insurance policy against the bad times and greatly extend an individual's range and chances of survival through participation in the construction and reproduction, in a social as well as a biological sense, of a human group. When the capacity to negotiate and maintain such complex webs is limited then what we see in the archeological record are, firstly, at the large scale, empty continents and, secondly, at the local level, temporary occupation on an ebb and flow basis into increasingly seasonal environments as they appear in each climatic cycle.

The stone age visiting cards of *Homo erectus*

The fossils of the Ancients in sub-Saharan Africa are few and far between, as I have already mentioned in chapter 4. To these finds must be added the many thousand stone tools usually attributed to the handiwork of *Homo erectus*. Due to the history of archeological research these are known as the Acheulean after the town of St. Acheul, situated on the Somme in northern France, where they were recovered during gravel digging in the nineteenth century.

The stone tools of this archeological culture display a wide range of repeated shapes and sizes (fig. 7.2). Implements made on flakes are particularly common, their edges often retouched by selective and careful flaking with the purpose of either strengthening or resharpening.[15] These have acquired a variety of names based on opinions about their likely function. Hence we find scrapers, knives, borers, points, and composite tools which means that more than one edge has been retouched to form, for example, a knife on a scraper. It must be remembered, however, that in most cases we have no idea what these implements were used for. Microscopic examination of unretouched and retouched edges for traces of damage due to utilization has shown that sometimes the latter were never used while frequently the unretouched or waste flakes were employed for cutting and scraping.[16] This is hardly surprising given

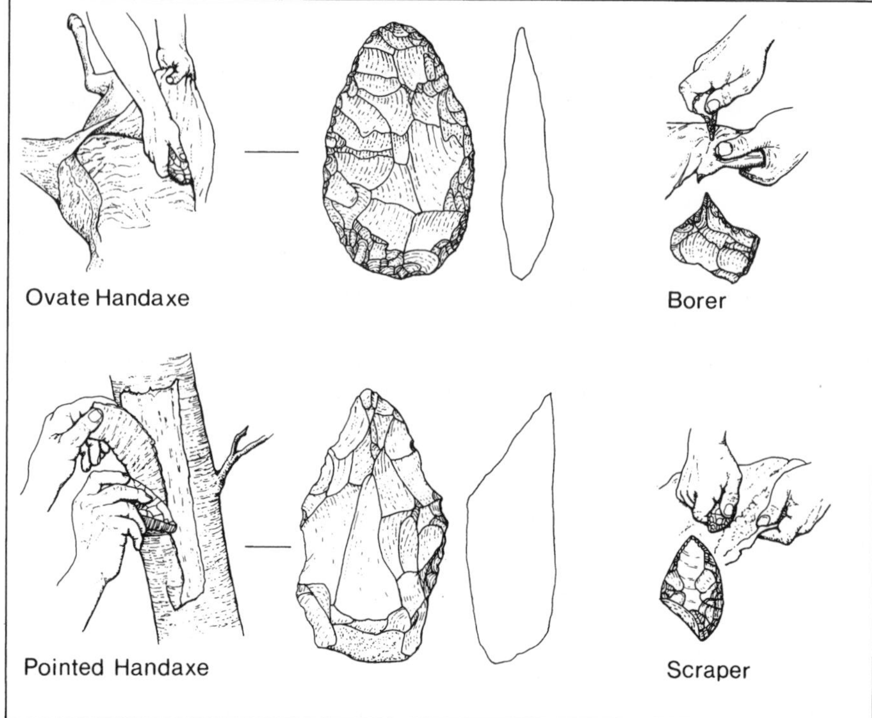

Figure 7.2 The major types of Acheulean stone tools and suggestions for their possible uses. The distinctive handaxes first appear in Africa about 1.6 Myr ago.

that a fresh stone edge is very sharp and can usually be held to avoid cutting the hand.

But by far the most distinctive and best known implements in assemblages of Acheulean stone tools are those variously described as handaxes or bifaces.[17] These have either been flaked down from a complete flint nodule or block of stone to form a pear-shaped object or else fashioned from a large flake struck from a nodule that has first been prepared into a core for just that purpose.

Handaxes come in all shapes and sizes—pointed, oval, triangular, thick, thin, elongated, and many more besides. They are worked on one or both faces, hence biface. We now know that what they look like and how well they were made is absolutely no guide to their age. This is in contrast to earlier views which held that materials were often arranged in ascending chronological order from crude to fine in a system which, as we have seen, owed everything to Lubbock, Pitt-Rivers, and Tylor. Handaxes have been recognized as antiquities since 1797 when John Frere perceptively identified some from Suffolk in eastern England as belonging to an age before the use of metals.[18] However, almost two hundred years later we are no closer to a solution as to their function and why they should be made in so many different ways and found over such large areas of the Old World. Nor are they the only large bifacially flaked tool. Cleavers with a long tranchet edge are also common in African assemblages of the Acheulean.

The exclusive association between early Acheulean tools and *Homo erectus* may not be as clear cut as once thought. At the Sterkfontein site Clarke has been able to show that the Acheulean is associated with *H. habilis* fossils.[19] Mary Leakey has evidence from her excavations in Bed II at Olduvai Gorge for two separate but contemporary stone working traditions—the Acheulean and a Developed Oldowan.[20] This contemporaneity is dated to 1.5 Myr in the middle of the Bed II sequence and marks the earliest appearance of the Acheulean.

Why should these differences exist? Louis Leakey originally thought the two industries were made by different species of *Homo*, but the Sterkfontein evidence casts doubt on such a crisp division. At Olduvai poorly made handaxes are found in the Developed Oldowan and their rarity has been put down to the availability of good raw material. Environmental evidence at Olduvai suggests that Acheulean sites were inland while Developed Oldowan were on lake margins.[21] Elsewhere in East Africa the raw material distinction does not hold up and the division between the two traditions is less clear cut.[22] The possibility, discounted for a long time in the discussion of human origins, that robust Apiths might also have been tool users must also be considered if for no other reason than that the association between some fossil species and some stone tool assemblages is undemonstrated (chap. 4). However, from 1.5 Myr at Olduvai we can trace Acheulean stone tools through sites such as Kilombe and Melka Kontuoré at 700,000 B.P. down through rich accumulations of artifacts at Olorgesailie 500,000 years ago and up to Isimila some 260,000 B.P. (fig. 4.1).[23] Excavated sites where only pebble tools have been found, which qualify for the label of Oldowan, are known from this span of time as at the FxJj50 site in Koobi Fora.[24] The general view is that patterns of stone tool use settled down into more than a million years of boredom. Local raw materials, usually picked up within 10 km of where they were eventually dropped, were invariably used.[25] The dating of many of these sites is poor since they are found as a result of gully erosion and often cannot be tied in to stratigraphic markers such as volcanic tuffs.

North Africa, the Middle East, India, and Central Asia

The track out of sub-Saharan Africa leads through the northern savannahs, up the Nile corridor or across the straits of Bab el Mandeb to the Arabian Peninsula (fig. 7.3). Even if fossil Apiths such as *A. afarensis* are one day found there—and this is highly likely given the proximity to the Hadar in Ethiopia and the tectonic history of the Red Sea—the colonization of the Ancients will, I believe, still be from the south. The lack of savannahs in Arabia during the critical period from 2 to 1.6 Myr rules out the *in situ* development of the Ancients in that region.

This first region of the Old World we shall examine is dominated by dry heat conditions and although plant foods are plentiful water is often restricted and the differences between seasons can be considerable.

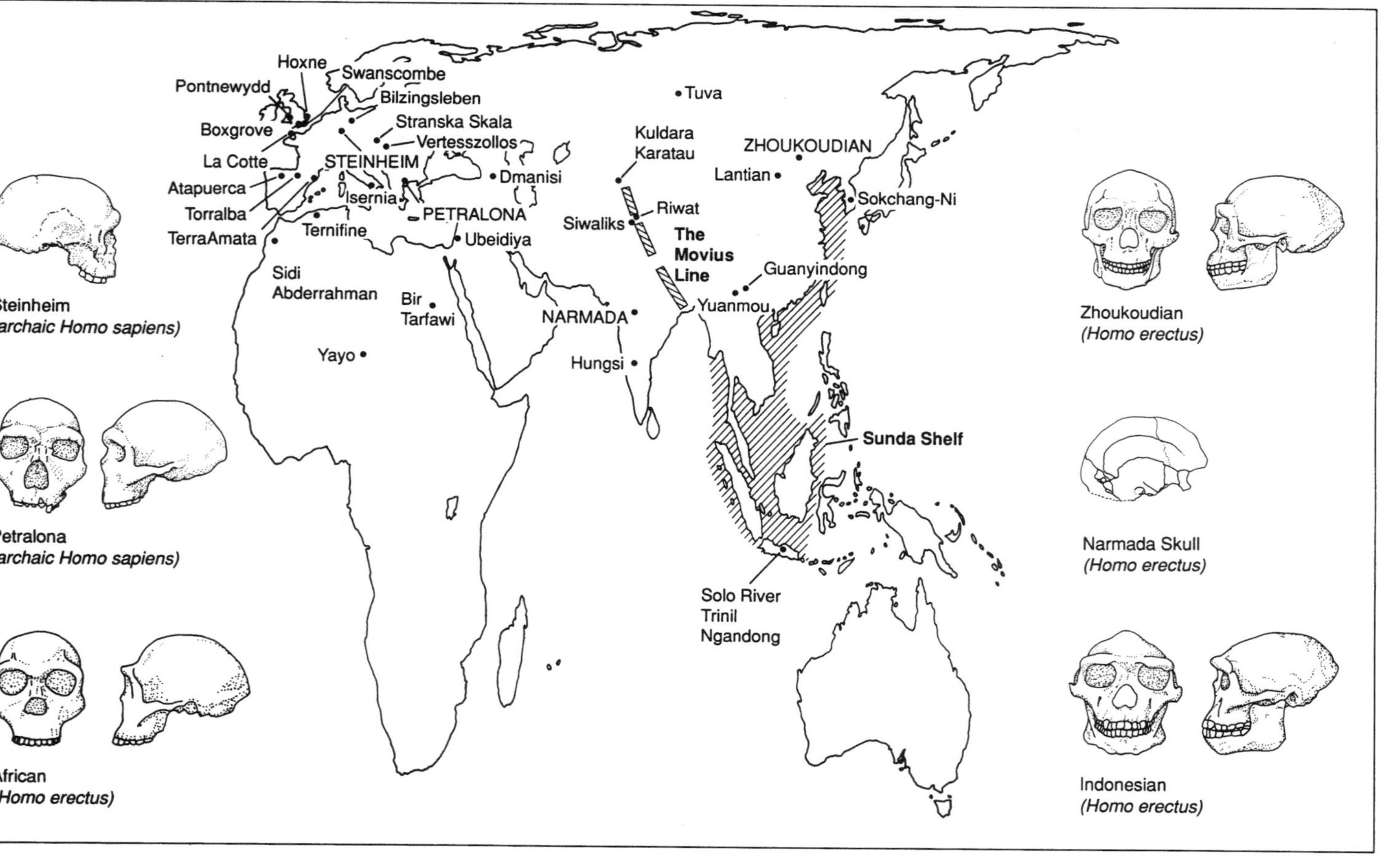

Figure 7.3 Location of the Ancients.

The fossil evidence
This great area of arid and semi-arid landscapes is so far poor in fossils and well dated sites. Important fossils come from Ternifine in Algeria where in 1954 three sturdy *Homo erectus* jaws and a piece of skull were found in an old lakeside setting.[26] One of their discoverers, Arambourg, drew attention to their close similarity to material found earlier at Zhoukoudian in northern China. The Ternifine fauna included camels, elephants, hippos, early zebras, antelopes, and rhinos as well as several carnivore species while the stone tools included handaxe forms as well as many retouched flake tools made on local raw materials.

Elsewhere in North Africa at the Sidi Abderrahman quarry southwest of Casablanca and at Salé, also in Morocco, similar material has been found.[27] The Sidi Abderrahman site has a complex series of deposits associated with changing sea levels and a long sequence of first pebble tools, then Acheulean type flake and handaxe industries.[28]

Apart from these sites the fossils are sparse. Parts of a face found at Yayo in northern Tchad are probably *Homo erectus* rather than *Australopithecus* as once claimed, while much farther afield a skull from Narmada in India is the only specimen of *Homo erectus* in the subcontinent.[29] At the moment the dating for all of these sites is poor. The Moroccan quarries have some absolute dates which suggest the long Casablancan sequence runs from 1 Myr to 300 Kyr.[30]

The archeological evidence
As in sub-Saharan Africa the Lower Paleolithic is poorly dated. Stone tool assemblages with handaxes or pebble tools are widespread. A key site in this respect is 'Ubeidiya in Israel. As Ofer Bar-Yosef points out, its position in the Levantine corridor makes it crucial for determining when *Homo erectus* left Africa.[31] Whether they reached the Levant via Arabia or the Nile is still unknown, although Bar-Yosef favors the Egyptian route. 'Ubeidiya is a very complex site, better described as a Paleolithic landscape, and lies between 160 m and 220 m below sea level in the Jordan Valley. Its many layers were laid down in a fresh water lake. The sediments were then moved as a block into a vertical position so that excavating this old landscape is like scraping a wall. The environment around the ancient lake can be reconstructed in some detail. In the valley bottom were open grasslands while the hilly slopes were covered with mixed Mediterranean forest. The lake surroundings provided a number of small microhabitats with pebble beaches and alluvial fans extending from wadis into the lake.[32] The artifact collections come from these areas. The stone tools were generally unstandardized and differences between assemblages in terms of the numbers of bifaces, picks, and choppers probably say something more about how these were swept off the landscape by natural processes and deposited in their final resting place. However, some raw materials were repeatedly used to make particular tools. Pebble choppers were made of flint, spheroids of limestone, while basalt and only occasionally flint and limestone were used for the Acheulean bifaces.

The sites in this landscape are old, but how old is still uncertain. A date of 1.5 Myr represents the oldest estimate while 0.7 Myr to 1 Myr indicates an upper range which I would support. The 'Ubeidiya material now has a wider context with the discovery in 1991 of a robust mandible excavated from beneath a medieval castle at Dmanisi in the Caucasus mountains of Georgia.[33] In an excavation measuring only 3 m by 1 m the excavators struck lucky. Lying beneath the skulls of two saber-toothed tigers[34] was the hominid mandible that preliminary studies suggest is *Homo erectus*. Simple stone tools and a rich fauna with ostrich and extinct bears, wolves, rhinos, and horses were also found. The archeology lies above a basalt plug dated to 1.8 Myr and the excavators favor a date of 1.6 Myr for their finds. Further work is planned and some downward revision of the age may be necessary.

However, the Dmanisi find supports Bar-Yosef's conviction that there was more than one migration of Ancients out of Africa. Dmanisi and 'Ubeidiya are very possibly the earliest sites in Eurasia, but he suspects at least a second influx into the Levantine Corridor, and again from Africa about 500,000 to 600,000 years ago. His evidence comes from Gesher Benot Ya'acov, also in the Jordan Valley, where the lowest level produced Acheulean tools made on local basalts. In an upper level the same tools were being made on flint which occurs between 2 km and 5 km away from the site. Flint is the raw material used in the hundreds of other Acheulean sites found across Syria, Lebanon, and Israel. Since basalt is the raw material of the African Acheulean he argues that this shows the arrival of another group who only gradually shifted their techniques to use the higher quality flints which were unknown in Africa. However, while flint fractures more predictably and cuts like a scalpel, the serrated edges from raw materials such as basalt and limestone provide, as Francis Wenban-Smith points out, a very efficient saw.[35] These might have been very useful to hominid scavengers when dismembering dessicated carcasses. Function rather than traditional preference might be a better explanation for the choice of raw materials.

Even though raw materials may have been carefully selected it is very noticeable that throughout the Old World both the chopping tool/flake and core tool/biface industries always employ local raw materials. Five to ten kilometers is the normal distance from sites scattered between China and southern Africa. In the small enclosed Hunsgi Valley on the Deccan plateau of peninsula India, K. Paddayya has systematically investigated the Acheulean use of the landscape.[36] It is noticeable that here the finely made handaxes are nearly all made from local limestones and are confined to the valley floor rather than to the low surrounding hills. During the excavation of an undisturbed floor Paddayya found a small piece of hematite (iron oxide), or red ocher. The closest geological source for this raw material lies 25 km northeast of the Hunsgi.[37]

In the Tadzhikistan loesses of Central Asia there are huge sections of wind blown sediments—loess—which contain horizons with pebble tools. The oldest come from Kuldara and may be as much as 750,000 B.P. The raw materials are once again local and in this area the distinctive Acheulean bifaces are absent.[38] At Karatau similar tools and flakes excavated by Vadim Ranov are well dated to 194,000–210,000 B.P., which just creeps under my chronological curtain for this chapter.

The environmental evidence

The major limiting factor throughout these arid regions would be water. The monsoon conditions of India would pose the problem of when it fell. Paddayya believes the Acheulean population of the Hunsgi Valley came together during the dry season around spring fed water sources. At this season less grass cover made it possible to hunt gazelle, blackbuck, swamp and spotted deer in this semi-arid season. Fission of the group occurred during the wet season which marked the major availability of fruits and vegetable foods.[39]

In the hyperarid eastern Sahara the total lack of water presented the biggest problems and Acheulean sites such as Bir Tarfawi, where no rain has fallen this century, are always found with evidence for former lakes. The excavators Fred Wendorf, Angela Close, and Romuald Schild estimate rainfall to have averaged up to 600 mm per annum when the area was occupied in the Middle Pleistocene.[40] Such changes in local conditions during the course of climatic cycles show that the Sahara was never, in the long term, a monotonous arid barrier to the extension of human settlement. The longterm mosaic of short lived but favorable conditions suggests yet again that the "melon pip" model of colonization is wrong (chap. 5). Something in addition to wetter conditions was needed for expansion to take place.

Similar problems are known in many parts of sub-Saharan Africa where a pattern of seasonal rainfall is found. The difficulty for the Ancients really seems to have been one of how the water was stored and distributed. Large bodies of water such as rivers and lakes presented no problems other than their degree of permanency. When sufficient rainfall was available in a local area, but unpredictable as to where exactly it was going to fall, then mobility played its part as the Ancients chased after it.

The major problems arose in landscapes where during droughts the only water supplies came in the form of small, permanent waterholes. In the first place these were the only water sources for other species, including carnivores, and so presented an unacceptable risk. Secondly, populations would end up tethered for many months to one spot which would place strain on the social group and on the local food supply. For these reasons early occupation of any arid area with highly localized water resources would not occur and the evidence bears this out.

Southeast Asia and China

The fossil evidence

Research in this region was inspired by Haeckel's claim for the Asian origin of humans which would be represented by a Pithecanthropine—literally Ape man—stage.[41] Although Haeckel's 1889 cradle actually rested in Pakistan, this proved no barrier to accepting Eugene Dubois' find of the top of a flat headed skull in 1891, at Trinil on the Solo river in eastern Java, as proof of the theory. More primitive in aspect than the few Neanderthal skulls which had by then been found in Europe, Dubois' find entered the scientific literature, at a second attempt in 1894, as *Pithecanthropus erectus*.[42]

Dubois had made a sensational discovery and at a stroke opened up the world to the study of human origins. For some time this would be the most primitive fossil available to physical anthropologists. The Taung child from South Africa was not found until 1924.[43]

The Solo river has turned out to be a prolific area for fossil material. From a number of clearly separated geological beds came distinctive animal faunas as well as further hominid material. Many of these were found by Ralph von Koenigswald between 1937 and 1941 and subsequently by Jacob and Sartono between 1960 and 1979.[44] In particular, von Koenigswald's discoveries in the Kabuh beds at Sangiran yielded a number of well preserved skulls.[45] More recently in 1973 in the upper part of this same bed another skull was found at Sambungmachan.[46] These are all regarded as being Middle Pleistocene in age and so no older than 730,000 years (table 3.2). However, some of the finds from Sangiran came from the older Pucangan beds which date to the Lower Pleistocene.[47] How old these finds might be is not known. Estimates vary widely from as much as 2 to ± c. 1 Myr old. A big problem is that absolute methods of dating are not easy. Either the volcanic horizons do not exist or else when they do they are not suitable for such techniques because their potassium content is too low. As a result dating is still done by relative means, where the changing fossil animal species are compared and ages estimated.

The hominid fossils have also been described under a bewildering number of names. Some were called *Australopithecus*, now discounted, while others because of their size were classed as *Meganthropus*. The majority view that gradually emerged is that they represent a regional group of *Homo erectus* fossils that show some variation through time.

Prior to von Koenigswald's discoveries at Sangiran archeological work had begun in 1921 at the Zhoukoudian cave complex outside Beijing. This rich archeological site produced its first hominid teeth in 1927 followed by a skull found by Pei in 1928. Eventually some fourteen skulls, eleven jaws and many skeletal elements were recovered as well as thousands of stone tools and an animal fauna with many remains of hyenas.[48] Sadly, the original skulls were lost during World War II.

The age of the Zhoukoudian material is also much debated. However, Wu and Wang have recently published a series of absolute dates for the various levels in the caves.[49] Three different techniques were used and are in good agreement with each other as well as with the stratigraphic position from which they were taken. Such internal consistency is important in confirming that the earliest levels are no older than the Middle Pleistocene in date. The base of the Zhoukoudian sequence corresponds to the boundary between the Lower and Middle Pleistocene, 730,000 years ago, while most of the archeological and fossil levels span the time range between 500,000 and 230,000 B.P.

At the open site of Yuanmou in southern China claims have been made that two large incisors belonging to *Homo erectus* were as old as 1.7 Myr. However, Wu Rukang and Dong Xingren have recently reexamined the evidence (which was found by others) and see no reason to place them any older than 600,000 years.[50] In 1963–4 Wu Rukang described a skull and mandible of *Homo erectus* at two separate findspots at Lantian, in Shaanxi Province. Based on the fauna they are also dated to the Middle Pleistocene.[51] A similar age has been given to two crushed skulls found in 1989 and 1990 at Yunxian in Hubei Province.[52] These important finds were uncovered *in situ* together with twenty-one stone artifacts and animals typical of the fauna which is to be attributed in China to the later Middle Pleistocene.

The fossil material from Zhoukoudian is regarded as a classic population of *Homo erectus*. The originals were treated to an exhaustive description by Franz Weidenreich in his monographs published between 1936 and 1943. The claims of cannibalism which Teilhard de Chardin and Breuil brought back from their visits to the site in the 1930s have now been replaced by identifying the hyenas in the site as the culprits for the broken skulls and bones.[53]

Elsewhere in China small collections of fossil Ancients have been made. Zhang Yinyun has dismissed the claims for *Australopithecus* from Jianshi and reclassified them as *Homo erectus*.[54] Very interestingly, they occur with the largest fossil primate known anywhere and aptly named for once as *Gigantopithecus*. So far this fossil primate is only known from its massive teeth which indicate a very large (some estimate its height as 4 m),[55] presumably ground dwelling animal, which coexisted during the Middle Pleistocene at 30°N with *Homo erectus*. Elsewhere *Gigantopithecus* is found in the caves of southern China with fossil bones of the orangutan.[56] When this evidence is combined with the much older finds of Siva and Rama apes in China we can see another widespread radiation of fossil primates in the period between 10 Myr and 1 Myr ago and prior to the arrival of *Homo erectus*.

The archeological evidence

From Zhoukoudian to the Solo river in Java covers an enormous transect through temperate, arid, subtropical, and tropical environments. Despite much work in central and northern China these two findspots still represent the largest and best studied fossil and archeological collections. It is common to find them compared as though they were side by side whereas in fact they are 5,400 km apart, which is like comparing Olduvai Gorge to the finds in the Dordogne as though they were neighbors.

Such unfair comparisons are frequently made when discussing the stone tools. After considerable controversy and many claims for their antiquity and authenticity it now seems that none of the early Ancients in Java is associated with stone tools. In all cases the artifacts have come from later deposits and are redeposited in the fossil bearing beds.[57] One of the reasons for accepting these artifacts as very ancient stemmed from their apparent similarity with the excavated material from Zhoukoudian. However, after careful reassessment Bartstra believes that the chopping tool and flake element was almost certainly introduced into Java at some time during the last 40,000 years by fully modern humans rather than by *H. erectus*.[58] The earliest stone tools now come from Ngandong where preliminary dates associated with later Ancients in a high terrace of the Solo river may be 100,000 years old. Instead of chopping tools, as at Zhoukoudian, many small, inconspicuous chalcedony flakes and cores were found, all heavily rolled in the gravel deposits.[59]

The earliest *Homo erectus* in Java therefore had no stone tool technology. This does not mean that they survived without artifacts and it has been suggested on many occasions that bamboo might have formed the main raw material for making knives, spears and digging sticks, and possibly containers. This argument has also been extended to the Chinese evidence where the simple chopping tools and flakes have been interpreted as a toolkit for fashioning more elaborate, but never preserved, bamboo and wooden artifacts.[60]

However, the invisible nature of the end product did not impress many archeologists. Hallam Movius, in a 1948 paper still cited by some as *the* authoritative view of southern and East Asian Paleolithic archeology, separated the chopping tool and flake cultures of the East from the handaxe cultures of the West. To the east of the Movius line (fig. 7.3) were "monotonous and unimaginative assemblages."[61] The area was generally regarded as one of cultural retardation which played little part in early human evolution.

Research in China as well as Mongolia and Korea since Movius's visit has revealed his scheme to be too simplistic. This is hardly surprising given the vast area he characterized from so few collections. The region of north China that has been most intensively studied shows geographical variation between stone tool assemblages specially in the proportions of particular types of retouched tools which they contain. The epithet of monotony is misplaced. Handaxes have been found in Mongolia as well as the neighboring Central Asian republic of Tuva.[62] The sites are undated. In Korea, handaxes have been found at Sokchang-ni and Jongok-ni, although whether they date to this early period is not yet established.[63] Zhoukoudian still provides the best studied site and fresh examinations show that flake tools made on local quartz dominate the assemblage. The majority are small, being less than 40 mm in total length and under 20 grams in weight each.

Considerable controversy has recently arisen over earlier claims at Zhoukoudian of thick ash layers interpreted as intentional burning and the control of fire. For many years this was accepted as the earliest evidence for the use of fire by hominids but of course has now been superseded by the African evidence. The Chinese excavators have continued to argue for intentional burning, while explanations of natural surface fires to produce the ash deposits have been put forward by a number of American archeologists, most recently Steven James.[64] Studies of the fire blackened animal bones show that they are not burnt but only heavily stained by manganese from the sediments. What is not in contention, however, is the total lack of any built fireplaces banked with stones and reused over several visits. As we shall see this is a very significant absence throughout the Old World at this early period.

The environmental evidence

What has almost a hundred years of intermittent searching revealed about colonization by these timewalkers? The picture is still far from complete and huge areas in tropical southern China and Indo-Malesia still wait to be investigated. Knowing what we now know, Dubois' discovery of the Trinil calotte in 1891 was a remarkable piece of paleontological prospecting.

The most significant finding from this transect is the lack of evidence for the use of tropical rain forests. In this respect the finds from Java are not fortuitous. Although Java lies close to the Equator the island itself is almost equally divided between year round high rainfall in the west and a marked seasonal climate in the east with several dry months during the winter. The effects on the island's vegetation are considerable. During the Pleistocene such climatic and vegetational differences would have fluctuated in the region but finds of *Stegodon*, an extinct elephant, in the Philippines, Sulawesi, Timor, and Flores indicate at various times and places a similar major division between closed tropical forest and broken, mosaic vegetation incorporating substantial open areas suitable for such megafauna. Even when the Sunda Shelf was exposed at moments of lowest sea level extensive savannahs never existed. The absence, as Geoffrey Pope remarks, of camels and horses from the ice age animal communities of southeast Asia bears this out.[65]

It might be claimed that finding fossil or archeological remains in the tropical rain forests is an impossible task. But archeologists working in the Indo-Malay Peninsula have been very successful at discovering the insubstantial remains of horticulturists' sites from the last 10,000 years.[66] Finds have also been made in the caves of subtropical southern China but here the earliest hominid traces are nearly all dated to after 40,000 years ago at a time when the rain forests were severely reduced. For the Ancients, Guanyindong Cave with big flake artifacts and Yuanmou at 26°N are among the most southerly sites so far known in China and, like the findspots in eastern Java, may also correspond to more open environments.[67]

A comparison of the paleontological faunas from the caves of southern and northern China provides an answer. There was quite simply more to eat in the north where large grazing animals dominate the lists of Pleistocene fauna. The southern grazers are often small in size such as the river, musk, muntjac, and tufted deer. Even when they are big, as are water buffalo, they only occur in small groups and are scattered through the forests. In the north of the country there are plains species such as ostrich as well as camels, horses, and gazelles while the deer are much larger and more gregarious. Some collections contain Arctic species such as mammoth and reindeer. Animal communities were very much on the move during the Pleistocene (chap. 5), producing many different and sometimes surprising combinations.[68]

The tectonic background

The East Asian evidence has also to be put into the context of Pleistocene tectonic history. As mentioned in chapter 3, the current forcing mechanism for the onset of glacial climates *c*. 2.5 Myr ago is seen by Ruddiman and others as stemming from the uplift of the Qinghai/Xizang (Tibetan) plateau. This massive block, the roof of the world, has an average altitude of 4,000 m and covers an area of 2.5 million km^2. It is fronted by the Himalayas where Qomolangma (Mt. Everest) is 8,848 m high and rising, as the Indian plate disappears under the Asian.

Such dramatic plateau uplift and mountain building is matched by the estimates for its inception. Figures are understandably rough and ready but all authorities are agreed that the Qinghai/Xizang plateau commenced its major uplift as recently as the beginning of the Pleistocene. Gansser has calculated that the block is moving upwards at a rate of 5 mm per annum, which means that in the last 1 million years it could easily have risen the required 4 km.[69] Moreover, as Wang and colleagues have shown, the Himalayas are rising 0.2 mm faster per annum than the Qinghai/Xizang plateau behind them.[70] Therefore, if we accept their figure of 1 mm per annum rise for the high Himalayas then the conclusion is that about 2 Myr ago they were actually lower than the plateau.

A rate of 5 mm per annum is about five times that for Alpine orogeny and Sharma, Liu and Ding settle for a figure between 1 mm and 1.5 mm per annum.[71] These figures still of course put the creation of the Himalayas within the timescale of hominid evolution. The fossil record of mammals shows that as mountain building isolated regional populations, some species on the Qinghai/Xizang plateau had, in classic vicariant manner, begun to differentiate from those to the south of the mountains in the Siwalik hills of Pakistan.

On this timescale the Ancients' colonization of East Asia by 1 Myr years and north China by the start of the Middle Pleistocene 730,000 years ago means that they would have evolved under very different conditions. Continuous mountain building would have intensified the monsoon climate and displaced the tropical and subtropical belts of vegetation in China. This would have created many more regional environments *after* the arrival of hominids in the area. In fact the opportunities for isolation either through high physical barriers or environmental zones would have been considerable.

In central China the effect of these trends on the landscape can be clearly traced in the massive accumulations of wind blown loess. Soils in areas with poor vegetational cover were stripped by the prevailing winds and deposited in a huge swathe between Beijing and Xian. The area covered by the loess increases during the Pleistocene, testifying to ever more arid, continental conditions that would have been further intensified by the exposure during glacial periods of the 600 km wide continental shelf in the East China Sea.[72]

One effect of these major events was the differentiation of East and West populations belonging to archaic *Homo sapiens* prior to 200,000 B.P.

Europe

Pleistocene Europe was often an arid place but droughts were rare. The cold, arid glacial phases also saw reduced evaporation which produced conditions similar to those found today in parts of Alaska and Siberia where, although the amount of rainfall is often less than in many hot deserts, lakes, pools, marshes, and mosquitoes are found everywhere. In western Europe the oceanic effect of the Gulf Stream, though shifted south by ice sheets, was nonetheless an ameliorating factor.

The fossil evidence

The earliest solid evidence for the colonization of this Asian peninsula comes from Isernia la Pineta southeast of Rome.[73] Here, in an old river and lakeside setting, a large collection of animal bones and many thousand small flakes and pebble tools were discovered in 1978.[74] The site is as well dated as any in East Africa. Just above the artifact level the section contains the paleomagnetic reversal that is the worldwide marker at 730,000 B.P. for the beginning of the Middle Pleistocene. Furthermore, a volcanic horizon at this same point is dated by absolute methods to the same age. Claims for much earlier colonization, up to 2 Myr ago, have been made in Spain, France, western Germany, but the evidence is never as convincing either in terms of the quantities of material or the security of the dates. On these grounds it can be reasonably claimed that timewalkers arrived in Europe at the start of the Middle Pleistocene.[75]

A major disappointment, however, is that we still do not know what these timewalkers looked like. Isernia has so far produced no human fossils. The majority of the early European fossils now have either relative or absolute dates which indicate ages of less than 400,000 years old. These would include the skulls from Steinheim in former West Germany and Petralona in Greece, the skull fragments from Swanscombe in England, Bilzingsleben in the former East Germany, Vértesszöllös in Hungary, the remains of more than twenty hominids from Atapuerca in Spain, and the thick rooted teeth from Pontnewydd cave in North Wales.[76] These are all regarded as more advanced forms, in terms of dental, facial, and cranial shape from *Homo erectus* and are lumped together as "archaic" *Homo sapiens*.[77] At the moment the 300,000 year fossil gap is possibly only filled by a very robust mandible from Mauer found in 1907 near Heidelberg. Estimates of its age could be as old as 600,000 years and its archaic features, including a receding chin, still divide opinion over whether it should be called *H. erectus*, claimed as an early, but archaic form of *H. sapiens* (*H.s. heidelbergensis*) or regarded as a separate species, *H. heidelbergensis*.[78]

What kept timewalkers out of Europe?

A recent suggestion by Alan Turner may go some way to explaining why we have such a slow start to the occupation of Europe.[79] In his study of the animal faunas, which accompany the earliest hominid fossils and stone tools, he has noticed that at about 500,000 B.P. there are some major changes. At this time a number of new species of deer, bovid, rhino, and horse appear. This development is accompanied by a major decline in some of the classic carnivores of the early Pleistocene such as the cheetah, saber-tooth tiger and dirk-tooth cat along with some giant hyenas. After 500,000 years ago the carnivores in Europe had a familiar, East African look to them. The changes affected the availability of food for timewalkers.

Turner argues that the saber-tooth cats would have left a good deal of consumable meat and marrow at their kill sites since the exaggerated construction of their teeth meant they could not eat all of a carcass. What they did not consume did not go to waste. It was competed for by the giant hyenas.

Indeed, between 1 Myr and 500,000 years ago the position may have been particularly bad for timewalkers who depended on scavenging opportunities as some of the saber-tooths (the carcass producers) became extinct and more hyenas (the carcass destroyers) appeared. Such competition may well account for the initially patchy and intermittent settlement of timewalkers over much of Europe in the period 700,000–500,000 B.P. The increase in hominid fossils and archeological sites dated after 400,000 years ago could well be indicative of the changes among the carnivorous competitors which produced more favorable foraging conditions for these early timewalkers.[80]

The archeological evidence

The artifact evidence is intriguing. The traditional schemes that organized material in a progression from crude to finely flaked handaxes and from poorly made flake tools to finely retouched types no longer stand up. From 700,000 B.P. to 200,000 B.P., but in no apparent chronological order, we now have well dated assemblages that range from the inconspicuous flakes and chopping tools at sites such as Isernia, Taubach, and Bilzingsleben in eastern Germany, Prezletice and Stránská skála in Czechoslovakia, and Vértesszöllös in Hungary.[81] Set alongside such expedient looking assemblages of tools and flakes are the Acheulean assemblages with a wide variety of handaxe shapes from pointed to ovate, occasionally cleavers, and great differences in their size and the amount of effort that has gone into their finishing. On average a handaxe weighs about 250 grams, but the size range is considerable. The largest weigh several kilograms and must have been used with both hands, the smallest are no heavier than the flake tools with a single retouched edge or an unmodified, so-called waste flake.

The numbers of handaxes to these flake tools vary greatly between assemblages. On occasion more effort and attention were paid to preparing the shape of the flake before it was struck from a flint nodule. Knapping nodules by such prepared core or Levallois technique (after the Paris suburb where the gravel pit is situated in which artifacts fashioned in this manner were first discovered) typically produced long parallel sided flakes, or blades as they are known, and ready made triangular points.[82]

Handaxe and non-handaxe assemblages are found throughout Europe. The classic Acheulean handaxes are, however, more common in western and Mediterranean Europe, particularly in Spain and Italy.[83] The frequency of material declines through central and eastern Europe until on the plains of the Ukraine there are no sites clearly dated to this period.[84] However, rich collections of this age are known from the Caucasus and in Soviet Central Asia the pebble tool industries of Karatau[85] and the handaxes from Tuva[86] define the northern limits of occupation under increasingly continental conditions.

Did they make a living by hunting?

In Europe it is usual to put the colonization of the continent down to two factors, hunting and fire. The first solved the problem that the continent is poor in plant resources for a hominid brought up in Africa, while finds of animal bones with stone tools in caves and open sites seemed to supply the evidence. Fire was vital, as popularized in the best selling *Quest for fire* published by J.H. Rosny-Aîné in 1911, for combating harsh northern climates, especially if you had no clothes!

Unfortunately, it is not that simple. Many of the associations between animal bones, especially the megafauna—mammoth, elephant, rhino, and bison—and stone tools are the result of collection and redeposition by the rivers and periglacial processes that slump material downslope. The interpretation of the Spanish site of Torralba as a fire drive to mire and kill elephants has now changed to one of a collection of bones and stones gathered by the local river.[87] This is not to say that the Ancients had no organized hunting parties—some cut marks on the bones show that they did[88]—but claims that they were big game hunters have to be tempered with what we now know about how our evidence was formed.

This has recently been the subject of intensive investigations at the granite headland site of La Cotte in St. Brelade's Bay on the island of Jersey.[89] While excavating the deep deposits in this cave-like fissure the excavator, Charles McBurney, uncovered two piles of mammoth and rhino bones that were stacked beneath the protective overhang. One pile contained the selected parts of at least 11 mammoth and 3 woolly rhinos while the other had 7 and 2 of the same species. Katharine Scott has interpreted these as two separate hunting episodes when small herds of these animals were intentionally driven over the cliff where the site is found.[90] Parts of the carcasses were then dragged into the site and abandoned after some butchery had taken place. These episodes occurred about 180,000 years ago when Jersey was joined to northern France.

Driving animals to their death is perfectly possible, although dangerous with such large beasts. However, the quantity of meat—at least 2 tonnes from every mammoth and 1 tonne from each rhino—smacks of an overkill which would have produced a food glut. Moreover, the seven mammoths in layer 3 consisted mostly of skulls which could be aged and sexed to a number of prime age females with their young. None of this suggests that hunting as defined today by the key elements of planning and information (rather than by the fact of killing animals) was taking place. "Dangerous driving by desperate people" might be more appropriate!

Where did they camp?

At the same time as a question mark appears over early hunting prowess a number of state of the art excavations in well preserved deposits have produced excellent evidence about what they did *not* do. The fossil beach sands at Boxgrove in southern England preserve *in situ* a landscape on which were dropped stone tools including handaxes, and the flakes and chips from their manufacture as well as animal bones, some of which carry traces of flint cut marks. This perfectly preserved landscape may be as old as 500,000 years.[91] Large flint nodules were being grubbed from a collapsed sea cliff, tested for faults, and then carried a few hundred meters where they were knapped into their final shape as oval handaxes. This involved producing many hundred small serviceable flakes. The flakes and bifaces were then used and thrown away. The overriding impression is of a short attention span, probably a "fifteen minute culture" when it came to the organization of stone technology within a life routine.

Elsewhere in England comparable good preservation can be found around the ancient lake margins at Hoxne and in the river loams at Swanscombe in a terrace of the Thames.[92] At the Netherlands site of Maastricht Belvédère[93] and in northern France at Biache-St.-Vaast preservation is also exceptional.[94] Elsewhere at Bilzingsleben and Ehringsdorf in former East Germany,[95] Kärlich in the former West Germany,[96] or the travertine quarry at Vértesszöllös in Hungary[97] the deposits are slightly disturbed, as is the case at Isernia, but not as geologically sorted as is apparently the case at Torralba in Spain.

At all these sites and landscapes we find the same evidence, or rather the lack of it, that we encountered in other parts of the Old World. Nowhere is there any evidence for well built hearths and fireplaces which we see today as the focus of a camp site and as a facility to be reused on a future visit. Construction in this instance acts as an indication of planning ahead. Neither is there any evidence for storage pits or the post holes which might indicate where a structure once stood. There are some enigmatic piles of stones at Hoxne which look as though someone emptied out a sack.[98] Neither do hearths appear in the caves. La Cotte has abundant evidence for burnt bone but nothing that passes as a hearth, despite careful searching for such features. Burning is also common at Vértesszöllös and other sites. In short all that excavation can show is where people sat, knelt, or stood to knap flint nodules into stone tools and on occasion where they scraped some meat off bones.

Larger claims have of course been made and one of the best known is the fossil beach site of Terra Amata at Nice. Here the excavator Henri de Lumley interpreted archeological horizons in a complex sequence of dune and beach deposits as a series of undisturbed living floors where hunters had repeatedly camped.[99] His reconstructions show oval brushwood cabins ranging in size from 7 m to 15 m in length by between 4 m to 6 m wide with people knapping flint and butchering animals around internal fires. Outside was found the earliest European coprolite. The site is now dated by absolute methods to 230,000 B.P.

However, this has proved to be one house that blew down. The coprolite, far from being the first example of Mediterranean beach pollution, has turned out to be a piece of limestone. More importantly, Paola Villa in a detailed study of the flint artifacts has been able to fit back together pieces that came from separate stratigraphic levels as recognized by the excavators.[100] This is not unexpected and simply shows how materials can glide through seemingly well partitioned sedimentary units. It does, however, cast grave doubts on the reconstruction of shelters which depended in the first place on demonstrating undisturbed, *in situ* materials.

In brief, the European evidence matches that from other parts of the Old World in pointing to a cultural landscape devoid of hearths, storage pits, and architecture of any kind. It leaves an overwhelming impression of spontaneous, highly episodic behavior where stone tools were made to do the job in hand before being dropped and their makers moving on. As we shall see in later chapters, this pattern of local and regional behavior is very different from what emerged over the next 200,000 and particularly in the last 40,000 years. What is lacking among these Old World Ancients is any indication for such modern practices as detailed planning, widespread contracts, or elaborate social display. There is no physical evidence for storage, raw materials all come from within a radius of 50 km, and usually less than 5 km of the sites where they were used and any form of art, ornament, jewelry, or decoration is entirely absent. Indeed, throughout the entire Old World of the Middle Pleistocene no Ancient ever made any tools from either bone, ivory, or antler.

What did they do?

So how did they make a living when there is no archeological evidence that suggests solutions comparable to the occupations of peoples living today in highly seasonal habitats? Taken altogether it seems remarkable that they managed to colonize as far north as Beijing, Tadzhikistan, Wales, and as far south as the southern Cape. In itself this is testament to the unimportance of technology in the process. They did it without toggle-headed harpoons, bows, sleds, dogs and even needles, hearths, and huts.

In many of these seasonal environments animals are going to hold the key to colonization. But what alternative ways of using animal tissues outside planned hunting can we suggest?

Let us return to Europe with its larger database, extreme limiting factor of what to eat in winter and examine how, in this part of their range, the Ancients survived.

One alternative would be to use the natural stores from animals which had died through old age, disease, at the paws of predators, and by accident. A famous example of the latter is the Beresovka mammoth which fell through the ice, broke its pelvis and was frozen in the Siberian permafrost.[101] When found eroding out of a river bank in 1900 its trunk was promptly eaten by wolves, in spite of it being 40,000 years old, before anyone could stop them. While hardly filet mignon—in fact Soviet scientists describe these thawing carcasses as smelling like old rubbish—such natural mortalities would have provided a substantial food reserve in the cold climates of glacial northern latitudes. Moreover, as Speth has shown, filet mignon was best avoided in the nutritionally stressful seasons.[102] Finding megafaunal remains would therefore be especially important because of the fat either in the carcass or still contained in the marrow cavities of long bones. The animal biomass on these high latitude steppe–tundras would have been considerable as measured by the megafauna and the main herd animals. Mortality would also have been high since these are highly stressed environments for animals as well as humans. My suggestion would be that in winter a niche existed in Europe for a mobile forager when many of the carnivores such as lions, hyenas, and wolves had migrated after the herds.[103] It involved searching for whole and partial carcasses scattered along watercourses rather than on the plains.

Obviously such a strategy could only work in areas where animal biomass was high so that scavenging such natural stores was predictable. Exploratory behavior would be needed to look for and locate resources for a larger group. The numbers of searchers would be critical, on the principle that with fifty people looking someone must find a carcass. Such behavior would need some limited skills and knowledge of the landscape. But rather than being specific to a particular landscape these could be employed anywhere that the density of natural mortalities was high enough to reduce risk to a level that survival was possible. Building survival information into a tribal encyclopedia was not needed. Information requirements both for foraging and social life would be low since the group was not under permanent pressure to split up as parts of large carcasses, when found, could feed many people.

From a landscape point of view I would expect groups and individuals to be highly mobile and for technology to be organized on an episodic basis, as resources were encountered, rather than founded on any depth of forward planning. Tools would be made from local raw materials for use on the spot and thrown away after use. In that sense the archeology of these sites tells us what went on at that location and nowhere else in the landscape.

Stone tools would be needed to cut, sever, and smash. Other implements might have helped to search and dig out the natural stores and it is intriguing that some wooden stakes, usually regarded as spears by those who favor the hunting hypothesis, have been found at Clacton and Stoke Newington in England. These might instead have been used as probes and digging sticks to get to carcasses buried in the snow. We are still waiting, however, for the first wooden shovel!

Over time the stone artifacts built up in different parts of the landscape and point to where, in the long term, such foraging behavior habitually took place. For example, in England the densities of artifacts along the Thames, the Ouse in East Anglia, and the rivers of the Hampshire basin stand out as core areas.[104] Population numbers in each of these areas may have ranged between 50 and 175 persons, with the higher number being sufficient to sustain a viable, longterm breeding population. Assuming modest birth rates and mortality such a number guarantees the constant production of sufficient mating partners as each generation reaches maturity.[105]

One aspect of technology was vital. Fire would be as important for defrosting parts of scavenged carcasses as for keeping warm. It gave European timewalkers a niche—frozen natural stores—that was not open to carnivores. Even a hyena could not use a carcass frozen like iron. But possessing fire did not necessarily make them human. The association of hearths today with planning ahead for another time, another visit and their focus for socializing and swapping information in conversation was not part of their function among the Ancients. For them both caves and rock shelters offered excellent opportunities for raising the temperature with a simple fire to get at the food. The lack of permanent slab built fireplaces can now be understood.

Finally in the spring and summer the herd animals returned and the natural stores diminished as temperatures rose and carnivores and other scavengers reappeared. During these months it would have been possible to scavenge fresh carcasses. Killing smaller herd animals such as horse, red deer, and reindeer as well as young bison and wild cattle for immediate consumption took place. Plant foods were also sought and eaten.

The main tactic of these early foragers was to position themselves in among the herd animals. In this way the predictability of survival stemmed from the numbers of animals in the landscape. The goal of security was met by the reliability of natural animal numbers—dead and alive. An alternative route to security through complex technological solutions and detailed planning where, when, and how many people to move around the landscape as food became available was avoided in these highly seasonal northern environments.

Fission of the larger group within a small area was the rule. People regularly met each other rather than lived in large residential groups. Both sexes foraged for themselves and gathered most of their own information. Even so extended absence, exploration, and the social skills to perpetuate, remember, and reestablish social links were under selection at both seasons of the year. The density of the Ancients' populations was critical for the functioning of their limited social skills. This in turn depended on the density of resources. When these declined then population would ebb from the local area and relocate in more secure surroundings.

Apiths and Ancients

This account of survival in Europe, framed in ecological time, does not of course cover all the possibilities and problems which faced hominids throughout the Old World. Natural stores in sufficient profusion would only be found in a small part of that vast range and would be of little use to groups in lower latitudes. However, what this case study points to is the scale at which such societies operated and how they were limited through the comparative simplicity of their social frameworks to particular resources and habitats. The barriers to occupation of either water poor or heavily forested regions lay in the way society was constructed rather than in the inability of limited technologies to overcome environmental obstacles.

What emerges from this account is a different understanding of the role of technology in both survival and colonization. I follow the view that the Old World Ancients were *tool assisted hominids*.[106] In that sense they were similar to chimpanzees who use stones to crack open nuts or sea otters who dive for rocks to use as anvils to smash shellfish. The Ancients required stone and wooden technologies to survive. These implements undoubtedly contributed to their longterm evolutionary success. But the toolkits and knapping technologies only assisted survival rather than formed, as we shall see in chapters 9 and 10, a medium for codifying information about peoples and places—in short acting as symbolic materials in a cultural system where change was expected. Assisted by tools these timewalkers colonized parts of the Old World. The use of technology did not by itself make this colonization possible, or determine where they could and could not go. While technology stayed the same for many hundreds of thousands of years social changes, which did not yet require the support of more elaborate material culture, undoubtedly took place. Variety came from the demands of local environments for the Ancients to cope with seasonality, drought, and the fickle abundance of food.

The foraging I have just described for Europe is one example of an alternative way to cope with seasonal environments by identifying those long winter months as the critical physical barrier to occupation. No doubt population ebbed and flowed in these northern latitudes; just as elsewhere along the Old World track, the importance of plant foods or water as a seasonal resource would have posed similar problems to local migration.

Consequently, if there was a universal strategy throughout the Old World in the Middle Pleistocene it was not as hunters, foragers, or scavengers that they made a living. Rather the universal strategy belonged to omnivorous specialists. They were specialists in the sense that their rudimentary planning, as shown by the lack of storage and the use of the landscape, limited where they lived and what they could eat.

But how did these Ancients differ from the Apiths and the earlier *Homo habilis* and *Homo* sp. in sub-Saharan Africa? They all had face-to-face societies for display, recognition, and negotiation. There is no evidence for either groupings or alliances being formed around external signs and symbols. However, what clearly marks the later, colonizing *Homo erectus* off from those first *Homo* sp. in sub-Saharan Africa is an ability to cope with seasonality and the increased absences, fission, and greater size of areas that will have to be exploited in order to make a living. Seasonal environments stretched social frameworks because individuals and small groups had to forage widely to find subsistence and therefore met each other less frequently. If the social networks could not cope with such stretching then colonization could not occur. Individuals would lose touch with the core group, social life unsupported by longterm memory would collapse, and the lack of incoming information would lead to poorly informed migration decisions and, inevitably, over time to local extinction.

What was needed to sustain colonization by the Ancients was an intensification of social life. The selection for this came initially from the advantages of calculated migration to a "K" selected, large bodied primate with bipedal locomotion. The sub-Saharan Ancients between 1.6 Myr and 1 Myr possessed limited social networks comparable to those of *Homo habilis* where information relevant to calculated migration was largely the result of subadult exploration on the periphery of a core group of males and reproductive females. This was the maintenance phenotype supported by a workable pattern of local migration that itself was finitely bounded by the scale and simplicity of social knowledge and relationships.

Selection to extend such ranges began to change these relationships. But was it an adaptive process? Were brains under selection to pursue that goal of constant expansion? I doubt it.

Higher intelligence as represented by much bigger brains was certainly necessary for holding longer term social knowledge. In this way the stretching of society in time and space took place. But there is no reason to look farther than the cooption of existing behavior to achieve such extensions. The fifteen minute culture as revealed in the manufacture and use of stone tools is a poor guide to the length of time over which social information could be retained. The occupation of seasonal environments provides a clear indication that such memory was now substantial.

Once the effect of seasonality was swamped by these migration patterns a new array of environments in the lower and middle latitudes was opened up. Using knowledge obtained for social contacts and selected by social contexts also produced the skills of memory, decision making, and evaluation for planning the use of the physical as well as the social landscape. By constructing society as a supra-organic niche the Ancients set in train the route out of their own circumscribed existence. Environmental extremes still posed problems since information on calculated migration could not be gathered. So mountains, deserts and closed forests where information requirements are that much higher were bypassed. No doubt over such long time periods there were occasions when non-calculated excursions occurred, but these have left no archeological trace.

8
Ancients and Moderns: what happened to the Neanderthals?

> There was the grisly thing again. It was running across an open space, running almost on all fours, in joltering leaps. It was hunchbacked and very big and low, a grey hairy wolf-like monster. At times its long arms nearly touched the ground.[1]

This was how H.G. Wells made the bones of the Neanderthals live in his short story "The grisly folk," seen through the eyes of the first "True men" drifting north into the lands of the Ancients. Through millennia of attrition they would eventually exterminate all such grisly people leaving only their ogres as shadows in the nursery and the stuff of nightmares, until Boucher de Perthes, John Lubbock, John Evans, and Edouard Lartet convinced the world of their archeological reality and its savage inheritance.

Wells wrote that "this restoration of the past is one of the most astonishing adventures of the human mind."[2] One such adventure, the meeting of the Ancients and the Moderns—the European Neanderthals and Cro-Magnons—has been celebrated in many illustrations, motion pictures, and novels. Perhaps the best known among the latter are William Golding's *The inheritors*, Bjorn Kurtén's *Dance of the tiger*, and most recently Jean Auel's *Clan of the cave bear*.[3] The outcomes to the story have been as varied as any science fiction. They reflect the use of the past to examine how different peoples today deal with one another. They reflect on our version of a nineteenth-century value, the psychic unity of mankind—better known as human nature.

Sons and Gardeners

The scientific stories are no less varied. At present there are two major groups which have pitched their tents either on the emotionally charged but unreliable ground of gut-feeling or on the comparatively secure base of dates and genetic, anatomical, and archeological data. Around these camps the protagonists roam the landscape like Old Testament prophets laying down the law.[4]

In one camp are those, including myself, who argue that the differences between the lifestyles and appearance of the Ancients, *Homo sapiens neanderthalensis*, and the Moderns, *Homo sapiens sapiens*, are just too great to be explained in any way other than by replacement of some sort. This does not mean that extermination took place in Europe and the Near East, where the data are still the richest and where the battleground is fiercest. Rather it is replacement either through the operation of genetic mechanisms or advantages conferred by superior behavior, for example spoken language. Existing populations were simply outcompeted, often in the nicest possible way, and usually with no direct evidence for conflict, by the new arrivals. The actual mechanism involved depends as much on a particular view of human nature as on any direct evidence from the past. The conclusion is that modern humans colonized the rest of the world from somewhere else. Many currently favor Africa, as we shall see below. This model of a single recent origin for all modern humans from a source population that was already modern has been called "Out of Africa 2" and most recently by its opponents "The Garden of Eden."[5] I shall call its supporters the Gardeners.[6]

In stark contrast is the interpretation known as "Regional continuity" or "Multiregional evolution," also christened the "neanderthal phase" by W.W. Howells.[7] To preserve a semblance of impartiality in the debate I will refer to its supporters as the Sons of Noah after the passage in Genesis much abused by polygenists and modern fundamentalists to explain the origin of racial differences (chap. 2). The scientific Sons of course claim no such thing. But they do trace the appearance of modern humans more or less independently from a number of regional populations founded by *Homo erectus* during their expansion into the Old World. Subsequently in parts of Africa, Asia, and Europe modern populations were derived by a process of parallel evolution from the Ancient pioneers. This did not necessarily occur in complete isolation. Indeed, it is difficult to see how modern humans could have popped up in different areas at roughly the same time without considerable gene flow between the regional populations. Selection leading to the independent allopatric evolution of *Homo sapiens sapiens* not once but several times would smack of latter day polygenism. While none of the Sons would now be such extreme isolationists, there is nonetheless a strong flavor of parallel immobilism, à la Croizat and his vicariance views (chap. 5), in their story.

Franz Weidenreich, who described the Zhoukoudian fossils, is credited with the creation of the multiregional model for modern human origins, and Carleton Coon in many books, but principally in *The origin of races*, presented it in its most extreme form.[8] He claimed five independent evolutionary events which gave rise to the modern races and which happened at different times, with the Africans last to cross the line and become modern.

Coon's chronology and insistence on strict parallel evolution has since been repudiated. But the fashion of the times which led in the 1950s to reassessments of some of the milestones in fossil research aided his argument for continuity. Firstly, there was the unmasking in 1953 by Oakley, Weiner, and Le Gros Clark of *Eonthropus*, the so-called Sussex "Dawn Man" from Piltdown whose remains turned up between 1912 and 1915.[9] Then came the reexamination in 1957 by Straus and Cave of the classic Neanderthal skeleton from the cave of La Chapelle aux Saints.[10] This was excavated in southern France in 1908 and described shortly afterwards in great detail by Marcellin Boule whose anatomical descriptions were duly converted by Wells into the "grisly folk." Cave and Straus pointed out that this, the classic Neanderthal, was an old man with hardly any teeth and suffering from grossly deforming osteoarthritis. The characteristic

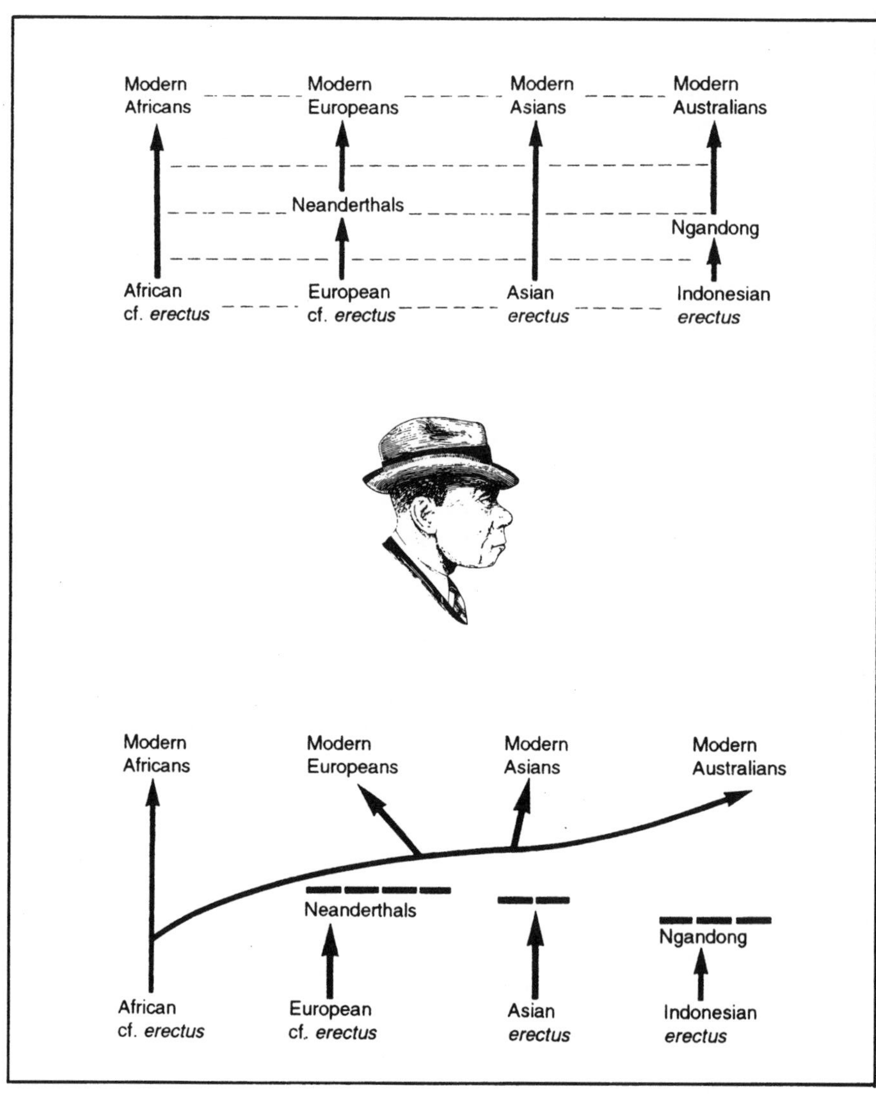

Figure 8.1 Carleton Coon's reconstruction of a Neanderthal in modern dress. His image suggests continuity in the evolution of the world's regional populations, as shown in the upper chart of multiregional evolution where the dashed lines point to gene flow between regional populations. The lower, "Out of Africa 2" chart interprets the fossil and archeological records as evidence for replacement; the view I favor. In this case the image is all dressed up with nowhere to go.

hunched posture with hands sweeping the ground was thus entirely misleading as the paradigm for this and all other Neanderthals. Their large brains, equivalent in size to the upper range of modern populations, appeared to provide further evidence for their close similarity to ourselves. So much so that Coon dressed one up in a hat to show that they would turn no heads today on the subway (fig. 8.1).[11]

Enter the archaics

In chapter 7 I showed that although timewalkers had reached Europe by 700,000 B.P. it is not until nearly 400,000 years later that we have our first well dated fossil evidence. By this time they were no longer *Homo erectus* but instead *"archaic" Homo sapiens*. They had changed considerably with larger brains (table 8.1) and more lightly built skulls and faces. By modern standards, however, such fossils as those from Swanscombe, Petralona, Steinheim, and Bilzingsleben are decidedly robust.[12]

Prior to 200,000 years ago it is also possible to see in the broadest terms an east/west division in the Old World. The European fossils called *archaic Homo sapiens* are sufficiently different from those of similar age in China and Java, while the small amount of material from African sites such as Kabwe in Zambia and Bodo in Ethiopia shows closer affinities with the much larger body of material from Europe.[13] Indeed, the Java material from Ngandong on the Solo river is generally regarded as late *Homo erectus* and has recently been dated by absolute means to *c.* 100,000 B.P.[14] In China the skulls from Maba, Yinkou, and Dali are certainly not advanced *Homo erectus*, although opinions differ as to whether they are transitional fossils within that region or if instead they have closer links with the European and western Asian material.[15]

One problem has always been the lack of dates. The critical period for the origin of Moderns between 200,000 and 40,000 B.P. has, until recently, fallen

Table 8.1 Head sizes in fossil populations. Measurements of braincase volume are in cm^3 (Stringer 1984, Day 1977).

	Number	Range	Mean
Homo habilis Africa	5	590–752	638
Homo erectus Java	4	815–943	896
Homo erectus China	5	915–1,225	1,043
archaic *Homo sapiens* Java	5	1,013–1,251	1,151
archaic *Homo sapiens* Europe	4	1,070–1,470	1,271
Neanderthals Europe and Near East	6	1,305–1,750	1,546
Modern humans Fossil material	13	1,116–1,659	1,475

between two tools for dating. The Lower and early Middle Pleistocene is dated by isotope decay methods using potassium argon (KAr) in volcanic rocks and checked against the stratigraphic sequence of paleomagnetic reversals. These are unable to date more recent material while radiocarbon methods can only provide an absolute chronology for the last 40,000 years. However, thermoluminescence (TL) dating of materials such as burnt flint is proving increasingly successful at plugging the gap. Here the "clock" is set back to zero by the heat of the burning and the radiation subsequently absorbed can be measured to establish when this zeroing took place.[16] Uranium series dates on stalactites in cave deposits and travertine spring material have also gone some way toward remedying the situation. Calculating the date depends upon predicting the uptake of uranium isotopes as the concretions are formed. It remains the case, however, that fossil material is still sparse from this time period and, as we shall see, the major impact of these dates has been to enhance our understanding of the archeological material.

What is a modern human?

There are at least three definitions of an archeologically modern human stemming from very different and not necessarily compatible lines of evidence. Modern humans have been defined anatomically, behaviorally, and genetically. It is also common to discuss them cognitively and psychologically. In recent years there have been several symposia bringing together the various disciplines to see what the modern human package consists of. Agreement has not yet been reached. Usually the anatomical evidence used by the Gardeners and the Sons sets the agenda, with claims that this is the prime evidence on the subject of origins. Furthermore, it is only recently that discussing our origins has shifted away from just being an investigation of European and Near East Neanderthals and Cro-Magnons to take in new evidence from a wider area.

But behind all this hue and cry the concept of a modern human may itself seem unnecessary. *We* know what a modern human is. As Paul Graves points out, our potential and variation are self-defined and defining.[17] It is an exercise that, as we saw in chapters 1 through 3, is under constant revision. But how are we ever going to find our equivalents buried in deep antiquity?

The problem is further muddied by the qualifiers that get put in front of "modern." Just as *archaic Homo sapiens* is an unhappy term—surely you are either *Homo sapiens* or you are not—so too is the epithet *fully* modern behavior with its implication that behavior which was only *partially* modern did exist and can be identified. This is like saying that chimpanzee behavior is *almost* human, which denies its unquestionable integrity.

Of course definitions like these lie at the heart of origins research, and we encountered similar problems in chapter 4 with the origins of hominids and how to classify fossil material into species. Without a contested definition to scrap over and confound with fresh field evidence there would be little origins research as we presently understand it. Defining ancestors is every bit as important, it appears, as setting down the rights and characteristics of modern humans. As a result, here I shall examine the evidence before presenting an alternative perspective in chapter 9. I hope to show that migration and colonization present an opportunity to examine the issue of our ancestry in broader terms rather than just reacting to new evidence.

Anatomically modern humans

The main features of anatomically modern humans are their large size, gracile skeletons, domed heads, small teeth, and lightly built faces.[18] That having been said, it must be immediately recognized that enormous variation exists within both living and prehistoric populations grouped under the *Homo sapiens sapiens* umbrella.

Nor does this modern variation cover all the prehistoric cases. A good example comes from Australia which was colonized at least by 50,000 B.P. and possibly earlier. There is no doubt that all the people who have ever lived on the continent would qualify as anatomically modern humans. Within such a large grouping modern Aborigines are distinctive for having some of the most robust skulls to be found anywhere.[19] This is shown in the thickness of the skull vault as well as the pronounced brow ridges and large tooth size. But even this generalization covers a great deal of intra-population variation where a range of values for these features is encountered.

However, collections of prehistoric remains from around the former Willandra Lakes in the semi-arid area of western New South Wales contain individuals that stand outside the ranges established by measurements on recent material. At Lake Mungo there are two very extreme Pleistocene specimens. A cremated female is dated to about 25,000 B.P. with a skull that is eggshell thin.[20] Whatever measurements are used this extremely gracile female falls outside the range of variation established by all other material. At the other extreme is a very thick walled skull known as WLH 50 with a preliminary date of 29,500 B.P.[21] Its cranial vault is so thick that most believe it to be pathological while others have combined it with skeletons from a 13,000 year old graveyard at Kow swamp in the central Murray river valley to argue for at least two Pleistocene populations—one gracile and one robust—representing separate colonizing episodes of the continent.[22]

Peter Brown, however, is unconvinced that the variation can be interpreted in this way. Apart from the tiny Mungo female the main feature of the Pleistocene collections is that they are large. His detailed measurements have shown that most of the robust specimens can be combined to form a single population with the recent skeletal evidence. Pleistocene Australians were large with males averaging 174 cm and females 165 cm in height. Recent populations from the Murray averaged 166 cm and 157 cm respectively. What he points to instead is that the postglacial populations are smaller in Australia and that this is a feature among populations worldwide.[23] In the Old World it is common to explain this by the adoption of agriculture—part of what C. Loring Brace calls the "culinary revolution"[24]—where new techniques for preparing softer food speeded up tooth reduction and jaw musculature and so led to smaller, more lightly built faces. But agriculture did not happen in Australia, the continent of hunters and gatherers. Brown suggests instead that the higher temperatures of the postglacial selected for smaller bodies which in turn affected cranial vaults and the face, particularly the size and shape of the nose, palate, and hence the size of teeth. These shorter modern populations may therefore have evolved between 10,000 B.P. and 7000 B.P. as part of an adaptation to continentwide climatic change involving, among other things, higher temperatures.

Plasticity is therefore a major feature of anatomical Moderns. To recognize populations which are not anatomically modern the differences will have to be very considerable indeed. This is of course exactly what was done by Boule in his description of the European Neanderthals. As Brace[25] and more recently Hammond[26] have pointed out, it was as if Boule went out of his way to make the Neanderthals as different as could be from finds such as the Cro-Magnon skeleton in order to remove them from human ancestry. Rehabilitation has since taken place to the extent of grouping them as the subspecies *Homo sapiens neanderthalensis* rather than as a distinct species *Homo neanderthalensis*, as Boule did,[27] while the "Regional continuity" model vigorously argued by Brace, Wolpoff and others places them squarely as the Upper Pleistocene ancestors of the Europeans.[28]

Not everyone would go so far. Chris Stringer is perfectly satisfied that the classic Neanderthalers with their large body size, jutting faces, big teeth, flat heads, and distinctive bulge at the base of the skull—the so-called occipital bun—are the descendants of the *archaic* Europeans such as Petralona and Steinheim, but not the ancestors of modern Europeans.[29] Even though their brain sizes are equivalent to and sometimes larger than those of anatomically modern humans (table 8.1), the differences in anatomy are considered too great. A key element in the discussion involves the role of biological adaptations to climate.

Big noses, long legs, and big heads

If the classic European Neanderthals, which date between 100,000 B.P. and 35,000 B.P. were, as seems agreed, descended from populations resident in the continent for over 600,000 years then it might be expected that some of the features which distinguish these Ancients from the Moderns were biological adaptations to climatic conditions.

The evidence is led by the nose, which among Neanderthals was very broad as well as being long.[30] The nasal passages were also separated from the brain by a large chamber which kept cold, inspired air away from it. Wolpoff has argued that the great breadth of the nose was an adaptation to warming the air and as a result the architecture of the face had to be massive. Climate may well be the cause of such an adaptation, as we have seen with the Australians. But if Neanderthals were ancestral to modern Europeans then why should there be a change to this time tested mechanism among anatomical Moderns who lived through the height of the last ice age 18,000 years ago?

Putting the head aside for a moment, another feature of the Neanderthal skeleton was its extreme power. The shafts of the long bones were very thick and the musculature of the arms and legs was massive. Both sexes were large and extremely strong while measures of sexual dimorphism show much lower values than we saw for the Apiths and *Homo habilis*. Sexual dimorphism was also slightly reduced from *Homo erectus* but almost the same as the Moderns. Where the skeletons of each sex are distinguished from anatomical Moderns is in their degree of robustness. Modern humans are taller, slender, less heavily muscled, and smaller boned. Both Neanderthal sexes weighed more than the Moderns which, as I showed in chapter 7, may reflect a biological adaptation to food shortage.[31]

The proportions of the skeleton are very important. Erik Trinkaus has shown that measurements in modern populations of the relative lengths of arms and leg bones (brachial and crural indices respectively) correspond well with the mean annual temperature where people live.[32] Hot environments generally select for tall people who dissipate heat by maximizing the surface area relative to body volume while the opposite—selection for short people—works best in cold environments. The values for the Neanderthals show that they had large bodies and short limbs, as would be expected. Interestingly, the earliest anatomical Moderns indicate just the opposite suggesting that they either came from hot climates or that these limb proportions reflect gene flow from those same areas.

Returning to the head, thermoregulation is also favored by Kenneth Beals, Courtland Smith, and Stephen Dodd to account for its size.[33] Their evidence shows that expansion of the braincase correlates most strongly with climatic factors. In a worldwide survey of modern populations they show that larger cranial capacities are found in colder climates. A small head is easier to keep cool and such thermoregulation accounts for variations of 100–150 cm^3 in cranial capacity between people living in the Arctic and at the Equator.

Their bioclimatic model predicts that head size will increase with distance from the Equator, from conditions of dry and wet heat through temperate to wet cold and finally dry cold conditions (fig. 7.1).[34] The comparatively recent appearance some time after 200,000 B.P.[35] of anatomical Moderns indicates to Beals that climatic factors were and still are the principal cause of variation. As an example they point to Native Americans who all share a common ancestry and a history of settlement of no more than 35,000 years, and many believe much less (chap. 10). However, cranial size varies in this hemisphere from the Equator to the Arctic just as it does on a similar transect through the Old World. Consequently, differences in cranial size between races are doing nothing more than reflecting climatic factors and have nothing to do with mental ability or racial character as once so fervently believed (chaps. 2 and 3).

This climatic trend helps explain large headed Neanderthals but does not account for the slightly smaller head size among prehistoric Moderns such as the Cro-Magnon populations (table 8.1) dated to 30,000 years ago. Surely they should have become larger still as climate declined to its coldest nadir in Europe 18,000 years ago? It is also interesting to see that differentiating Neanderthals from Moderns by their limb proportions applies equally well to the Eskimo and the Native North Americans of Arizona and New Mexico. As expected from the bioclimatic argument, the latter have long and the former short limbs. What Trinkaus found, however, is that in Europe the later Cro-Magnon populations which lived through the period of maximum cold at 18,000 B.P. show no crural change. They were not apparently adapting to cold stress through their anatomy.

This contradiction in expectations serves as a warning that correlations between climate, size, and stature among modern day populations are not infallible predictors for the anatomical adaptations of past populations. Longterm biological adaptations to cold stress by European Neanderthals and Eskimos could refer to 200,000 years for the former and only a few thousand for the latter (chap. 10). Searching for a functional correlation to explain variation is not always worthwhile. Adaptive explanations of biological variation are not always easy to support. Exaptive effects relating to the history of a population, its degree of dispersal, migration, and colonization could be as important as the climatic stresses they met.

All we can conclude from these bioclimatic studies in the Americas and Australia is that anatomically modern humans can undergo significant changes on relatively short timescales. In the case of Australia the changes have taken place over the past 6,000 years while in the Americas the figure may be three to five times as long. A short timescale for the development of the pygmy physique as a rain forest adaptation is also now indicated (chap. 9).[36]

Transitional forms and sleeping partners

The Garden of Eden model predicts that transitional forms, in the best Darwinian sense, will be found, but that these should only appear in the continent where the transition took place. In Africa these are few and far between but a strong candidate is a skull found in 1961 in a quarry at Djebel Irhoud in Morocco (fig. 8.2).[37] It is a robust skull with a rather flat vault which suggests to some that it is a North African Neanderthal while others have placed it with more modern forms from the Near East and Europe.[38] It has recently been dated to between 106,000 and 190,000 years old.[39] In Tanzania Laetoli hominid 18, also known as the Ngaloba skull after the geological bed from which it comes, is dated by uranium series to 120,000 B.P.[40] At Omo in Ethiopia two incomplete skulls have been recovered from the Kibish formation which has been dated by the same method to 130,000 B.P.[41] They show very well those modern features of an expanded vault and a round base to the back of the skull.

These skulls are certainly very different in shape from their Neanderthal contemporaries in Europe. It is also interesting to see how these features are picked up in a group of anatomical Moderns from Israel whose respective ages have been debated ever since they were found. These are the important skulls and skeletons from Mt. Carmel excavated by Dorothy Garrod in the 1930s. In the Tabun cave she found two Neanderthals and a few meters away in the Skhul cave the remains of ten individuals. It was eventually demonstrated to most people's satisfaction that these were separate and anatomically different populations.[42] However, as we shall see below, they both made the same types of stone tools which ran counter to the orthodoxy that stated that anatomical Moderns

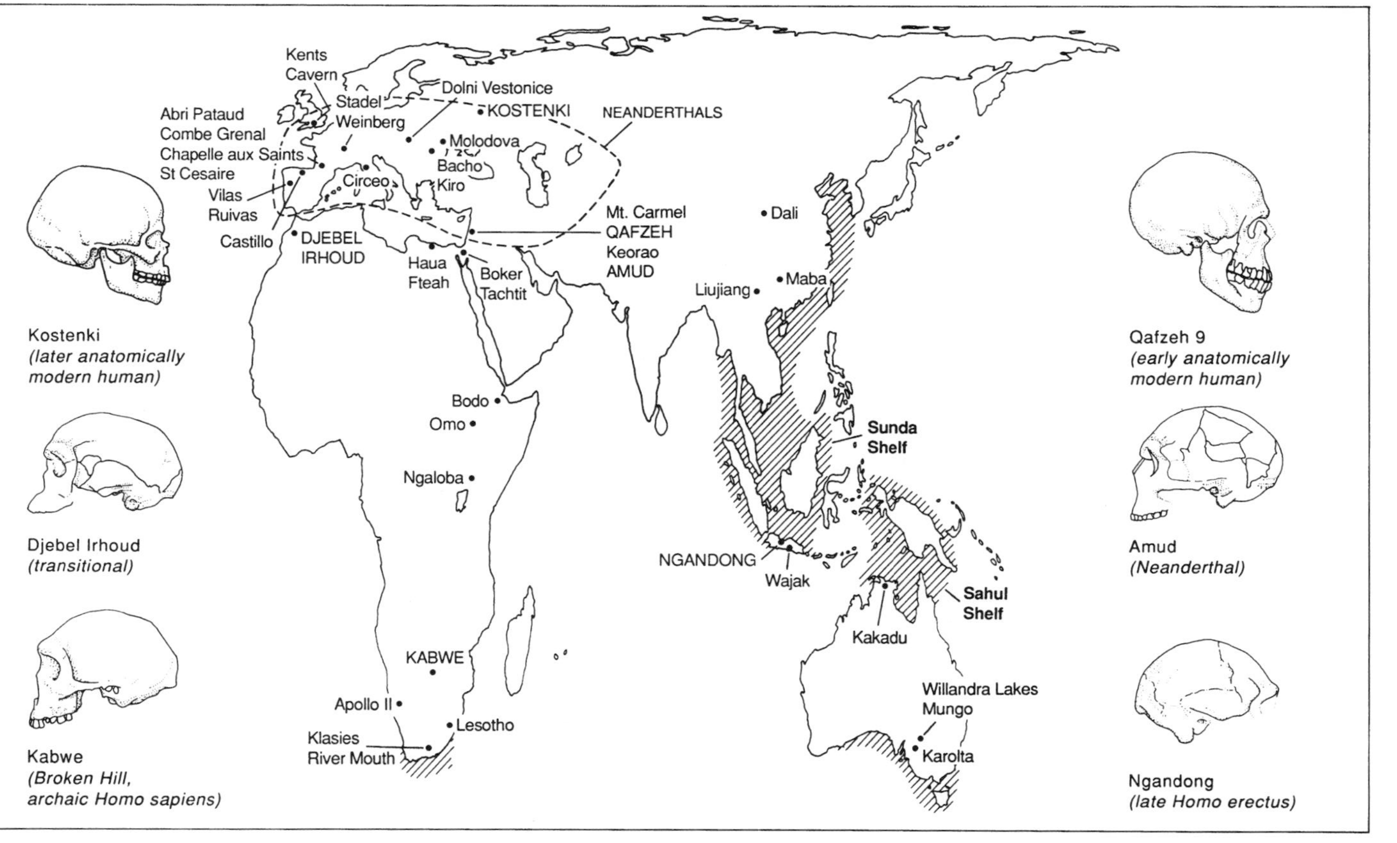

Figure 8.2 Location of Pioneers and Moderns.

manufactured Upper Paleolithic and not Middle Paleolithic toolkits. Much of the dating debate revolved around resolving this conundrum with the majority opinion favoring a late date for the Skhul remains and an early one for the Neanderthalers from Tabun. This debate was extended to later finds from other cave sites including Amud[43] and Kebara[44] where Neanderthal skeletons were found and Qafzeh where eleven modern looking individuals associated with Middle Paleolithic stone tools were discovered.[45]

Absolute dates using TL methods now exist for Qafzeh and Kebara. They reverse the archeological expectation where a little late Middle Paleolithic toolkit might have been made by a very early Modern. Far from it. The Qafzeh dates are 92,000 B.P. while Kebara is only some 60,000 years old. Moderns preceded Neanderthals in this instance.[46] This is confirmed by further absolute dates from Skhul which bracket the hominid remains there between 101,000 B.P. and 81,000 B.P. and so are very comparable in age to the similar looking material from Qafzeh.

These dates lend considerable support to the Gardeners model, although the Sons insist that in such a crossroads region their preferred pattern of continuity will be more difficult to detect. Instead they make their strongest case on the less well dated samples from China, Java, and by extension to Australia. As developed by Alan Thorne and Milford Wolpoff the model distinguishes the center from the edge.[47] In the center gene flow is greatest and variation in phenotypes and genotypes most marked. Investigating the center of each regional population would reveal this variation which makes simple evolutionary schemes impossible to trace since blurring and mixing have taken place.

At the edge decreased gene flow results in less variation. Separation, founder effect, and allopatric speciation are all too possible. In the long term, edge populations will appear very stable but any changes will be dramatic and easily recognized.

Where are these centers? Thorne and Wolpoff regard them as areas of optimal conditions with high population densities and contiguous but variable local populations.[48] East Africa is one of their central areas, at least for the original Ancient, *H. erectus*; Java and China were once an edge to this center but then at some unspecified later point in time either became a center in their own right or an edge to a new regional center somewhere in eastern Asia. The similarities between the Solo river fossils of all ages from Sangiran to Ngandong is one example of a longterm edge, while continuity from *archaic Homo sapiens* to Neanderthals and anatomically modern humans in Europe is also put forward.

What seems necessary for the model of multiregional evolution is the appearance of transitional forms not just in Africa but in each of their major regions. This is claimed for Africa using the fossils mentioned above while in China the Dali and Maba crania are strong candidates as transitional forms.[49] The unbroken line would therefore run from Lantian to the oldest fossils from Zhoukoudian then on in time, but in fact undated, to the *archaic Homo sapiens* crania from Dali and Maba finishing up with Liujiang and the Upper Cave of Zhoukoudian with an undisputed anatomically modern skull and skeleton.[50] But while Wu and Wu see the Upper Cave specimen as ancestral to the modern Chinese others are less convinced.[51] Indeed, the origins of the Asian races particularly the Chinese and Japanese are not currently well understood from the late Pleistocene fossil record (chap. 10). Colin Groves, in a detailed study of sixteen anatomical features which Weidenreich thought linked the Chinese *Homo erectus* material to modern populations in northern Asia, thinks there is little evidence to plead the case. A key feature such as shovel shaped incisors occurs with great frequency among some Mongoloid peoples in Asia but was also the norm among Australopithecines and occurs at frequencies as high as 30 percent in some European populations. He concludes that it is nothing more than a primitive trait, in a cladistic sense, which has, for whatever reasons, been retained more commonly in some rather than other populations.[52]

Rightmire has also challenged the view, favored by Weidenreich, that the Trinil and Sangiran fossils form an unbroken line either with the later Ngandong material (and from there to the Australians),[53] or that a skull from Wajak found in Java in 1889 leads to the modern Javanese.[54] Instead he argues that the Ngandong skulls with a possible maximum age of 100,000 B.P. are much closer to the earlier *Homo erectus* populations. If his view is accepted then replacement seems more likely as an explanation for the appearance of both the Australians and the populations of modern southeast Asia. As with the shovel shaped incisors, which are said to demonstrate continuity, considerable store is set by the flattening of the frontal bones in the forehead among Pleistocene Australians. This has been called "the mark of Ancient Java" since it is found on the Ngandong skulls.[55] On closer inspection Peter Brown was able to show that for some of the Australian skulls such a feature was due to the cultural practice of head binding rather than being a morphological feature indicating where they came from.[56]

Attitudes to change

Should we be surprised at this diversity of views? Hardly, since everyone agrees that trying to determine biological species from fossil data is at best hazardous and at worst impossible. Ian Tattersall, for example, has shown that just using primate skeletons to determine how many living species there are leads, inevitably, to an underestimate.[57] Selecting mates and maintaining species integrity is not done with X-ray vision to assess the finer points of skull shape but is instead based on recognizing aspects of the phenotype—hair, color, markings, horns, tails, scent, calls—that do not survive.

Furthermore, most parties are willing to admit that in some areas "Regional continuity" currently best explains the evidence as in Central and Eastern Europe.[58] But the replacement hypothesis is more persuasive in the western arm of the same continent and in the Near East where the classic Neanderthal skeletons have so far been found.

What probably divides the two models more than the details of their arguments concerning gene flow, isolation, and the longterm stability of populations is, however, the issue of speciation. The Sons favor unbroken lines and no speciation events from *Homo erectus* to modern humans. Gradients, or clines, in anatomical features are part and parcel of larger geographical processes which in the center and edge model keep variation on the boil in some regions and on the back burner in others. Consequently, the definition of a fossil species can be very elusive. Illuminating the process by discovery and description is all. To that extent I would summarize the approach as a longterm view of the maintenance phenotype (chap. 6).

In my opinion this highlights a weakness of the multiregional model with its assumption that genetic populations are maintained at equilibrium. As Cavalli-Sforza and others have pointed out, this cannot explain the single most important fact which is the very rapid colonization of modern humans to the whole earth where the equilibrium of regional populations has clearly been upset.[59]

I have shown how habitat changes encountered by *Homo erectus* to *Homo sapiens* over their Old World range were very varied in both time and space. They witnessed the elevation of the Qinghai/Xizang plateau, the exposure of the Sunda Shelf, the changed rhythms of ice age climates, the radiation of other mammalian species and yet according to the Sons they buffered these changes in other ways than by seizing the opportunity to speciate through colonization, either by exaptive or adaptive radiation. As a result they only underwent relatively minor tinkering to the basic anatomical structure that hotfooted it out of Africa.

The Gardeners, on the other hand, define discrete species rather than an unbroken continuum of regional forms. This is done through the careful enumeration of their shared and derived traits. While disagreement might arise about the tempo of change, be it gradual or punctuated, the view is very much that speciation involves dispersal even though the process might best be understood by vicariance within a larger geographic region rather than by movement from a center of origin. I would therefore expect any buffering behavior to be constantly swamped by the magnitude of habitat changes, to the extent that selection for change to the genome, and its anatomical reflection, became probable. It is also obvious from the recognition concept of species (chap. 5) that the opposite could happen. Very little might change even under severe environmental provocation but a reproductively isolated species could still be produced. Selection might be relaxed so that exaptation via a macroevolutionary effect takes over. According to Turner and Chamberlain only the fertilization system must change at speciation.[60] That may have nothing to do with the anatomy that survives in the fossil record.

If this dichotomy is allowed then we see that the two models are fundamentally at loggerheads over the mechanism of adaptation to environmental change. The Sons are telling us that behavior buffers so that speciation is unnecessary. Gene flow will bring about change according to where populations are located to the demographic center and edge. The Gardeners put the onus for adaptation via speciation onto biological solutions to selection pressures from the environment. This occurs either at a regional scale with, for example, the appearance in the Old World of east/west populations of *archaic Homo sapiens*, or internationally with a possible second African center, this time for the origin of modern humans.

Undoubtedly my conclusion will be disputed by both camps since neither necessarily see behavior as their main preoccupation.[61] By pleading the case for a Neanderthal phase in human ancestry, the Sons are committed to a Darwinian position of slow, cumulative development.[62] Cultural behavior is just a convenient method to explain away major and apparently sudden changes in the fossils that would contradict this view. Brace and Wolpoff have both argued that an increase in culture rendered biological solutions such as big teeth and large noses unnecessary in the face of labor saving stone technology and adequate clothing. They claim the Upper Paleolithic was a culinary revolution that had far reaching and dramatic effects for the rapid evolution of the European face. Rough flake knives were swapped for precision stone blades while earth oven cookery now tenderized meat to the point where tearing it with big incisors became passé. The Gardeners, by comparison, are avowedly migrationist in either a genetic or population sense. Any jumps in the archeological record are used as support for their case, with peoples replacing peoples, their new cultural behavior ousting existing behavior. Nor does this mean that they all support a punctuated tempo to human evolution or that rapid change is more likely to occur in a peripheral population. In this regard both models are remarkably similar: they both have centers and edges; they both see evolution as gradual. The main difference is between the role of colonization versus *in situ* development to account for any regional changes. The adherence is to the process of adaptation, irrespective of whether it led to longterm continuity or new species.

Behaviorally modern humans

Is it at all astonishing we should discover that anatomical Moderns were intelligent, tall, and good looking? A "finely chiselled head poised on a well balanced vertebral column" as Grahame Clark put it almost fifty years ago.[63] Changes in the fossil evidence have been interpreted in behavioral terms with cultural adaptations replacing biological solutions. Big, robust, and very strong Ancients using simple tools such as short stabbing spears tipped with relatively heavy triangular stone points became gracile Moderns who used their brains rather than their muscle to equip themselves with more efficient weapons systems. Even when the hunt was done the sounds around the camp fires of the Ancients tearing at their tough food with their massive incisors was replaced by the delicate chewing of food, cooked until tender, and sliced with purpose made stone knives.

For the Gardeners the differences are judged to be too great, the timescales too short, for any other interpretation than replacement from outside. By contrast, the Sons welcome everyone aboard their Ark. They complain that to leave any of the Ancients on the shore is to commit hominid catastrophism, where scientists act as God in deciding who is and who is not worthy of inclusion in human ancestry, with Neanderthals the all time losers.

Their case for behavioral continuity has recently been put by Geoff Clark and John Lindly. In the long run, they argue, cultural solutions to universal problems of survival gained the upper hand. Strength and endurance were replaced by new behavior and linked technological dependencies.[64]

This brings us straight back to the problem of how we think ourselves into the so-called primitive mind and the existence, or not, of something like *partially* modern behavior. What evolutionary significance can we attach to Neanderthal differences? Was it, as Grahame Clark also supposed, a quality of mental evolution between the time of the Ancients and Moderns that was reflected in their respective achievements in technology and material culture? To what extent can this be regarded as *the* human revolution?[65]

Stone tools

It has long been noticed in Europe that the appearance of anatomical Moderns coincides with new forms of stone tool manufacture and different types of retouched implements. The Upper Paleolithic implements made by the Moderns stands in contrast to the Middle Paleolithic fashioned by Neanderthals. The change took place between 35,000 to 40,000 years ago.

As research has unraveled the sequences in other parts of the world a broadly similar pattern has been traced. Throughout sub-Saharan Africa the Middle Stone Age (MSA) is followed by the Late Stone Age (LSA) with a very similar suite of techniques and types of stone tools to those in the European Upper Paleolithic. Recent evidence from China supports the pattern while in India and the Near East abundant material also exists for similar technological changes about 40,000 years ago. I will show later that this was not the case in Australia.

How big was the change? The main difference revolves around the way that stone nodules were prepared for flaking. The Levallois and disc techniques were superseded by a form of core preparation which trimmed nodules so that many light, slender, parallel sided stone blanks could be detached. These elongated forms, known as blades, varied in size from large down to microlithic but, by definition, are always twice as long as they are wide. While such blades are found in Middle Paleolithic and MSA assemblages they are usually thicker and broader and never the dominant blank which, with retouching, was turned into a variety of implements such as end scrapers, awls, points, and knives. The Upper Paleolithic/LSA had an obsession with blades while large implements such as handaxes vanish altogether.

The explanation for this development in stone working is usually put down to the practice of combining several stone components to make a tool. Blades are both lighter as well as forming segments which can be mounted as long continuous cutting edges in knives, arrows, or as end scrapers and borers at the end of wooden or bone hafts and handles. These may well have been the most important part of the tool. The advantage of blades, as opposed to flakes, lay in the ease with which a new segment could be made and fitted. Greater attention to repairing equipment when it was broken rather than starting again from scratch was the hallmark of such maintainable technologies.[66] Investing more time in acquiring and selecting better raw materials as well as in the preparation and manufacture of tools reflected the importance and usefulness of that implement in reducing risk.

Such investment in materials and tools will be curated so that repair and reuse will be the order of the day for hunting trips. Lewis Binford considers this to be another aspect of planning ahead, where selection to reduce risk hones such behavior to help survival.[67] In that sense the technology of behavioral Moderns is also under selection from the scale and complexity with which they exploit their habitats. The implication is of course that the Ancients' technology, although under the same selection pressures, was geared to different scales of exploitation and organizational complexity and hence the differences in tool shapes and knapping skills.

The Pioneer phase from 60,000 B.P. to 40,000 B.P.

On closer examination this neat division between the Middle and Upper Paleolithic or MSA to LSA becomes decidedly fuzzy (table 8.2).[68] In order to examine what is going on I have nominated a crucial twenty thousand years as a Pioneer phase. You will see from table 1.2 that it is peopled by Ancients of various types including archaics, Neanderthals, and Moderns. It overlaps in time with the Moderns, as defined through a spectacular burst in colonization that occurs after 50,000 B.P. in some parts of the world such as Australia. The Pioneer phase sees only limited geographical gains, hence my choice of name. It also shares some artifact and fossil traits with the real Ancients and the thoroughly Moderns (table 8.2). Such special status focuses on how archeologists expect change to manifest itself in the materials of the past.

For example, hafting precedes the Upper Paleolithic. In North Africa there are some obvious projectile points with a large shaped tang from assemblages known as Aterian. Desmond Clark confidently regards these as the earliest evidence, possibly as old as 100,000 years ago, for a bow and arrow technology.[69] The rest of the assemblage is typical Middle Paleolithic which means flakes often prepared using the Levallois technique and many retouched pieces, typically side scrapers.

In fact it is possible to see from a number of places that large, hafted projective points using the Ancients' approach to stone knapping are common between at least 60,000 and 35,000 years ago (fig. 8.3). In Europe they come in all shapes and sizes.[70] In southern Germany at the Weinberg caves in the Atlmühl valley, a tributary of the Danube, a Middle Paleolithic assemblage of flake tools contained almost a hundred large leaf-shaped points.[71] They had been made on thin tabular blocks of flint that occur locally and had been carefully worked on both faces. At their base is a clear hafting notch while their size, up to 12 cm long, means that they must have been hafted on a spear rather than as arrows.

Variations on this leaf point theme have turned up in small collections of stone tools from England, Belgium, eastern Germany, Poland, Bulgaria, Greece, Hungary, Czechoslovakia, and along the Dnestr river in Russia.[72] Sometimes the rest of the collections are made on flakes and therefore called Middle Paleolithic, while elsewhere they occur with flints knapped according to the best traditions of the Upper Paleolithic, as in England at Kents Cavern 38,300 years ago.[73]

As a result these assemblages with leaf points have been treated as transitional industries between the two European traditions of stone knapping. But even in the time bracket 60,000 B.P. to 35,000 B.P. they do not always appear. Standard

Table 8.2 The changes in behavior between the Ancients and Moderns, as captured in the archeological record. Clearly the period 60,000–40,000 years ago, which I refer to as the Pioneer phase, was a crucial one, for it is only after this point that so many of the elements associated with modern behavior are to be widely found.

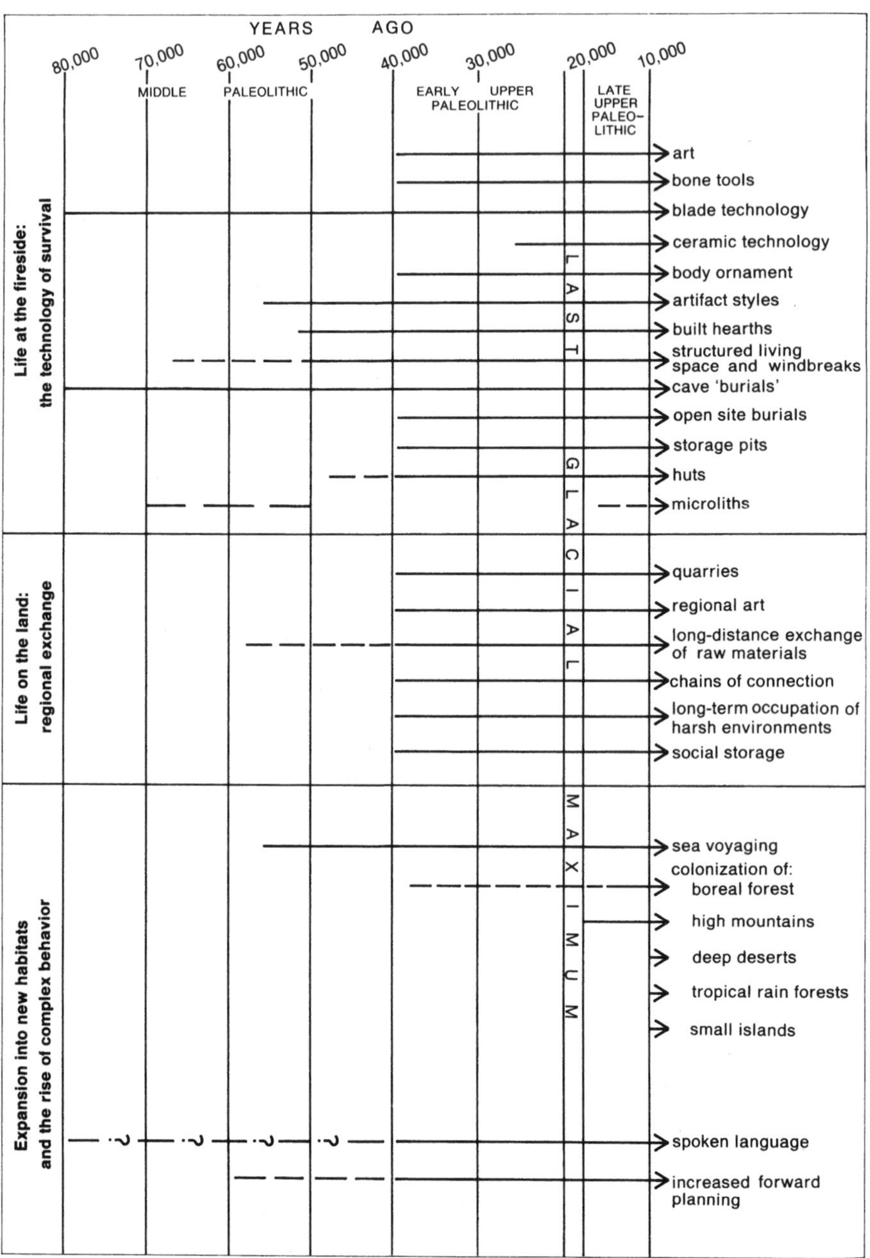

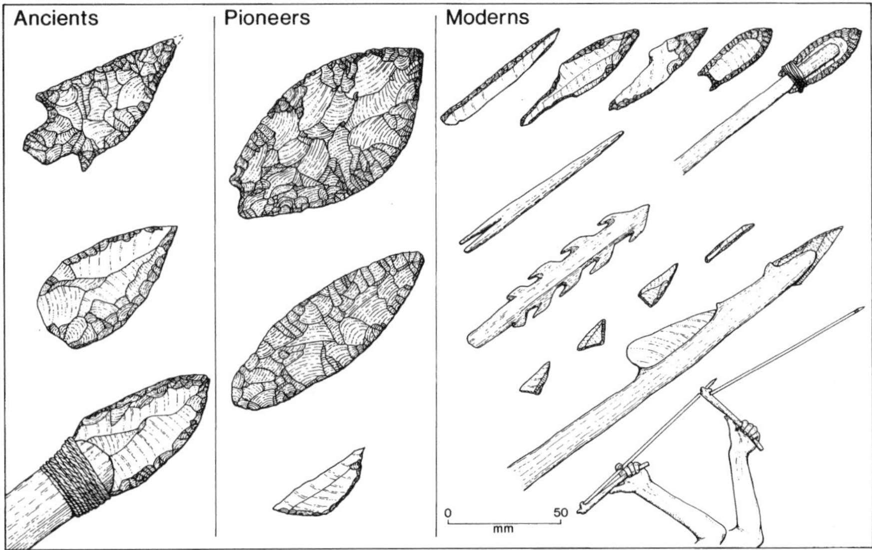

Figure 8.3 Technological changes in hunting weapons. Tanged Aterian points (possibly the oldest arrows) and simple Mousterian points were hafted by the Ancients. During the Pioneer phase we find bifacially worked leaf-shaped points from Germany, England and across much of Europe. These were probably mounted as spear tips. We also find geometric-shaped microliths (an element in the Howiesons Poort industries) from sites such as Klasies River Mouth. The Moderns, who were marked technologically by the appearance of the Upper Paleolithic, introduced further changes in weapon size and materials. Using blades as blanks, we now find blunted-backed, tanged, shouldered and fluted (shown hafted) points. Lightweight bone points, with bases split for hafting, and antler harpoons are also found. A variety of geometric-shaped microliths and small backed blades are now mounted as segments in knives and arrows. Finally spearthrowers made of antler are found for the first time.

Mousterian assemblages are still common while at the Bacho Kiro cave in Bulgaria, Janusz Kozlowski has excavated a blade assemblage, without leaf points, dated to >43,000 B.P.[74] At the other end of the continent in Cantabrian Spain, the deep stratigraphy of the El Castillo cave, currently being excavated by Victoria Cabrera-Valdés, has a blade industry, closely linked to the earliest widespread Upper Paleolithic culture known as the Aurignacian, and dated to 40,000 B.P.[75]

There is no doubt that after 35,000 B.P. Upper Paleolithic type industries sweep the board, not only in Europe but across much of the Old World, with Australia forming a notable exception (see chap. 10). Prior to this was a long period when assemblages varied considerably in both tools and techniques.

The leaf points of Europe generally come from very small collections, just a few hundred flints at most. We know most about them from caves along the upland arc that lies to the south of the North European Plain. Irrespective of the debate over their transitional status it seems more appropriate to consider them as part of a

Pioneer phase in using these upland areas during the cold, but not coldest, conditions of the last ice age. For the first time we see small hunting parties moving into the hills, supplying themselves, and repairing broken equipment around small hearths lit in the entrances to caves. The implications from these meagre traces of very distinctive flint points are that groups were better able to cope with social fission and hence the scale and logistic complexity at which they exploited landscapes had increased (see chap. 9).

In the Near East and Africa the Pioneer phase is even more complicated. At the cave sites of Klasies River Mouth (South Africa), Tabun, Amud (Israel), and in Cyrenaica at the Haua Fteah there are industries regarded as Pre-Upper Paleolithic, PUP for short.[76] This means that they contain blades and, in the case of the Howiesons Poort assemblages from Klasies River Mouth and a number of rock shelters in Lesotho, microlithic crescents and lunate segments, made on non-local stone and obviously hafted.[77] Yet far from these innovations carrying all before them, they appear then disappear. At all these sites the PUP levels are followed by the return of flake based Middle Paleolithic/MSA industries. What appears to be a considerable time elapses before the Upper Paleolithic/LSA appears and stays for good.

The monolithic division of stone technology based on making flakes and blades between the Middle and Upper Paleolithic is not the force it once was. The PUP may come and go, as in Israel, while in the same area classic transitional industries are found. Nowhere is this better demonstrated than in Anthony Marks' excavations at the Negev Desert open site of Boker Tachtit.[78] From a single horizon he was able to fit back together the flakes and blades which had been struck from a large nodule. The knapped end product was a particular type of projectile point. From one end of this nodule the knapper had been making them using a Middle Paleolithic Levallois point technique. As the size of the nodule dwindled he/she switched to an Upper Paleolithic blade technique that made identical points. One lump of flint, two different traditions which should be unrelated and one common end product. Overall, the four levels at the site, the earliest of which is dated to 47,000 B.P., show the technological transition from Middle to Upper Paleolithic techniques while the end product of triangular Levallois points remains the same.

Further blurring of the Middle to Upper Paleolithic cultural transition comes from the association of the anatomically modern remains of Skhul and Qafzeh with Middle Paleolithic tools, while at St. Césaire in southwest France the reverse happened and a classic Neanderthal, now dated to 36,000 B.P.,[79] has been excavated in a level containing the earliest Upper Paleolithic industries in the region.[80] Such seeming role confusion, at least by standard archeological thinking, is not surprising when we remember that the stone tools of the Pleistocene Australians show that *Homo sapiens sapiens* was making what was essentially a Middle Paleolithic stone toolkit when they first arrived.

Ethnic groups and the land

The conclusion must be that no crisp boundary exists for the Upper Paleolithic revolution and that behavioral Moderns are not best described by the techniques and types of stone tools they made. An alternative is to treat different assemblages of stone tools as representing tribal groups. The eminent archeologist of the Paleolithic François Bordes argued this for the material from the classic sites in southwest France. Here he identified through detailed analysis of many collections excavated from caves and rock shelters in the Dordogne and Charente five basic types of Mousterian assemblage.[81] This Middle Paleolithic industry was named after the rock shelters above the village of Le Moustier where early excavations produced a ton of flints and later some important Neanderthal skeletons.[82] The five assemblages contained different numbers of side scrapers, points, finely made triangular handaxes, and crudely flaked pieces known as denticulates because of the gap-toothed look to their edges. Bordes interpreted this patterning as the material culture of five Neanderthal tribes who occupied the rock shelters of the Périgord region and made and left behind these stone age visiting cards as proof of their ethnic identity. Writing in his *The old stone age* he described the process as follows:

> Man is more ready to exchange his genes than his customs, as the whole history of Europe demonstrates. If a woman from the Quina-type Mousterian was carried off by an Acheulean-tradition man, she may perhaps have continued to make her tribal type of thick scraper . . . but after her death probably no one went on making them.[83]

The iron hand of tradition which dictates that Frenchmen wear berets and Englishmen bowler hats was, according to Bordes, honored in deep time. Instead of hats, Neanderthals, and then Upper Paleolithic populations, said it in stone. In the archeologically rich caves and rock shelters around the Dordogne, these five assemblages are found one on top of another in deposits which span many millennia. Of these the most important is the rock shelter of Combe Grenal, dug by Bordes, and producing no fewer than fifty-five separate levels with these different assemblages of tools.[84] To explain this longterm patterning Bordes turned to an older tradition which had been strongly put by Denis Peyrony, one of the great excavators of the Paleolithic.[85] For both men the Dordogne and particularly the Vézère valley was a privileged corner, a Garden of Eden, which was actively competed for by Paleolithic tribes in both the Middle and Upper Paleolithic. Those who succeeded got to live in the rock shelters such as Le Moustier, Combe Grenal, La Ferrassie and later in the Upper Paleolithic at Laugerie Haute and La Madeleine.[86] If their star waned then there was always another group eager to push them out and leave a record to the fact. This consisted of assemblages made up of different proportions of tools from the basic list of types, as well as some distinctive type fossils such as handaxes, and in the Upper Paleolithic projectile points and tools made of bone and antler.

Strong identification with the region was, and still is, a hallmark of the approach. Peyrony lived in a house that sits inside the Laugerie Haute rock shelter and on top of Paleolithic deposits, while Bordes wrote many science fiction novels with a Paleolithic twist, under the nom de plume Francis Carsac, a name taken from the Périgordian village where he lived. Regional and group identity emerged in their works as a continuum from the Neanderthals through to the present. Bordes argued the case that the earliest Upper Paleolithic in the region evolved directly from the local late Mousterian. The St. Césaire find from the Charente-Maritime would be just the sort of evidence to clinch his argument.

The sort of assemblage patterns which Bordes brilliantly distinguished for the French Neanderthals have now been claimed between regions in Africa and China and are regarded by Desmond Clark as distinguishing MSA behavior from the Early Stone Age.[87] Compared to France the African regional groupings are at the moment more inter- than intra-regional. More research will almost certainly change the picture.

But alternative views have forced a reassessment of the cultural and ethnic explanations for differences between assemblages of stone tools. The Binfords argued twenty-five years ago that an area such as the Dordogne had to be seen as a region over which people moved on a daily, seasonal, and yearly basis.[88] As the location of their settlement changed and the tasks which needed carrying out varied this would be reflected in the manufacture and discarding of stone tools. Over long periods of time what was left behind in the rock shelters were not name cards but worksheets.

Later Lewis Binford expanded the explanation to consider very different patterns of land use between the Neanderthals and the Moderns. The former, he argued, made, used, and threw away tools when and where required. By contrast Upper Paleolithic tools of the Moderns were made with future use in mind. They carried them around the landscape and repaired and recycled them when they broke. He called this *curation* and contrasted it with the immediate or *expedient* making, using, and discarding of stone tools. Where tools in a curated system were finally thrown away they might bear no relation to where they were actually used or broken.[89] What separated the two technologies, and by inference the Ancients from the Moderns, was an increased depth of planning among the latter.

The typological basis of Bordes' scheme has also been questioned. Harold Dibble has shown that different types of side scraper are more economically interpreted as stages in resharpening and reuse (fig. 8.4). Michael Barton, working with Spanish material, stresses this point by showing that the edge rather than the whole shape of the piece, or tool, is what counts.[90] According to his regroupings there are no longer five tribes but just two types of assemblage distinguished by the edges found on scrapers and those on the irregular, notched denticulates.

Bone tools and raw materials

The stone tools of the Pioneers now reveal much more complex patterning than what was once seen as a simple cultural revolution. But for all this some other common Upper Paleolithic tools do not appear at all in the Pioneer phase. Artifacts made of antler bone and ivory only appear with the Moderns (table 1.2).

The LSA/Upper Paleolithic is the first time, worldwide, that any bone and antler tools are found.[91] Since wooden artifacts are always rare because of lack of preservation, the use of bone, antler, and ivory assumes greater significance.

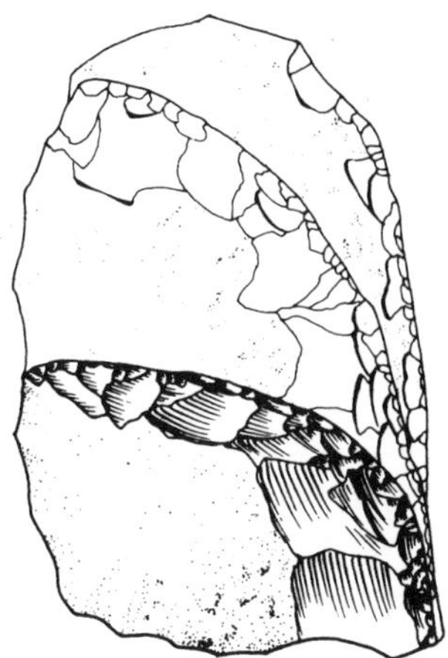

Figure 8.4 The biography of a stone tool. Dibble illustrates a cautionary tale for stone tool typologists. Instead of conforming to discrete types, this scraper, which he made, passed through three differently recognized types as it was used and resharpened.

Their appearance in the Upper Paleolithic/LSA alongside blade based technologies fits in with notions of hafting and weight reduction. Initially these organic materials were used for making simple hafted points and only much later eyed needles and elaborate harpoons. But from their first use they include carved figurines, ornaments, and jewelry.[92] Moreover, in many parts of the world it takes some time for these to appear in any numbers. In Europe bone tools and ornaments are only common after the last glacial maximum at 18,000 B.P. and are therefore separated from us by almost as much time as they were from the earliest Upper Paleolithic in the continent some 40,000 years ago.[93] Janette Deacon has made the same point for the numerous LSA assemblages excavated in South Africa and where it is not until the Post Glacial and after 12,000 B.P. when the climate picks up, that such objects become common.[94]

Raw materials provide another means to look at the revolution. In a recent survey of the Middle Paleolithic, Wil Roebroeks, Jan Kolen, and Eelco Rensink have shown that in Europe the use of higher grade, exotic stones does in fact predate the Upper Paleolithic.[95] Some raw materials were being obtained from distances of up to 300 km although most came from within 50 km. Distances increase during the Upper Paleolithic when it is common to find flints moving 400 km and fossil shells, for use as ornaments, as much as 700 km from their source. Interestingly, the break is not 40,000 years ago with the Upper Paleolithic revolution but rather with sites older than 200,000 B.P. Before this only very local stones were used as we have also seen in the much older Oldowan and early Acheulean of Africa.

Hearths and burials

By now we are beginning to see that at best the earliest Upper Paleolithic/LSA is divided from the Middle Paleolithic/MSA by a dotted rather than a solid line. Either side some features are present/absent, such as stone blades/bone tools, but even more elements which differ by degree such as blade technology and the use of exotic stone. These shared traits are defining my Pioneer phase (table 8.2).

Well built hearths are rare on the Pioneers' sites. Luís Raposo has excavated a stone-built hearth at Vilas Ruivas in Portugal. It was found with a Mousterian assemblage dated to the last glaciation, possibly 50,000 years ago.[96] Hearths of similar age, but less substantial, are known from the multilevel open sites of Molodova on the Dnestr in Russia. The significant fact is that they appear at all since the older sites of the Ancients do not contain them (chap. 7).

With the Upper Paleolithic well built hearths become common. At the Abri Pataud, by the Vézère, Hallum Movius excavated a row of large stone-lined Aurignacian hearths running parallel to the back wall of the rock shelter.[97] The earliest date to 34,000 B.P. The appearance of built structures in rock shelters is matched, as Jan Simek has shown, by spatial patterns within them as areas are used for special purposes rather than as haphazard dumping grounds.[98] Dwelling and storage pits are found dug into the loess on open sites in southern Russia as early as 32,000 B.P.[99] These are arranged in a regular fashion around hearth settings. This investment in camps is strong evidence for modern behavior where building for the future, reuse, and planning are taken in our stride.

More contentious is the evidence for burials. The issue has recently been reopened by Robert Gargett who casts considerable doubt on the claims for Neanderthal burials, many of which were dug up in the early years of the twentieth century.[100] However, the response to his article has been to exhume the certificates signed and witnessed by archeologists at the time of discovery saying they saw what they saw, rather than what they wanted to see.[101] This involves grave pits and claims for grave goods with some of the skeletons. The latter usually comprise only stone tools, animal bones, and the occasional lump of red ocher. Whether these are corpse disposal or burial is another matter. Burial is a highly variable rite. It can involve either complete interment in a pit, the redistribution of parts of the skeleton around a region, or cremation.[102] A femur placed in a crevice in a rock shelter is as much a burial in one society as a full body, lowered into a pit, is in another. We have already seen the dangers of interpreting stone tools as cultural and ethnic markers. The same caution has to be applied to these complete Neanderthal skeletons. Continuity cannot be assumed between the remote past and the present just because the burial practices are superficially similar.

I would stress instead that for Europe and the Near East the interesting thing is the recovery for the first time after 100,000 B.P. of complete skeletons. Indeed nearly all the Neanderthal bodies so far dated lie in the range 60 Kyr to 40 Kyr B.P. and which I have taken as the limits to my Pioneer phase.[103] It is very interesting to see how bodies of either Neanderthals or Moderns first turn up in those parts of Europe and the Near East where remains of bear, hyena, lion, and wolf are poorly represented in those same caves. This suggests to me that at earlier times and in other places these carnivores were using caves for hibernation and as dens during which they dug the floors and so disturbed any skeletons. Finding complete bodies is not necessarily the signal for the appearance of spiritual thought, the glimmerings of humanity as many claim, but rather the first opportunity for whole skeletons to survive because, for some reason, the carnivores changed their behavior. Elsewhere at the Circeo cave in Italy the carnivores continued to leave behind their unmistakable traces on an isolated Neanderthal skull.[104] With these natural processes contributing to the evidence there is clearly more work needed to unravel what went on.[105] But even before this work is completed one pattern stands out. It is only after 35,000 B.P. that any complete skeletons from open sites are found in the Old World. These include the remarkable triple burial from Dolní Věstonice in Czechoslovakia dated to 27,640 B.P. with two males flanking a gracile skeleton thought to be female.[106] The Mungo remains at 25,000 B.P. in southeast Australia consist, as already mentioned, of a cremated female and, slightly later, the complete burial of a man with a diseased elbow and molars worn down by use as a vice to twist fibers, possibly for making fish nets.[107]

Art and ornament

The Pioneer phase is also notable for its lack of art and ornament. The lack of such items as grave goods alongside Neanderthal bodies might, some claim, be evidence that the Pioneers and Ancients had rites and ideologies that did not require such objects. The sudden appearance of art after 40,000 B.P. marks a great watershed in the prehistory of material culture. John Pfeiffer has aptly termed this the "creative explosion" when on a worldwide basis we find evidence for ornament and jewelry and engraved stone plaques, carved figurines, figures pecked on rocks and iron oxide pigments daubed on cave walls and ceilings.[108] Is such a creative explosion better evidence for a human revolution and a worldwide replacement of existing ways of life? Or is it, as some claim, further evidence for continuity in the potential of peoples to produce art but a discontinuity, for whatever reasons, in terms of their need to realize that potential?

What does the evidence reveal? When the rock engravings from Karolta in South Australia, now dated directly to as much as 32,000 B.P.,[109] are compared with the colored images on two small stone slabs dated to 26,000 B.P. from the Apollo cave in Namibia or the lion headed anthropomorphic figurine from the Stadel cave near Ulm in southern Germany at 31,000 B.P., what is impressive, but not surprising at such a scale, is the variety of mediums.[110] The Mungo male burial also shows the practice of evulsion where some of the front teeth are knocked out, perhaps during initiation.[111] This raises the strong possibility that for a similarly wide sample, bodies were also being painted, scarified, and pierced while no doubt hairstyles were quite the fashion just as they are today.

Art is a broad term and like burials and hairdos, embedded in our own preconceptions of what it is and what it is not. This should make us cautious in our interpretations. Moreover, Pleistocene age art varies so much in time and space that it is not possible to do it adequate justice here. However, I would emphasize that its content and mediums were far more diverse than is commonly appreciated by a brief perusal of the coffee-table books devoted to the spectacular and, to us, easily accessible painted images of mammoth, horse, bison, and ibex from the French caves. These are once again late, dating from about 16,000 B.P. to 9000 B.P.[112] The earliest rock paintings may well be Australian; George Chaloupka has made the case for the oldest painting style in Kakadu, east of Darwin, to be in the order of 35,000 years old. A first radiocarbon date of 20,320 B.P. for blood protein, incorporated in the pigments of a painting at Laurie Creek in the Wingate mountains of the northwestern Northern Territory, indicates that he may well be right in arguing for a very considerable antiquity.[113]

Art in its broadest sense is universal, but limited to humans. Artistic ability is not faithfully capturing a likeness but rather decoding, interpreting, and acting on the messages in this medium through an appreciation of style. This in turn is derived from social contexts such as story telling, ritual, and history, the entries in Pfeiffer's tribal encyclopedia. Art expands and enriches such social knowledge. It can act as a sign to chunk enormous quantities of information into manageable proportions. As a result memory is expanded either in a practical sense, for example the signs and information in a map, or ideologically by substituting action with signs, where a simple object such as a crucifix can stand for a complex system of belief. It is unthinkable that modern humans could be modern humans without such constant reminders from material culture and their stylistic and artistic codes. At the moment there is no good evidence anywhere that the Pioneers, Ancients, or anyone else for that matter produced or utilized these properties of material culture prior to 40,000 B.P. A few older but contentious bones with zig-zag marks[114] are best accounted for with prosaic explanations, as Philip Chase and Harold Dibble have recently documented.[115]

The situation can be summed up for the Pioneer phase as follows:

60,000 to 40,000 B.P.:
Middle Paleolithic and MSA: stone tools—no art, no bone tools
Pre Upper Paleolithic: stone tools—no art, no bone tools
Pioneers/Ancients (e.g. La Chapelle aux Saints)—no art, no bone tools
Moderns, older than 40,000 B.P. (e.g. Qafzeh)—no art, no bone tools

after 40,000 B.P.:
Upper Paleolithic and LSA—art, body ornaments, bone tools

This does leave one major area still open—stone tools. Were those leaf points, or indeed earlier some of the handaxes, refined through flaking that was not just directed by functional requirements?

The question is how does material culture, stone tools included, do more than just perform mechanical functions? An ethnographic example will help to define the options. Polly Wiessner has convincingly shown how the styles of modern iron arrow heads and decorated headbands and aprons correspond to the different goals of group and individual identity.[116] The Kalahari !Kung San among whom she conducted the study are like any other peoples negotiating their corporate and individual positions in a flexible social world. The arrowheads are visible when hunters from different groups meet. Interestingly, the styles correspond to the language groups from which the hunters come. The bead work shows no such correlation. Here individual skill and taste is expressed as the women seek their own social goals. Social context, message, and medium are thus intertwined. Style as an element of human cognition works through comparison. Through this process, as Wiessner argues, it can renew, alter, and create social relationships.

But was extra information prior to 40,000 B.P. added to some tools through their design and form? Does the similarity shown in some longterm stone working traditions also imply comparison and hence a modern route to negotiating social relationships? Martin Wobst has argued that if art is being used as a communication system, then it is an all or nothing situation.[117] Either everything is put into context and has meaning, so that leaving a bone-point blank rather than carving a design on it in fact carries information, or nothing is. Only in this way can ambiguity leading to misinterpretation be avoided.

It therefore seems improbable that the Ancients would be so selective as to devote their artistic outpourings solely to stone tools. They may, for example,

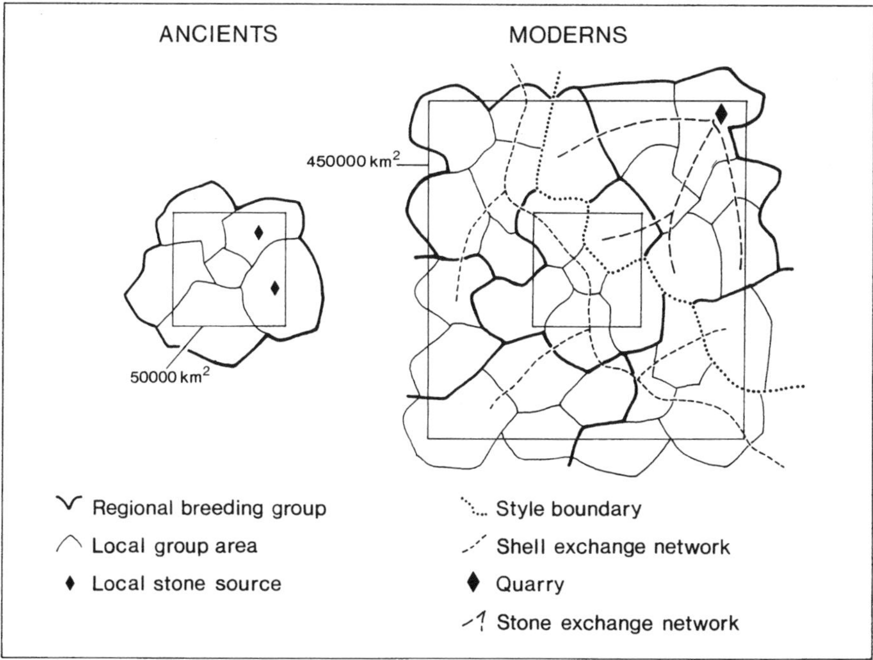

Figure 8.5 The social revolution. The main difference between the societies of the Ancients (including the Neanderthals) and the Moderns was one of scale and complexity. I interpret the archeological evidence for increasingly far-flung exchange and social interaction as an indication that the emphasis of survival shifted 40,000 years ago from local self-sufficiency among the Ancients to a system of storage based on social ties and alliances, thus reducing the risk of starvation. With this leap in scale came the unexpected opportunity, through exaptation, to expand territories and acquire raw materials (especially shells, amber and other important social "trinkets") over much greater distances. Colonization could then proceed rapidly as, perhaps unintentionally, social solutions were found to survival problems in most environments.

have tattooed their bodies and arranged their hair into elaborate displays. This seems to me unlikely for the simple reason that they probably did not need to. The absence of art is, I feel, significant in the context of the scale of their societies. Lower levels of fission and the simple alliances outlined earlier would not have needed reinforcement by material culture produced specifically to engage participants in the multiple realms of social life that existed, simultaneously, in their past, present, and future. The scales of social memory and geographical exploration during migration were exaptively linked (fig. 8.5).

The lack of art associated with all the Ancients and the early Moderns such as the hominids from Omo Kibish, Qafzeh, and Skhul does, I believe, put them outside even the broadest definitions of what a behavioral Modern might be. Adding to the range and shapes of stone tools is not enough by itself. Imitation of actions rather than comparison between objects and their social context is still an adequate answer to the question of why some stone tools such as handaxes and stone knives appear to be so "stylistically" similar.

The information content of all types of material culture—stone tools, figurines, body painting—rather than the ability to imitate would eventually be needed as a major element in constructing society at new scales and levels of organizational complexity. As a result elements of the full Upper Paleolithic package such as modern skulls, blade technologies, hearths, even, I would suggest, undecorated bone tools, will be found over wide areas and throughout the period from 200,000 B.P. to 40,000 B.P. This does not mean that changes to the scale and style of society go undetected until Pfeiffer's creative explosion. During the Pioneer phase those hunting parties with leaf point armaments, found in the uplands of Europe, represent a move towards the creation of larger social systems. Art and ornament, on the other hand, will never be much older than 40,000 B.P.

Language and the evolution of speech

As might be expected, a further human universal, spoken language, has been added to this part of the equation. Is the late appearance of art in human evolution somehow explained by necessary cognitive developments, the quality of mind Clark spoke of? Is its appearance linked to similar patterns of cognition that support articulate speech? Or are they completely separate so that speech and language could have been produced from the very earliest times?

At the outset I would distinguish between speech (the production of sounds), communication (which can involve gestures and speech), and language which, even after the success of teaching chimps and other primates sign languages, is still regarded as unique to humans.

Speech as we know it is only possible with a modern voicebox and, as Philip Lieberman has shown in a series of pioneering studies, depends very much on the position of the back of the tongue in relation to the vocal tract since this regulates the air passages and controls the range of vocalization.[118] In his original study, published with Crelin, he claimed that a reconstruction of the voicebox of Neanderthals could not have produced some key vowel sounds—

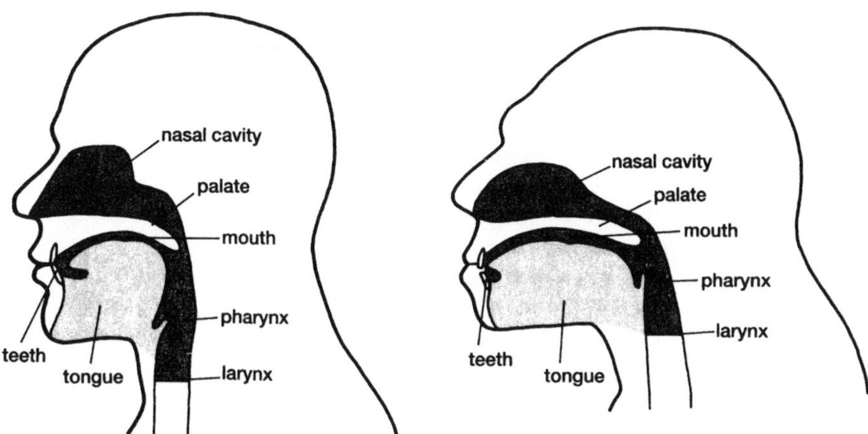

Figure 8.6 The Neanderthal voice box differed significantly from modern humans (left). Sounds are produced in the larynx, but to form words the tongue must vary the size and shape of the mouth and pharynx. The Neanderthal pharynx is limited by the larynx, which sits higher in the throat than in modern humans. Furthermore, since the tongue is long and rests almost entirely in the mouth rather than the throat, it can only alter the size of the mouth. This single-chamber acoustical system may have restricted Neanderthals to a slow and limited form of speech.

a, i, u—and would have had difficulties forming their g and k. To put a curved, modern tongue into a Neanderthal supralaryngeal vocal tract means that their larynx ends up in their chest (fig. 8.6). This is impossible. Modifications have to be made to reconstructions of their tongues and larynx and hence the restriction in vowels and speech production follows.

This does not mean that Neanderthals could not speak but rather that they spoke slowly thus missing out on the high transmission rate of human speech. In other words, the listener might forget the beginning of the sentence before it had been finished. What his provocative reconstruction suggests is that a key element of speech production, without which language would be less than best, was under selection to evolve. The voicebox was not a perfected mechanism whose time had come. It had to undergo evolution to its present state in the same way that babies' vocal tracts are similarly undeveloped for full speech and must undergo "evolution" during normal growth. It was always a voicebox but became adapted to a more sophisticated function. Indeed, the completely different function and quality of speech points to the process as one of exaptation.

The selection pressure for these changes in the voicebox comes from the advantages for speech in the form of phonetic output. As Lieberman remarks, this selection was so strong that other disadvantages were shrugged off. Among these is the deficiency of the human supralaryngeal tract for swallowing. We choke when we eat if food falls into the larynx and blocks the passage to the lungs. No other animal is so prone to killing itself by eating. Perhaps we would have done better to keep our mouths shut.

His reconstructions have been challenged. It has to be admitted that reconstructing voiceboxes from the angle of skulls on vertebral columns, basicranial morphology, and the shape of mandibles is bound to lead to controversy. More recently the hyoid bone, which forms the Adam's apple, has been studied in detail in the Kebara Neanderthal.[119] There is no significant difference from modern examples, and it is suggested that the larynx beneath the hyoid has also been unchanged for at least the past 60,000 years.

Speech and language require more than a voicebox. Lateralization of the brain is also a necessary but not sufficient requirement, with Broca and Wernicke's areas controlling speech in the dominant hemisphere. In particular Broca's area, as Passingham shows, controls the sequencing of the vocal cords and directs them according to context.[120] This adds up to extreme vocal skills. The earliest brain cast showing Broca's area comes from the 1470 *Homo habilis* skull from Koobi Fora.[121] Indications of asymmetry, and hence lateralization in the brain, have been traced through preferred handedness in the manufacture of stone tools as well as endocasts of the brain of *Homo erectus* to 1.5 Myr.[122]

But does this help us to distinguish between communication and language? Passingham maintains that primates, and indeed all animals, can specify locations, as with the dance of the honey bee, but that they lack the means to *describe* the outside world. In his terms a system of communication is a language only if it can be used to communicate about the outside world, which it does through symbols.[123] In my model in chapter 6 such a description is provided by the contrasting phenotypes of individuals returning from foraging trips. But this sort of communication together with facial gestures, calls, and grooming rituals, is clearly quantitatively and qualitatively very different from the sort of information that can be stored and transmitted though spoken language. Naming and counting things emerge as the critical factors in assessing language skill since they describe the outside world. But how and when did these skills originate?

Once again there is a long and short chronology. Gordon Hewes has consistently argued the long chronology for the gestural origins of language.[124] He sees tool use and manufacture as critical evidence for gestural languages, comparable to modern sign languages, so that when spoken language took over relatively late and as a result of cognitive developments it appropriated rather than invented a language system.

This view is not widely favored at the present time although judging by its past history it will no doubt come back into fashion. Instead, Jerison sees language stemming from environmental selection for more cognitive capacity rather than improved communication skills.[125] If communicating about the outside world had first taken place with a gestural language he wonders why Broca and Wernicke's areas are where they are in the brain instead of being close to those areas which control the hand and thumb. He sees language as a substitute for the loss of olfactory senses among early hominids who now produced "maps" of their territory through sound and shared these cognitive constructions with others.

Talking pictures

A recent approach by Iain Davidson and William Noble puts the case for a short chronology to language as we know it.[126] Language, they maintain, transforms humans into modern humans. It is a system of recognizable meanings arising out of shared and repeated signs. For them the critical transition between communication and language can be traced through depiction, which brings art into the process of language evolution. What depiction does is change the rules. Previously communication depended on the immediate context to interpret and act on the signals which were being picked up by the eyes, ears, nose, and skin. Depiction through signs, freezing the image of people and animals, indicates a different system at work, one which is independent of context and so capable of reflection and narrative. Depiction is based on the memory of social and physical environments. What is signed and depicted only has meaning, defined narrowly as the ability to communicate to others about the outside world, if it is also subject to reflective language where an attitude of reference is adopted to the perceived world. Rather than just monitoring the environment and solving these migration equations, as many animals do, our unique feature is that we reflect upon what we perceive. We do not just register how hot, cold, wet, or dry it is and how much and where the food is. In that sense Davidson and Noble state that with language we *realize* what we perceive. Using language is to engage in myriad constructions of *reality* based upon this reflection and on sharing it with others.

The link they argue for between depiction and reflective language puts them firmly in favor of a short chronology for the origins of language. Indeed, they go as far as to suggest that "There can be no such thing as culture without language and the socially determined sharing of meaning and value. It will therefore be misleading to talk of culture for any hominids before fully modern humans."[127] They clearly have in mind that the earliest art, broadly defined, marks the origins of language as we know it at about 40,000 B.P.

Not everyone sees the break quite so starkly as Davidson and Noble. Robert Whallon has examined language ability in the context of innovative social contexts which show up in the archeological record as those major demographic expansions into Australia and the Americas.[128] I shall discuss these in chapters 9 and 10. He follows the line that some environments, which can only support humans at low densities, put pressure not only on getting food but also on the social system and the contacts, marriages, and necessary alliances between local groups. Humans are of course found today in such extreme environments as the Arctic, the deserts, and the tropical rain forests. What Whallon proposes is that the barrier to settlement of such habitats depends on the possession of memory and the ability to plan forward. Language confers efficient memory, as Randall White puts it,[129] and Whallon, using Bickerton's model presented in *Roots of language*,[130] suggests that those elements dealing with tense—modality—aspect, which allow discussion of the future to be part of conversation, were the last major elements to evolve—what Davidson and Noble refer to as displacement. Restricted memory and planning depth therefore acted as constraints to the settlement of those environments where foraging and social life required an increase in such capacity. The innovative social contexts required to allow colonization to proceed were, according to Noble and Davidson, made possible by language.

Whallon does not, as do Davidson and Noble, see life prior to the Upper Paleolithic as precultural. In his opinion protolanguages existed that communicated well enough but lacked the expression of, and reflection on, the future and the past so critical to the elaboration of "extended kinship networks, communication beyond face-to-face encounters and exchange of information beyond the here-and-now, the organization of logistical economic strategies, and the extension of the time depth of adaptation to environmental fluctuations."[131]

I find much to support in this view. Communication existed among all timewalkers and the social context provides the exaptive process to upgrade this to language. The link between, on the one hand, exploration and the use and knowledge of space, and on the other the scale, depth, and complexity of social life is irrefutable, as I have shown in earlier chapters.

Viewed exaptively the origins of language escape from the adaptive problem of the chicken and egg. Which came first, social relationships requiring memory and the ability to juggle time and people? Or language, making possible intricate socializing interrupted in time and place and complicated by fluctuating numbers? Neither aspect sprang up unannounced or even evolved gradually to its present function. What good would 5 percent or 50 percent of a modern voicebox, Broca's area, or an extended alliance partnership have been? They were adapted/exapted to satisfy current needs and not to meet an unknown future goal.

Language as we know it has to be understood in these contexts where its main impact has been in the exaptive elaboration of social life, as might be expected in a species that as a result colonized almost the entire world. The construction of society coevolved with the constant need to interpret the environment to create an ever more sophisticated niche, which in turn placed selection on all timewalkers for the evolution of modern human attributes in existing behavior.

Consequently I would seek the "origins" of language in the colonization of those environments which required its unique properties to elaborate and maintain social life rather than basing the argument on the appearance, anywhere, of art. If the two coincide then so much the better, but don't be surprised if they won't.

Genetically modern humans

The first thing to note about genetic Moderns is that the differences between populations when measured genetically are, as Chris Stringer and Peter Andrews show, so very small.[132] These are much lower than among other primates while differences are often greater within human populations than between them.

Such minute differences emphasize to many the recent origin of the genetic populations they measure and the constant gene flow between them. The short chronology for the appearance of genetic Moderns has been made by Luca Cavalli-Sforza and his coworkers where they forge an explicit link with language.[133] Like many others they regard fully developed language as the hallmark of modern humans. They compare the pattern of modern world languages with a genetic tree based on 120 alleles (forms of a gene that arise by mutation and occupy the same place on similar chromosomes) in 42 human aboriginal populations. This tree is drawn by measuring gene frequencies between populations to produce a dendrogram that summarizes genetic distances. The clusters this produces are interesting to both the Gardeners and the Sons of Noah.

The six sub-Saharan African populations that were sampled form a distinct group separated by the greatest genetic distance from other populations in the world. These in turn form two superclusters. The north Eurasian supercluster subdivides into the Caucasoids and a large group containing the northeast Asians and Amerindians. The second, southeast Asian supercluster subdivides into the southeast Asians proper, the Pacific Islanders and a cluster with the Australians and Papua New Guineans.

What is the significance of such a genetic tree for human evolution? This depends on the evolutionary models used to draw up such trees and where a constant evolutionary rate is assumed as populations split and continue to evolve independently. Small populations have to be avoided since including them might result in demographic "bottlenecks" which might upset the assumption of constant rates. To test this they match the genetic distance to a timescale for the appearance of modern humans such as at Qafzeh at 92,000 B.P. and the timing of global colonization provided by archeological evidence. This gives reasonable agreement that population splitting has occurred at a constant rate, although it has to be admitted that this falls well short of being an independent test.

The interpretation of the clusters favored by Cavalli-Sforza is that the greater genetic distance between sub-Saharan and all other populations is indicative of a longer period of evolution of modern humans in that region and their expansion from this nuclear area. He therefore comes out strongly in favor of the Gardeners' model of where the Moderns come from.

But why should his assumption of constant evolutionary rates be correct? Question this and the Darwin/Wallace implication of dispersal from a center of origin at steady rates also becomes an issue. Matching the major language families to the genetic clusters can be expected to produce some correspondence due to nothing more than the scale at which the analysis is expected to produce significant results. Correlations of this sort prove little and still beg the question of timing, rate, and origin. Genes and language are not necessarily correlated, just as we saw in earlier chapters that shapes of stone tools and skull size have no necessary relationship. Demonstrating either common factors of selection or linked macroevolutionary effects is needed to make the case. What we might consider instead is how long genetic Moderns existed before their genes, by whatever mechanism, started to colonize the world. The assumption that they would start to spread as soon as they first appeared only works if, in some way, those modern genes are controlling geographical expansion. This was Parsons' question put earlier in chapter 5. Since dispersal is about social behavior, and this is constructed rather than inherited genetically then the chances are very strong that nuclear DNA could undergo considerable local changes without any major extension of range. It will only serve as a measure of colonization if genetic changes were needed to survive in novel habitats. The generalist character from the gracile Apiths onward argues against any necessary changes in the genome before colonization could occur.

The correlations between genetic trees, major world language groups, and migration while tantalizing is not yet proved. But this is not the only genetic approach claiming a second African origin. Nuclear DNA studies, by Jim Wainscoat and his team, have compared eight population groups and shown once again that a significant genetic distance exists between African and all other populations.[134]

Equally striking is the similar genetic separation using mitochondrial DNA (mtDNA). Mitochrondia are present in all cells but are not part of nuclear DNA. They reproduce independently and hence are not believed to be under selection as is the case with nuclear DNA. They also evolve more rapidly with a mutation rate of up to four times that of nuclear DNA and are only inherited through the maternal line. A male's mtDNA, inherited from his mother, dies with him.

Because mtDNA is believed to be neutral and outside environmental selection a constant evolutionary rate is assumed. A small sample of 147 people has been analyzed by Rebecca Cann, Mark Stoneking, and Allan Wilson.[135] When plotted as a similarity tree this again points to Africans as separate from other populations and with greater diversity of mtDNA than other populations. They believe this is consistent with a longer history for the constant evolution of mtDNA at a rate of between 2 percent and 4 percent every million years. It has also been suggested that the mtDNA which with slight, but measurable, variation characterizes all modern humans had a point of origin in a single African female whose mtDNA has subsequently colonized the world. "Eve's" age, estimated by calculating the divergence rates, is put between 166,000 and 249,000 years ago,[136] with the first move of genetic Moderns out of Africa happening sometime between 90,000 B.P. and 180,000 B.P., depending on whether you accept the high or low estimate.

But all is not well in this genetic Garden of Eden.[137] Doubts have been cast on the way the tree was traced using the mtDNA data. The result has been to question where the "lucky mother" (as Eve is now referred to) should be located. As a result the estimates of the lucky mother's age are only suggestive and should not be given too much credence. Moreover, as Wainscoat points out, "Eve" is only our mitochondrial common ancestor and might have contributed nothing to our nuclear DNA. Since mtDNA is not supposed to be under selection this genetic hitchhiker cannot correlate with any aspect of either the anatomy or behavior of the Moderns. It would be more significant to pin down the timing of the demographic bottleneck that provided future populations with only a single mtDNA lineage. This would at least mark a migration event for genetic Moderns if not the advent of modern humans.

Modern bees and ancient humans

Here, then, are three different ways to classify a modern human in the past and two models that attempt to marshal all the lines of evidence to prove their position. Somewhere between of course lies a host of alternatives, borrowing from either and molded to the circumstances as they appear in one part of the world rather than another. The key issue once again comes down to colonization, and before we move to the evidence for what the Moderns did after 40,000 B.P. I want briefly to step outside to look at two parallel cases that neatly summarize many of the problems touched on in this chapter as they relate to colonization and change.

My cases come from work on that miniscule workhorse of evolutionary experiment the fruit fly, *Drosophila subobscura*, and African honeybees. Both have been rapidly expanding in the New World having been imported from elsewhere. In the case of the fruit fly, 44 percent now contain the mtDNA from a single ancestral form.[138] In a thousand years time it will appear that they were all descended from the fruit fly equivalent of "Eve," but this does not mean a single ancestor. How could it, since just under half of all living female fruit flies could currently claim that honor? As Graham Richards has pointed out, it is a bit like being surprised that so many Welshmen are called Jones or Evans.[139] It does not mean that each named lineage had an ancestral "Eve," but only that a population bottleneck, no doubt related to social factors, imposed its own founder effect on the inheritance of surnames. If names were like mtDNA then we can see that there would be many Joneses and Evanses contributing to the nuclear DNA, which is what matters. The bottleneck that produced such restricted surnames in subsequently much larger populations could be an important indicator of the timing and tempo of migration and colonization but not of dispersal leading to speciation.

But by far the most detailed study of genetic change and colonization comes from the recent spread of the neotropical honeybee through two continents. The parallels with the Ancients and Moderns and between the Sons and the Gardeners is instructive.

The African honeybee was imported into Brazil in 1956 and subsequently twenty-six of the forty-seven queens escaped from controlled apiaries to establish feral colonies. At that time the bee population in Brazil consisted of two main types, both *Apis mellifera*, but derived from European and now African stocks. While cross fertile they are very different bees. The Europeans are small, easily controlled in apiaries but poor producers of honey in the tropics. The Africans are large, fierce (these are the African killer bees that have become a significant public health hazard), difficult to keep but marvelous honey producers. Africans swarm more often and this is how they colonize. They produce more drones and fly farther than the Europeans and, as might be predicted with such mechanisms, have a shorter period of development and a briefer lifespan.

This behavior has yielded spectacular colonization results. Since 1956 and after approximately 150 generations they have colonized from Brazil to Mexico, a distance of 8,000 km and at their peak were extending their colonization front at the rate of 500 km every year. They have outcompeted the "native" Europeans. It is estimated that before the Africans arrived Venezuela had between 30,000 and 40,000 colonies of European bees. Now it has an estimated 1 to 2 million colonies of feral Africans. This has happened since 1976.

But what is highly significant is that in less than thirty years these conspecifics (closely related species) have produced evidence for highly asymmetrical gene flow between the expanding Africans and the resident Europeans. As Africans establish feral colonies so their drones invade the peaceful European apiaries. Within a period of two to three years the apiary is Africanized. This Africanization leads to changes in behavior, morphology, and nuclear DNA in what were once European bees. This does not necessarily occur through takeover of the hives but rather as a result of the abundant African drones mating with European queens. Here, then, is one route to replacement between conspecifics and the disappearance of native populations as defined morphologically, behaviorally, and genetically.

But the rapid dispersal of the Africans is not just a factor of drone behavior. The queens do eventually disperse having handed on their hive to a daughter. What this means is that as the European/African queens leave their hives, which have been Africanized, and join the feral population, which is entirely African, they bring to it European mtDNA since none of these genes can be inherited from the African drones. We should therefore expect substantial evidence for European mtDNA in these feral African populations, a factor that would reveal the demographic history of bees in South America. But this is not the case. There is in fact very little European mtDNA in the feral swarms. Gene flow between these Africanized queens and feral populations is minimal. Instead it seems, as Deborah Smith, Orley Taylor, and Wesley Brown conclude that "the mitochondrial genomes of a small number of African females have rapidly colonized most of a continent. . . . Thus, an essentially African population is expanding in neotropical habitats through migration and colonization of new territory by African females."[140]

An explanation for this might be that mechanisms of mate recognition and sexual selection are operating which in fact have nothing to do with mtDNA but result in the preferential favoring of some maternal lines.

On occasion such preferential selection has resulted in the takeover of hives by African queens, and Glenn Hall and K. Muralidharan in their studies of mtDNA have concluded that unless actively preserved in the tropics these European maternal lines may vanish through attrition.[141] Their explanation for the failure of some maternal lines to enter the feral populations is that mtDNA may not be neutral after all. In other words it *is* under selection and, as they speculate, "European mitochondria may limit a metabolic capability needed for long distance dispersal."[142] The aggressive, fast growing, shortlived, long-distance flying African has a different metabolic structure congruent with rapid dispersal and effective colonization. What is under selection in the competition between these two populations is therefore the complete package of colonization. The result is expressed both geographically and genetically, although it will not be long before those interested in apiarian history will have to unravel the arrival of the Moderns and the demise of the Ancients by other means.

9
Pioneers and diehards in the new lands of the Old World

The dance of the honeybee that shows where the best pollen is to be found bears some resemblance to the conduct of research into human origins. The most vigorous dancers attract the largest followings as Roger Lewin shows in *Bones of contention*.[1] At the moment the door of the hive is buzzing with controversies over the number of hominid species we can recognize and the fate of the Neanderthals. This is as it should be, but beware of stings.

The dance, however, runs the risk of becoming a ritual rather than an exploration. There is a dispute over definitions, as we saw in chapter 8 with the current controversy over how to assign Moderns and Ancients to their rightful place.[2]

These are just the new steps in a very old dance. It is apparently deep rooted that we should classify the world and the humans within it into a myriad categories. I showed in earlier chapters how civilized and savage people were once distinguished and we have just seen the present solution to the same question, with the savage Ancients thrust back in time rather, than as before, into some other part of the world. The intention, however, is the same—to render prehistory acceptable and harmless to support our contention that as a species we are different and special. From that platform all sorts of action and opinions flow.

Discarded notions

This may seem part of the agenda inherited from Darwin and especially Huxley who argued for man's place in nature. Well, so it is, but what can we offer from all the archeology that has been undertaken in the past 100 years that is novel and informative about this dialogue with a nature we defined?

To begin with two deeply embedded notions can now be discarded. Suitably amended these are:

—don't judge anyone by their stone tools; and

—don't be mislead by their heads.

As I have repeated throughout this book, stone tools can tell us a great deal about the organization of technology and those all important longterm survival strategies, but precious little about either intelligence or its potential, however that is defined. Technological progress is neither the rule nor a yardstick. For example, apparently sophisticated hafting technologies appeared then disappeared in the long cultural sequence at Klasies River Mouth. Elsewhere, Australia was peopled by modern looking humans with a stone toolkit that was rudimentary by the standards set in other parts of the world 50,000 years ago when this process of colonization began.

In Australia the stone tools and other excavated technology used in foraging challenge our view of how the world works. Nowhere is this more apparent than in Tasmania, which was joined to the mainland until 12,000 years ago when rising sea levels at the end of the ice age flooded Bass Strait. Since then the island population has undergone the longest period of isolation of any group of *Homo sapiens sapiens*. Communication between the mainland and Tasmania by boat or raft did not take place because their watercraft, built from rolls of bark, usually stayed afloat for about thirty minutes—just long enough to reach inshore islands. When the archeological and ethnographic toolkits are compared it is obvious that during this period of isolation they became simpler. Tools made from wallaby bones, such as points and awls, were present 20,000 years ago, but disappeared about 3,000 years ago.[3] The first Europeans to visit Van Diemen's Land could only list about two dozen items of material culture, among which the stone tools seemed particularly rough and ready and were used as modern examples of the Lower Paleolithic finds then being made in Europe. The world's simplest material culture apparently became simpler.[4] Had this been known in the nineteenth century it would of course merely have confirmed the views of those, such as the Duke of Argyll,[5] who argued that the state of modern savages resulted from degeneration, particularly those at the utmost ends of the earth. Today Rhys Jones uses the archeological discovery to make the point that the Western view of unrelenting technological progress and continual change toward greater complexity is our shortterm view rather than a universal longterm norm.[6]

But the Tasmanians should not be judged just by their stone tools and simple shelters. In common with the rest of humanity they engraved rock surfaces, decorated their land by carving trees, built cairns, painted and scarified their bodies with elaborate designs, and festooned themselves with jewelry. All this transformed their land from a "natural wilderness" into a highly charged landscape of metaphor, religious knowledge, and sacred meaning based upon the enactment of myth and the observance of ritual. These custodians of the Dreamtime, as elsewhere in Australia, are as complex as any other society in their transformation of nature. To ignore this and dwell instead on their leaky rafts and simple stone tools misses the point about human diversity.

The Tasmanians also know a good deal about the second discredited notion—don't judge people by their heads. With the death of Truganini in 1876 the Tasmanians were declared extinct, which understandably still deeply offends many modern Tasmanians, who can trace Aboriginal ancestors through the troubled social history of the island. The fate of Truganini is well known. Her worst fears of being dissected and put in a museum were realized since her body, in particular her skull, was considered too important to science to be left in the ground. She was dug up in 1878 and put on public display from 1904 to 1947 in the museum of the Royal Society of Tasmania. The relationship between cultural status and physical appearance was all too plainly made, although her skeleton was not studied scientifically until 1971. After many years of protest Truganini's skeleton was returned in 1976 to the Aboriginal Tasmanian community, cremated, and cast on the waters of D'Entrecasteaux Channel.[7]

But the legacy which Truganini's history illustrates continued. In Europe the link between modern skulls and advanced stone tools is the clearest example. The discovery elsewhere in the Old World of modern skulls with Middle Paleolithic and MSA stone tools at Qafzeh, Skhul, Border Cave, and Klasies River Mouth has seriously dented the archeological arguments. However, it is still widely held that the Neanderthals of the Near East, Central Asia, and Europe were responsible for the thousands of finds of Mousterian or Middle Paleolithic stone tools, even though their fossil remains are only found with them on very rare occasions. This willingness to ascribe lithic authorship to an unseen primitive exposes our bias to regard the shape of skulls as significant.

Our innate bias applies equally to the production of art. Those same Paleolithic sites with Modern heads but Ancient flints show very clearly that art in the form of ornament, engraved slabs, figurines (not to mention painted cave walls) is *not* invariably associated with modern looking skulls. The irony of the Tasmanians should, once again, not be missed. Once held up as living representatives of the Lower Paleolithic, where there is no hint of art, and yet themselves prodigious producers of art in many different mediums.[8] But then, of course, in the early decades of this century there was art and then there was Art. The celebration of African and Polynesian sculpture by French and Spanish artists and American collectors still had to make its general impact in redefining the limits of good taste. The general view was that

> The decorative art of a people does, to a certain extent, reflect their character. A poor, miserable people have poor and miserable art. . . . The finer the man the better the art, and . . . the artistic skill of a people is dependent upon the favourableness of their environment.[9]

Modern looking humans, like those from Qafzeh, capable of collective survival in hard environments, yet without art and very conceivably lacking a spoken language, are very difficult for us to countenance. How would we classify and cope with them had they survived in some small corner of the Old World? Would they have been treated to the same fate as the chimpanzee in laboratory, circus, or zoo?

Exaptive colonization

Where does this leave us if heads and stone tools are dismissed as reliable signposts to what is modern? In chapter 8 I might have perched too much on the fence when talking of the Sons and Gardeners, recognizing strengths and weaknesses in both their positions. But when it comes to a decision I side with the Gardeners' view of replacement, if not with their mechanism. The anatomical, cultural, and genetic changes that are magnified in some areas of the Old World, minimized in others, do suggest change from outside. But by themselves these are necessary but not sufficient evidence for determining the appearance of Moderns.

Moderns are modern because of the societies they construct and live in. Social context is all. More particularly, the *exaptive* social contexts demonstrated by the burst in colonization starting 50,000 years ago. From this I infer new patterns of organization which led to global humanity in a short time, at least judged by prehistoric standards. This involved only minimal further changes in anatomy and genes.

The archeological case rests not with the shapes of tools or heads but instead on the massive colonization of new environments that began some 50,000 years ago (table 1.2). It is quite clear that without global colonization we would not be discussing any of these questions. As I have maintained throughout this book, the changes in behavior required to complete this process are what made us human, even though that behavior had no such goal in mind. We were not adapted for filling up the world. It was instead a consequence of changes in behavior, an exaptive radiation produced by the cooption of existing elements into a new framework for action. Any selection in this process would have been local and, of course, took place in ecological time.

This radiation was not achieved without significant cultural developments of which art, appearing after 40,000 years ago, ranks as one of the major contributions. While art is a human universal it must be seen as an effect of this colonization process rather than as its cause. In other words, inventions such as art, boats, and sewn skin clothing did not come first and then help one group rather than another achieve the journeys into inhospitable lands or across vast ocean distances. Instead they were all devices whose time had come. They were required, so they were coopted from existing materials. Once they had served their purpose they were, on occasion, just as Tasmanian bone tools or the boats which brought people to Easter Island (chap. 10), dropped.

Hard habitats

The pulse in global colonization which began 50,000 years ago took two forms. Firstly, there was the occupation of new areas in the Old World which previously had only been occupied under very favorable climates or not at all. The second aspect involves people moving into Australia, the Pacific, and the Americas (chap. 10). With the former we can trace the timewalkers' route through the relative ease of making a living in many of the major habitats (fig. 1). With the latter continents they were faced by barriers that needed surmounting. These were water, distance, ice, and cold (fig. 7.1). Such obstacles to colonization have ranked as important as the Sahara for the Apiths and early *Homo* in explaining the late appearance of timewalkers in fertile lands. We shall see again that very often the biggest barriers have been our ideas about "primitive" abilities and motivations.

Old World

In the Old World there are four major habitats that need examining—plains, deserts, mountains, and forests (fig. 9.1). In all of these areas the evidence from earlier periods is, at best, a regular ebb and flow of occupation to the rhythms of the ice ages, described in chap. 2, and at worst, as in the equatorial rain forests, a human desert.

Figure 9.1 Location of Old World Moderns.

Plains

The earliest hominids were of course adapted to the high plains and plateaux savannahs of East and southern Africa. In the Old World the major plains environments not used extensively until after 40,000 B.P. were those vast areas that start with the North European Plain and expand through eastern Europe into the steppes of the Ukraine and the Russian plains west of the Urals. Under all types of glacial and interglacial climate these areas had low evaporation so that water was never a major limiting factor. It was always present in lakes, rivers, and in standing pools during even the coldest and most arid periods. The amount of forest cover depended not only on climate but also on the degree of continentality that produces ever greater seasonal differences on a west to east transect. These variations in temperature, snow cover, and rainfall patterns affected the degree of frozen ground (permafrost), the depth of the active layer above it, the extent to which solar radiation was reflected (albedo), and the growing season. The result was to thin out resources along that same transect.

Most of these plains saw no trees, outside of well protected valleys, during either early or maximum glacial conditions. But despite this bleak picture, for much of any single climatic cycle these were productive environments (fig. 1.1). During the ice ages these lower latitudes produced a rich herb and shrub steppe tundra that supported herds of woolly rhino, reindeer, horse, bison, and other large mammals such as steppe ass, musk ox, hyena, wolf, and bear.[10] Many agree, however, that in the more continental areas of these plains the key animal was the mammoth. With its short inner and long outer hair coats, huge fat reserves, small ears, sensitive grasping trunk and curved tusks, which could sweep snow from vegetation, it combined a number of physiological adaptations to extreme cold and a grazing environment. In eastern Siberia, the dry land paleocontinent Beringia, which linked Asia and America, is described as part of the much wider mammoth steppe in recognition of their importance (fig. 9.2).[11]

Plant foods for humans were rare on these plains so that the density and distribution of animals were critical for survival. Sites dated well before 40,000 B.P. are found only in their western part. Gerhard Bosinski has remarked on the use of the North German plains for the first time possibly as early as 200,000 years ago and therefore by the Ancients (table 1.2).[12] The open site of Salzgitter Lebenstedt, near Hannover, has exceptional preservation of organic material and a rich stone assemblage with handaxes.[13] The animal bones are dominated by remains of reindeer with mammoth, bison, horse, and woolly rhino also present. The archeological remains, although not *in situ*, are typical for these early periods with no hearths or structures.

Even allowing for subsequent destruction similar sites of this age are rare on the northern plains. In particular there is scant evidence for the use of the Russian Plain, centered on the rivers of the Dnepr and Don, until much later.[14] Open sites dating to the Pioneer phase (60,000 B.P.–40,000 B.P.) can be found at Khotylevo on the Desna, a northern tributary of the Dnepr, where spectacular

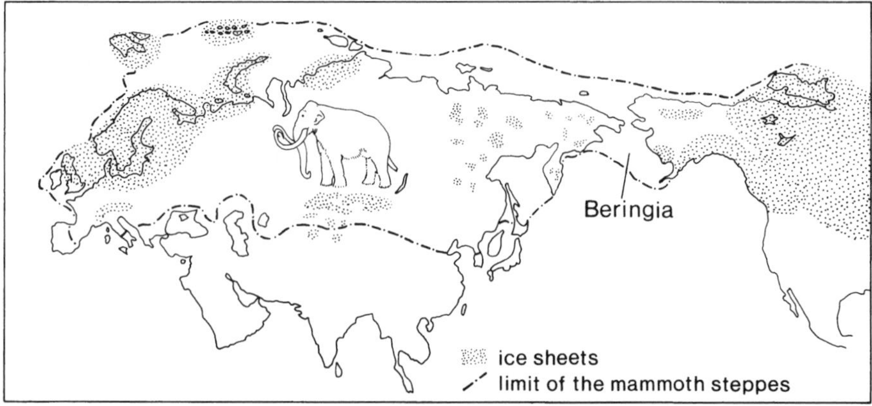

Figure 9.2 The mammoth steppe at the time of the last glacial maximum 18,000 years ago (Guthrie 1990).

finds of large projectile points and rich flint workshops with classic Levallois cores and flake blades have been excavated.[15]

The situation is different along the Dnestr which forms the border between Moldavia and the Ukraine. Here in the massive loess profiles of Molodova, Korman, and Ketrosy multilevel open sites with Middle and Upper Paleolithic occupations have been found.[16] This concentration, which is continued in the rolling hills of the Prut valley along the Roumanian border, points to the conditions which the earliest plains populations preferred. Plateaux and topography broken by these major valleys had the effect of concentrating resources by providing more varied ground vegetation and feeding opportunities. This is a pattern also seen in southwest France, repeated throughout the uplands of central and eastern Europe through to the comparable plateau country of the Crimea and Caucasus.[17]

The plains to the north of the rolling uplands were certainly well stocked with food but at densities that would have needed storage, mobility over large distances, and much forward planning to secure a living. Moreover, and perhaps of greater importance, these conditions would sorely tax the rudimentary systems of individual and group communication which these Pioneers were capable of. As a result we find them in those areas, such as the incised plateaux, where animal resources were concentrated, predictable, and came in the right package size. No doubt forays were made on to the plains in the Pioneer phase from 60,000 B.P. to *c.* 40,000 B.P., and possibly even earlier. But they soon ebbed, as can be seen in Donbass north of the Azov Sea, when climate exerted a strong control.[18]

After 40,000 B.P. the change is subtle rather than dramatic. Drawing a simple line is not possible (table 1.2). The Moderns can be traced at the edge of the inhabitable world in Pioneer sites such as Khotylevo and through sparse occurrences in the caves of England, Germany, Poland, and the Carpathian arc (chap. 8). In fact after 30,000 B.P. the North European Plain was left unoccupied for over fifteen thousand years. Occupation continues in the uplands and protected basins to the south.[19] What is significant is that there should be any occupation at all as the climate plummeted and the ice sheets moved out from the Alps, Carpathians, and Scandinavia to reach their maximum extent 18,000 years ago. Much of central Europe was caught—literally—between the sheets, and at the height of the last glacial maximum settlement traces become very rare in parts of Germany, Czechoslovakia, Austria, and Poland.[20] But before this happened there was a remarkable plains adaptation which linked these areas with sites along the Don river over 1600 km to the east. Here at the small village of Kostenki, named after the thousands of mammoth bones and tusks dug up over the years, a hundred years of investigation have uncovered twenty-one major sites many with multiple occupation horizons.[21] None of them, however, has been dated to more than 38,000 B.P. At Kostenki site 1 level 1, more than 1,500 m^2 has been excavated and is still under investigation. The settlement is radiocarbon dated to 23,000 B.P. at a time when global climate was moving rapidly to the glacial maximum and when the protection afforded by the steep

Plate 9.1 Pit house filled with mammoth bones at Kostenki on the River Don.

chalk ravine in which the site sits would have been appreciated. Pit houses were dug in the loess loam sands down to the top of the permafrost, a depth of 110 cm, roofed with tusks and no doubt insulated with turf and moss. These are grouped around a line of massive dumps of bone charcoal dug out from open hearths. Mammoth bones supplied the necessary fuel in this largely treeless landscape while in many smaller pits bones with meat on were stored. Bones of fur-bearing animals are very common in the pits although, when you visit these sites, it is difficult to concentrate on anything else but the remains of mammoths which cover everything like a fallen wall of huge, misshapen bricks.

Kostenki is rich in decorated bone and ivory objects. Some of these are identical to those found at the site of Avdeevo 200 km to the northwest and of the same age.[22] Several large flat bone wands, up to 50 cm in length, have distinctive hand grips with a round handle pierced by four "eyes" while their business ends are smoothed. Their function is unknown; they could be wands, daggers, spatulas, or snow knives. The stylistic links they establish between the two sites is quite clear. Equally distinctive are the many female figurines made either of stone or ivory from both sites. There is great variation in shape and size. Some are thin and elongated, often with the knees pressed together and their feet splayed, while others are fat and squat. Large buttocks and breasts form the hallmark of the ensemble. Faces are rare but belts, bracelets, and sometimes necklaces are shown.

Figurines of the same age and within the same general style have also come from Willendorf by the Danube in Austria and from Dolní Věstonice 80 km northeast across the border in Czechoslovakia.[23] This site, and its extension at Pavlov, rivals Kostenki in size and complexity of structures, pits, and layout. A further link is provided by the appearance at Willendorf, Pavlov, and a site dug in Krakow's Spadzista Street in southern Poland of highly distinctive stone projectile points with short tips and long hafts formed by trimming a shoulder on one edge. This same type also comes from Kostenki and Avdeevo.[24]

I regard these similarities in both non-utilitarian and functional items as indicating the scale of social systems at this time.[25] The intensity of interregional interaction they point to was vital to such largescale, longterm survival. Further evidence is supplied by the source for the flint used at Kostenki which comes from 200 km away while shells in the sites come from the Black Sea, some 600 km to the south.

It is unlikely that a single group was responsible for all these sites and journeys. The overall similarities point instead to the vital importance of pancontinental alliances reinforced no doubt by ritual, art, and tradition. These alliances provided insurance policies between communities and individuals for bad times on the mammoth steppe. They allowed the unimpeded flow of people at such low population densities to contract marriages and negotiate partnerships of social and economic value.

Nowhere is this intensity of social life more vivid among these plains peoples than at the northern site of Sunghir some 200 km east of Moscow and hence, 24,000 years ago, very close to the approaching ice sheets.[26] Another massive settlement but this time without pit dwellings and mammoth bones. In among the occupation, as indicated by the density of flints, the excavator O.N. Bader found two graves. In the first an old man, possibly sixty years old, was covered in drilled beads that had been sewn on to a cap and cloak. In the other pit a young boy of between twelve and fourteen years and a girl of seven had been laid head to head. They too had been dressed in clothes richly decorated with many thousands of ivory beads. Ivory bracelets, animal figurines, pendants, and a wide variety of other grave goods including a thigh bone (due to its robustness thought to belong to a Neanderthal) were found with the bodies. By their sides were ivory wands and spears of straightened mammoth ivory, the longest of which measures 2.40 m.

It seems no coincidence that the appearance of sites which justify being called villages should also coincide with elaborate burial practices and a plethora of material culture. The recent discovery at Dolní Věstonice of a triple burial, dated to 27,640 B.P. underlines the point made later at Sunghir. At least two interpretations are possible. Firstly, they represent a "golden age" or, secondly, as Martin Wobst has argued, a form of Arctic "hysteria."[27] I favor the latter explanation. This explosion of art and ornament and the burials and villages where they are found, tells us that these peoples had a hard life where every cultural trick and means of support had to be marshaled and brought to bear on the problem of social and economic survival. This was no epoch of leisure leading to great art, whatever that might mean. Instead every additional means of exchanging information and minimizing risk was tried. The fact that this was achieved as the last glacial maximum crept closer is evidence that the ebb and flow pattern of earlier settlement on the plains had been transcended.

After 18,000 B.P. the climate began its slow upturn. On the Russian plains at Kostenki there is good evidence that settlement continued through this climatic nadir. Elsewhere on the Dnepr, south of Kiev, Olga Soffer has shown how this central portion of the plain was only used intensively as the climate improved.[28] The village site of Mezhirich has four circular dwellings made of mammoth bone with low walls built either by stacking mandibles and flat shoulder blades or fixing long bones in an upright position like fence posts. Outside hearths and the contents of storage pits allowed Soffer to conclude that this site was occupied for six months during the winter by up to sixty people, twelve families. It is dated to 15,000 B.P. Similar complex villages are known from Mezin, Yudinovo, and Khotylevo in the same area. Mammoth dominates the faunas. Controversy still exists over whether they were hunted or their bones just collected for raw material and fuel. I believe life would have failed without the use of their meat fat and marrow reserves. How they were killed is still an open question.[29] Mammoths are extinct on these plains by 12,000 B.P.[30]

To the west we find repopulation of the plains between the ice sheets as soon as they began to retreat. Large camps of reindeer hunters, but without clearly defined houses as in the Ukraine and Russia, are known from the sites of Pincevent, Verberie, and Les Etiolles in the Paris basin dated between 10,600 B.P. and 12,300 B.P.[31] A rich scatter of sites is found to the east on the north German plain. At Stellmoor, near Hamburg, reindeer was again the dominant food animal,[32] while horse and reindeer provisioned the site of Gönnersdorf in the Köln basin. Dated to 12,400 B.P. this site has a large oval tent structure floored with flat schist plaques many hundreds of which are engraved with images of humans, mammoth, horse, rhino, wild ox, and some birds and carnivores.[33] Reindeer were not drawn.

What is striking about this flow of settlement is the speed with which the northern fringes of the then inhabitable world were repopulated and the high density of settlement traces which have been recovered. Even allowing for problems of preservation this contrasts strongly with the comparatively few traces from earlier periods.

I have dwelt at some length on the plains evidence since our chronologies are better, but far from perfect, and the pulse of settlement well documented. In the other environmental zones the same general pattern can be seen with marked increases in either permanent or temporary occupation only occurring after 40,000 B.P.

Deserts and arid lands

The arid zones, where evaporation is high so that water becomes a major limiting factor, may not have such spectacular artistic remains, thereby supporting Wobst's point about Arctic "hysteria," but nonetheless the same general pattern of colonization can be seen.

Richard Klein has made a detailed study of climate and animal economies in southern Africa and concluded that the earlier Paleolithic, as marked by

Acheulean tools, is never found in conditions which are either as dry or drier than historic ones.[34] This is well shown by the extensive site of Elandsfontein in the western Cape, where in 1953 the robust Saldanha skull, generally regarded as archaic *Homo sapiens*, was found in the shifting sand dunes. When hominids patrolled this Middle Pleistocene landscape, perhaps 200,000 years ago, local conditions were much moister. It is littered not only with handaxes and large flakes but also bones of megafauna—elephants, buffalo, and rhino. These animal bones carry tooth marks from the large carnivores but no stone tool traces. Klein concludes the hominids were very ineffective hunters.

Fieldwork by Myra Shackley in the central Namib Desert in southwest Africa produced handaxes as well as flake assemblages comparable to Middle Stone Age assemblages elsewhere.[35] Although undated all the sites are closely tied either to pans and other traces of former water sources or to the Kuiseb River itself. This is hardly surprising but preliminary findings also indicate that occupation occurred during wetter phases than those at present.

In the aptly named Wadi Arid, 200 km east of Lake Nasser in Egypt, no human occupation exists today. However, handaxes have been found with deposits indicating wetter conditions at 45,000, 141,000 and 212,000 B.P.[36] The expansion of population as we saw in chapter 7 at the equally arid Bir Tarfawi was entirely limited by one dominant environmental factor—water.

The same pattern is repeated in the dry loess steppe of Soviet Central Asia where, at Karatau, tools have been dated to 230,000 B.P. and at Lakhuti to 130,000 B.P. Both dates coincide with warmer, moister climates as indicated by the deepsea core chronology (chap. 3). The well developed soils from which they were excavated indicate occupation of the area during the climatic optima of interglacials.[37]

The effect of aridity on settlement is also demonstrated with Middle Stone Age evidence, i.e. pre-40,000 B.P. in the interior of southern Africa. The frequency of Late Stone Age sites after this date is still sensitive to the availability of rainfall, as Janette Deacon has shown, but even at the last glacial maximum at 18,000 B.P. some settlement traces can now be found.[38] Klein believes that hunting only approaches any sort of modern competence at this time.[39] These interior traces are, as John Parkington argues, the tip of the settlement iceberg. Population would have been densest on the now submerged coastal plain that added a further 100 km to the Cape.[40] As with the Pioneer phase on the plains of the Northern Hemisphere, the important point to make is that there was any occupation at all in these dry inland areas. It may be significant in this context that at the Apollo 11 cave in Namibia painted slabs have been recovered from an archeological level dated as early as 27,500 B.P.[41] However, it is only after 12,000 B.P. that ornaments made from ostrich eggshell become common in southern Africa. Once past the climatic minimum colonization of the upland interior and the arid basins is rapid and widespread, as Shackley discovered in her work in the Namib Desert.

Elsewhere in the arid lands of the Old World the last glacial maximum 18,000 years ago draws attention to some important changes. These allowed human populations to cope with being tethered to permanent water and exploit highly seasonal but small-size resources. Angela Close and Fred Wendorf working in the Wadi Kubbaniya on the west bank of the Nile, north of Aswan, have evidence for the repeated use of the wadi between 19,500 B.P. and 17,000 B.P., hence all through the hyperaridity of the last glacial maximum.[42] During this time all settlement clung to the river. The key to survival was the massive harvesting of catfish in a brief period from early to mid July. Many hundreds of thousands of fish bones indicate an intensive period of processing and storage for use in other lean seasons. Some settlement movement was forced by the Nile flood. When this receded nut grass and club rush tubers were gathered. The latter are nutritionally best in December and January, but have to be carefully processed by grinding and cooking in order to render the volatile toxins harmless, as well as to make the fibers more digestible. Close and Wendorf have found many grindstones in the Wadi Kubbaniya sites. The use of abundant but small resources such as tubers and catfish and the extra costs in terms of processing and storage that go with them nicely point up differences in organization between these Modern groups and those before them. The latter only managed to establish settlement during very favorable times.

The Nile sites also show the social consequences of being tethered to such a limiting resource. Intensification in subsistence can pose other problems for survival especially when it involves living together for longer periods. Hunters and gatherers usually cope with conflict and disagreements by walking away from the problem. However, this is not always possible. Wendorf and Close have uncovered a good deal of evidence for violence along the Nile which is not found elsewhere at 18,000 B.P. in the more fertile North African refuges of the Maghreb and Cyrenaica. One skeleton from the Wadi Kubbaniya is presumed to have died from the stone bladelets found in the region of his abdomen, while in the Jebel Sahaba graveyard at least 40 percent, regardless of age and sex, had met a violent death.[43]

The desert and arid zones of the Middle East have not produced such dramatic evidence, probably because no feature comparable to the Nile determined the pattern of occupation. In the Negev, Marks and Freidel have shown how the Middle Stone Age sites occurred *c.* 60,000 years ago during colder and more humid conditions.[44] Deterioration in the food supplies leading up to the arid conditions of the last glacial maximum was solved by greater mobility and seasonal shifts between areas. In the semi-arid northern Negev the site of Far'ah II has been investigated by Caroline Grigson and Isaac Gilead. Middle Paleolithic artifacts date to a humid phase between 45,000 B.P. and 40,000 B.P. and a range of animals was exploited. These included steppe ass, wild cattle, gazelle, goat, and camel while hartebeest and even hippo were possibly present.[45] It is interesting that although dating from the same time the Negev site of Boker Tachtit yielded completely different stone tools based on blades, usually indicative of the Upper Paleolithic, rather than flakes as at Far'ah II. The elision between Pioneers and Moderns (table 1.2) is very apparent with these data. Archeological evidence for the occupation of hard habitats is the only way to resolve their status. They cannot be judged either by the shape of their stone tools or, if anyone ever finds any, the shape of their heads.

Seasonal patterns are obvious in the Azraq basin in the Jordanian desert where Brian Byrd and Andrew Garrard have been carrying out intensive research around this famous oasis.[46] Earlier artifacts, including handaxes, are confined to the oasis and elsewhere to spring deposits indicating generally wetter conditions. Radiocarbon dates confirm, however, that in Wadi Jilat and Wadi Uwaynid settlement continued throughout the last glacial maximum. These sites were no doubt used on a seasonal basis as groups moved in the dry summers either to the oasis or east to the Jordanian plateau where conditions were slightly moister. The animals they depended upon were primarily gazelle and wild ass while some wild cattle could be captured around the oasis.

The desert sites are very impressive. Deflation of the land surface by the wind has left behind huge mounds of worked flint. In Wadi Jilat these flint domes reduce further erosion and preserve the earlier levels beneath. One of the excavated sites in Wadi Uwaynid contains the staggering amount, for an open site, of up to 22,000 pieces of chipped stone per cubic meter. In other sites the figure drops to a more manageable 400, but the average for all the sites from both wadis is 12,500 per m^3. The flint comes from the surface of the desert and at 18,000 B.P. is mostly knapped into hunting armatures. However, in Wadi Jilat basalt grinding stones are also being brought in from the edge of the Black Desert 50 km to the north. In both wadis there are finds of beads made from shells transported 200 km to 300 km from either the Mediterranean or Red Sea

Plate 9.2 Walking on a mound of chipped stone at one of the Wadi Jilat sites in the Jordanian desert. Occupation here spanned the height of the last ice age 18,000 years ago.

and lumps of red ocher (hematite) came from the western highlands 70 km from Jilat and more than 100 km from Uwaynid and the Azraq sites.

As might be expected the arid zones are highly variable. In the Jordan Valley 120 km west of Azraq exceptional preservation in Wadi al-Hammeh has allowed Phillip Edwards to trace adaptations at the last glacial maximum as well as the upturn in climate.[47] At 18,000 B.P. gazelle and tortoise were eaten in considerable numbers while wood charcoal from the hearths indicates that oak, pistachio, almond, and hackberry were all in the area and may have been exploited for food. Later sites in the wadi form part of a much wider Natufian complex dated to 12,000 B.P. Here grasses, legumes, and other seed bearing plants have been recovered. At the site of Ein Gev I in Israel reaping equipment, including sickle hafts made of bone and flint blades with edges glossed by silica sheen caused from cutting grass stems, are thought to date back to 15,700 B.P. Such items are very characteristic in the later Natufian and further point to an intensification in using resources, allied to climatic opportunity, that is a general hallmark of the last 40,000 years. Natufian sites are unknown from the arid landscapes of Wadi Jilat and Wadi Uwaynid and are only found at the Azraq oasis.

Similar variation is found in the arid regions of India where D.P. Agrawal has been studying the evolution of landscapes in the Thar Desert.[48] Settlement at 18,000 B.P. is highly variable and still depends on local rainfall patterns, which in India means the appearance of monsoons. However, a feature of the Indian sites from 40,000 to 10,000 years ago which closely follows the evidence from the Middle East and southern Africa is the paucity of art objects. They are rare or absent forming a major contrast with the Arctic "hysteria" of the northern plains and uplands where most sites have some carved bone and ivory objects. At a few, like Gönnersdorf, the Volp caverns in the French Pyrenees, and Parpalló cave in southeast Spain many hundreds of stone slabs have been found engraved with drawings of animals. Finally there are several hundred sites dated rather loosely between 17,000 B.P. and 9000 B.P. that contain painted and engraved walls. Apart from Kapova cave in the Urals these are only found in southwest France and northern Spain and include Lascaux, Pech Merle, Altamira, and Tito Bustillo.[49]

The arid zones are much more restrained. Elaborate burials are unknown until the Natufian when graveyards appear.[50] The date of 12,000 B.P. in the Near East strikes a chord with the increased importance after this date of ornament and art in the sites of southern Africa.[51] By this late date there had been a tradition for over 20,000 years on the northern plains of making and using ornaments, figurines, and other such paraphernalia in the many rituals and practices of life. So far the only hint that the pattern may be older and more complex comes from the painted slabs at the Apollo 11 rock shelter in southern Namibia.

Mountains

This type of environment raises some interesting questions about earlier occupations. The main problems mountainous areas present to colonization are the fragmentation of plant habitats, the effect of altitude on tree foods such as nuts, and the response by most animals to be small in both body and herd size.

The wild sheep and goats are very typical of the type of resources available.[52] Mountains can be rich habitats but very demanding in terms of planning and scheduling to decide which species to take, when and where and how to move people around. Compared to large herds of large bodied animals in the diverse communities on the plateaux and plains these can be expensive environments to exploit.

What has to be borne in mind is that altitude is no guide by itself to these problems. Chapter 4 looked at the high-level plateaux of sub-Saharan Africa where at 1,400 m in the Sterkfontein Valley Apiths lived. Much later the pebble tools at Karatau in Central Asia lie at an altitude of 1,700 m. Furthermore, the problem of archeological remains surviving at high altitude is very pertinent given that most peaks, at whatever latitude, had local ice caps during glacial periods. However, taking that into consideration we still see a two stage process: between 40,000 B.P. and 18,000 B.P. some settlement, which is remarkable for being there at all; and afterwards when the climate starts to improve, the demonstration of what they can really do.

The colonization becomes interesting when we consider very broken mountainous country. Within these areas it is possible to make high altitude finds which are usually dated by typological comparisons. Hence Middle Paleolithic artifacts are claimed to occur above 2,000 m in caves in Switzerland, Austria, and the Caucasus. Closer inspection shows, however, that these are invariably small collections associated with bones of hibernating animals such as cave bear. The Drachenloch in Switzerland lies at 2,445 m. The site was poorly excavated and most if not all the quartzite artifacts were probably formed by stones being crushed in the sediments due to ice action.[53] The dating depends entirely on their coarseness and similarity to a Middle Paleolithic typology. By comparison rich, well dated cave sites such as Shanidar in Iraq (alt. 765 m) or Houmian in Iran (alt. 2,000 m), both much older than 40,000 B.P., also turn out to be more in the nature of high level plateau sites rather than incidence of occupation in broken country. This is also the case with a series of recent discoveries made by Sergei Astakhov in the Tuva Republic of Central Asia to the west of Lake Baikal.[54] At elevations of between 1,200 m and 1,400 m there are finds of pebble tools as well as handaxes from stream deposits. The date of these is difficult to determine since later sites in the lower Yenisey also have pebble tools and advanced blades in the same toolkit. Astakhov has also discovered a large number of Mousterian type artifacts in the same area and which he tentatively places in a warmer interstadial between 60,000 and 40,000 years ago.

But there is one very interesting site which I cannot at the moment explain away and which may date to the Pioneer phase. This is the cave of Teshik Tash dug by the Russian archeologist A.P. Okladnikov in 1938–9.[55] It is near Karatau to the east of Tashkent at an altitude of 1,500 m. The photographs and description of its location convey just how rugged its situation is with deep gorges and high peaks above the caves. This is not a plateau site. The stone tools are very large and technically simple—scrapers, cores, points, and large flakes. The fauna is dominated by Siberian mountain goat. Most significantly, bones of carnivores are extremely rare indicating that this was not a den. On the first day of the excavation Okladnikov discovered the partial skeleton and skull of a Neanderthal child, originally regarded as a boy but now thought to be a girl. It was covered by a protective cage of goat horns.

Now this evidence is most interesting; five levels with a total chipped stone inventory of almost three thousand pieces, those with retouch equivalent to the Mousterian of western Europe, a Neanderthal to confirm the expected association, and yet in rugged topography at high altitude with extreme continentality. The several levels and quantity of material argue against either a single or chance visit. Moreover, there are hardly any carnivores in the site to account for the bones of the mountain goat. Hunting seems the only explanation. Is this modern behavior as recognized by the colonization of habitats which were previously off limits?

Living in such broken rough country, even if only on a seasonal basis, requires specialization and there have been no shortage of claims. The excavator of the Drachenloch cast his alpine Paleolithic finds as specialist cave bear hunters and threw in some fanciful evidence for bear cults among the Ancients to support his case. These have proved difficult to dislodge both from the archeological and general literature but, however reluctantly, the much illustrated stone "cupboards" with bear skulls stacked inside have to be dismantled. Bears frequently die during hibernation and other sleepers in the cave will push bodies and skeletons out of the way and in doing so create all sorts of interesting patterns with skulls and bones.

The quantity of bones is often impressive. In the Drachen cave near Mixnitz in Austria at an altitude of 950 m no less than a quarter of a million bones and teeth of cave bear were dug up representing between 30,000 and 50,000 animals. There were also a few flake and blade tools which, with similar finds in other lower altitude sites such as the Tischofer cave (600 m), point to a limited presence sometime before 18,000 B.P.[56]

It is, however, difficult to find clear evidence for the specialist hunting of mountain species. Well dated Middle Paleolithic sites such as the Hortus cave in southern France[57] and the early Upper Paleolithic site at Bacho Kiro in Bulgaria[58] both have animal faunas dominated by bones of mountain goat. However, it is unclear how many of these bones are really the end products of hunting since at both sites carnivores, which would be major predators on ibex, such as leopard, wolf, and lynx are also common. This raises the distinct possibility that the human record is once again being swamped by the archeology of other species. Further, this suggests a use of mountain resources in an opportunistic and sporadic fashion that falls well short of what we would expect from specialist hunters.

This is what makes the evidence from Teshik Tash so interesting. Here carnivores are almost absent and those goat bones, which are still housed in St. Petersburg, show no trace of animal gnawing whatsoever. This is a pattern recently unearthed in the Klithi rock shelter in northwest Greece by a team, of which I am a member, led by Geoff Bailey.[59] The site lies at a height of 500 m in the narrow steep sided gorge of the Voidimatis river at the foot of Mount Gamila. The peaks rise above the site to heights of 2,497 m in this Epirus section of the

Plate 9.3 Living in the mountains. The Klithi rock shelter in northern Greece was only occupied by Moderns after the local ice sheet had retreated. The arrow points to the site.

Pindus chain. The Klithi deposits have been cored down to bedrock and no trace has been found of any occupation older than 16,000 B.P. Elsewhere in Epirus, sites have produced older material but these are found either closer to the coastal plains and hence at lower altitudes or on flat limestone plateaux in the uplands. The site of Klithi is well away from either locality. Furthermore, the peaks above the site were, at 18,000 B.P., some of the most southerly to be glaciated in Europe.

This mountain valley provides an excellent opportunity to see what humans could now do once the climate began to turn in their favor. The answer is that by 16,000 B.P. colonization commenced. Hunters moved into the valley as soon as the ice sheets receded and plants and animals returned. By 14,000 B.P. and for the next four thousand years the cave was intensively used during the spring and summer for a limited range of tasks. The animal bones are dominated by remains of ibex with small amounts of chamois. Very few bones in this rich collection are gnawed and carnivores are almost absent. The only possible explanation is that here we have the remains of human meals. The third most numerous species in the deposits is beaver, indicating the refuge for trees provided by the deep Voidimatis gorge.

The stone tools and animal butchery all point to very repetitive activities. There are a few drilled beads made either from seashells collected along the Adriatic 100 km to the west or from the canine teeth of red deer which was a food staple in other Epirus sites. There is no art but several needles and plenty of tools which could have been used to make garments or repair equipment which incorporated skins and binding materials made from animal sinew. The result is a dense midden of bone and stone which accumulated in the deep rock shelter over 4,000 years of regular and intensive use. This is the archeological record of specialist ibex hunters. What made it possible to visit Klithi with a small hunting party on a regular basis, kill game perhaps as it forded the river and before it moved to the high summer pastures, butcher and process it for storage and packing out of the valley, was of course planning. This applied to the personnel as much as to the equipment, timing, and decisions on how long to stay. Forward planning made it possible for the Klithi task group to meet up with other members later in the season. The number of times this was repeated in 4,000 years speaks volumes for the way in which the problems posed by the distribution and nature of mountain resources had been surmounted so that colonization of a new habitat took place.

Identical timing for similar specialist ibex hunters has also been substantiated by Lawrence Straus at the other end of Europe.[60] He has noted that the high mountains of the Picos de Europa in Cantabria, northern Spain, are first used after 18,000 B.P. The faunas have few or no carnivores and at altitudes between 600 m and 1,400 m the sites must be seasonal in nature. The most likely reconstruction is once again for a pattern of individuals and small hunting parties fissioning for some time from the bigger group, provisioning themselves and possibly taking back some surplus meat. As Straus points out ibex hunting is a specialization. It is not associated with villages or long occupations. If we want to find an example of modern behavior then the organization, backup, information, and planning required to make ibex hunting a worthwhile and secure part of a wider strategy would fit the definition. The absence of older sites at these elevations and the extreme rarity of ibex in them points to fundamental differences in behavior.

As the climate moves into the postglacial warming the hills become alive with hunters. The Alps are rapidly colonized with hunting camps, as at Colbricon in the Dolomites, situated around small lakes at over 2,000 m.[61] Through the great mountain arcs of Soviet Central Asia the same timing can be seen. Richard Davis and V.A. Ranov report camp sites at Oshkona at an altitude of 4,000 m in the eastern Pamiyrs and dated at 9300 B.P. to the early postglacial, while at Shugnou in eastern Tadzhikistan at the lower elevation of 2,000 m there are five artifact levels, the second from the top of which is dated to 10,700 B.P.[62]

Further east on the semiarid plateau of northern Tibet, An Zhimin and his colleagues have found eighteen sites at altitudes from 4,500 m to 5,200 m.[63] The artifacts are all surface scatters and so undated, but the typology is advanced with small bladelet cores and microliths. These are typologically closest to material from northern China. In the Shenja region today there is no settlement and An Zhimin stresses that all the sites are found near traces of former marsh and river deposits which may date to the early part of the postglacial warming.

Finally, but by no means least, the high level cave sites excavated by Pat Carter in the Drakensberg mountains of Lesotho show how important it is to appreciate local factors.[64] At altitudes above 1,000 m the rich plateau grasslands and clear winter skies of this zone of tropical convergence produce exceptional grazing conditions. The plateau sites from the Middle Stone Age onward contain the bones of large herd animals such as eland, horse, and wildebeest. Such grasslands provide timewalkers with some of the easiest living (fig. 1.1) in contrast to the difficulty of exploiting species normally found in high, broken elevations.

Tropical and deciduous forests

With the tropical forests we encounter a paradox. In the first place, are we not popularly supposed to have come down from the tropical trees? Secondly, and more significantly, these are the richest environments on earth since they receive the most solar energy. They have the highest figures for biomass and productivity as well as the greatest diversity of plant and animal life. And yet, the conclusion from both historical and archeological evidence is that occupation occurred very late indeed. Appropriately enough the Sanskrit word *jangala*, from which English derives jungle, means desert. No wonder Tarzan was pleased to meet his Jane.

The main problem for humans in tropical forests, where rainfall is usually more than 1,000 mm per annum and temperatures never drop below freezing, is that much of this energy and productivity is turned into inedible woody tissue as trees battle for light and nutrients. Proportionally there is not much left for humans to eat and that is often high in the canopy. Moreover, trees and plants are widely spaced rather than clumped and animals are often arboreal, nocturnal, small, and lean which makes them inefficient, inaccessible energy sources. Finally, seasonality does exist within these forests so that predictability is not as high as once thought and movement during the year is essential for successful foraging.

Let us now turn to the ethnographic evidence for forest dwellers in Africa, India, Malaysia, including the Andaman Islands, and New Guinea. As Robert Bailey and his coauthors point out, the pygmies of Zaire are frequently presented as the original inhabitants of the African rain forests who have only recently come into contact with encroaching horticulturists and village life.[65] They show, however, that there is scant evidence to support such a view from either archeology or ethno-history. They argue instead that colonization of the tropical rain forests depended upon agriculture. By planting clearings with domestic crops such as cassava, rice, and bananas it became possible to harness the tremendous productivity of these environments. The pygmies and all other tropical forest groups have a special relationship with their horticultural neighbors. They supply forest products such as honey, monkeys, and other small protein packages and in return receive the all important seeds which make existence in the forests possible.

Archeological evidence is difficult to obtain and assess in such environments. Alison Brooks and Peter Robertshaw in reviewing the problem in Africa stress the fact that as the forests retreated into their refuges at 18,000 B.P. there would have been many more opportunities for hunting along the junctions between savannah and forest.[66] Two sites which date to this last glacial maximum contain

Plate 9.4 The Nukak collecting food in the Colombian rain forest. Could it be that such forests were only used by foragers when agriculture was nearby?

evidence for open conditions in areas once thought to be the site of forest refuges. As the forests rapidly expanded between 9000 B.P. and 5000 B.P. there may well have been many such habitats. The introduction of domestic crops into the area is also an open question and yams may be the earliest at 6500 B.P. If this is the case then the timescale involved in the evolution of the pygmy physique, which is often put forward as evidence of great antiquity for forest hunters, would have to be revised.

The answer to the question of whether the rain forests were used prior to agriculture cannot be satisfactorily answered with the current evidence from India. In China Late Paleolithic sites cluster around latitude 25°N. There are no earlier sites in this area and although only a small part of China falls within the true tropics only one site, at Changzuo in Guangzi province, has so far been identified. Earlier occupations are largely confined to Shanxi province and northern China in what would, during much of the Upper Pleistocene, have been temperate environments. Under glacial conditions these would have extended into the tropics, as already mentioned in chapter 8.

In Sundaland Peter Bellwood has concluded that hunter gatherers did not use the ever wet equatorial rain forest areas.[67] Today sites found in such settings usually contain evidence that when occupied more open, broken conditions existed as at Niah cave in Sarawak. The core regions of equatorial rain forest in southern Malaya, eastern Sumatra, and much of inland Borneo show no traces of occupation at all until after 5000 B.P. when agriculture enters the region. Hoffman's demonstration that today's forest hunters, the Punan and Kubu, are very closely related to the surrounding cultivators supports such a finding.[68]

Outside the core area, occupation of rain forest and monsoonal areas goes back to 10,000 B.P. with the widespread Hoabinihan settlements based on hunting and shellfish gathering as well as limited horticulture.

On balance much of the tropical forests appears to have been true human jungles until comparatively recently. They provide a mirror image of the colonization of humans into the deep Sahara and other arid deserts. Agriculture apparently held the key. The controlled planting of carbohydrates solved the problems of nutritional stress in the most productive environments on earth. Herded animals such as camels opened up the arid deserts. Even so, much of central Borneo is uninhabited today as are large parts of the Sahara.

The deciduous forests of northern latitudes provide another example of how unattractive forests could be. The oak mixed forests were once held to be optimum environments for the Ancients during warm interglacial periods. The genial conditions in the forests of western Europe were thought by Charles McBurney in 1950 to account for the preponderance of handaxes in that part of the continent.[69] Favorable environments apparently encouraged higher levels of culture as represented by finer artifacts—a tradition of thinking that goes back to Montesquieu and the Enlightenment and underpins cultural chauvinism everywhere.

The reality is somewhat different. Nowhere is this more clearly seen than in Britain where oceanic conditions during the last interglacial 130,000 years ago allowed hippos to graze in the Thames while the rich oak forests resounded to the footfalls of straight tusked elephants, narrow nosed rhino, and other forest species. This was an exceptional interglacial with the deepsea cores indicating the smallest ice to ocean volume in the entire Middle Pleistocene. Warm temperate conditions for a large island well stocked with a diverse animal community should surely suit the Ancients and their big game hunting lifestyles that are so commonly portrayed? If there was going to be a Paleolithic Garden of Eden for such a lifestyle then this was surely the time and place.

Despite a great many finds and much fieldwork it has proved impossible to pin down any of the British sites to this interglacial stage.[70] Hippo are found across the country from Victoria cave in Yorkshire to the Barrington gravels near Cambridge, where complete skeletons were unearthed, to beneath the lions in Trafalgar Square. And yet no flints are found with them. The same picture is repeated in other geological and paleontological contexts. Britain 130,000 years ago was a human desert.

This, of course, clashes with the usual picture of life in northern latitudes. Here the cold of the ice age is seen as keeping out the scantily clad Ancients (even though in our present interglacial few of us would linger long, unclothed, in the open, on a February day on the north European plain!) But the story, at least in this corner of Europe, has to be reversed. The extreme forest conditions may have many resources but they come in the wrong size package and at densities with which the organizational strategies of the Ancients were ill equipped to cope. Elephants and acorns both require a great deal of processing and even under interglacial conditions storage for winter was vital for survival. It is this which gives these habitats a low ranking in terms of accessibility (fig.

1.1). The alternative natural stores discussed in chapter 8, which were available during the more open conditions of the early glacial, were simply not there due to higher winter temperatures and smaller herd sizes producing fewer mortalities.

It might be argued that the forests have nothing to do with this pattern and that instead it was Britain's island status at the time which led to it being a human desert for 10,000 years. But Britain was also an island during earlier but less extreme interglacials. Under cooler, temperate conditions timewalkers were present.[71] It is also instructive to jump forward over a hundred thousand years and look at the occupation of Britain between 9000 and 5000 B.P. by Moderns. A patchwork of sites and stone tools covers the country. Deciduous forests and open environments were used, the former controlled to some extent by burning. The toolkits of these fisher–gatherer–hunters include axes and many small hafted armatures for improving hunting efficiency on the low density, open forest animals such as elk, red and roe deer, pig, and rabbits. Comparing these tools to those of the Ancients, 120,000 years previously, brings home the point that while there is a living to be made in such forested environments this is only achieved through increased levels of planning, organization, and a consequent gearing up of toolkits to improve hunting success. The selection to make technology more complex and hence better than before comes not from the type of resources or the nature of the environment but instead from the principles within those societies which govern the pattern and process of colonization.

Coniferous forests

Nowhere is this concept of a society's capacity to adapt the available resources to its needs better illustrated than in the last environment to be examined, the coniferous forests. While widely distributed the coniferous forests I will deal with here are those which now cover Siberia in that vast area east of the Ural mountains. Three main divisions can be drawn. The west Siberian plain drained by the Ob is a basin of marshes supporting Siberian fir and larch, spruce, silver fir, and stone pine. It is bounded to the north by tundra and to the south by wooded steppe grading into full steppe. The eastern boundary is formed by the Yenisei river. Between the Yenisei and the Lena lies the central Siberian plateau while the third region of high, folded mountains encircling the Kolyma plain above the Arctic Circle lies to the east of the Lena. These two regions form the eastern Siberian forests with Siberian fir, eastern larch, and stone pine. Tundra fringes them to the north while along the southern margins are the more diverse coniferous forests of western Siberian type, as well as high altitude steppes.

Today these are sparsely populated and inhospitable lands. The cold and drought resistant forests are the most difficult environments for hunters and gatherers to extract a living from. While these estimates are naturally relatively coarse, the table brings home the inhospitable character of the vast tracts of Siberian forest which even during glacial phases, when they became tundras, were hard environments. This does not mean that humans are prevented from living in either habitat. The ethnography of subarctic boreal forest adaptations of North America shows just the opposite.[72] However, in Siberia the evidence is

Plate 9.5 The inhospitable boreal forests of Siberia and Alaska provided a challenge to timewalkers.

overwhelming that use of these forest environments and the open tundras was a late phenomenon.

Much of Siberia remains unexplored. Archeological research has concentrated on the rectangle formed by the upper Yenisei and its major tributary the Angara in a mountainous region west of Lake Baikal. The earliest evidence now comes from the open-air site of Kara-Bom in the Ob drainage basin. The stone tools show a classic change from Middle to Upper Paleolithic types and techniques.[73] The shift has been dated to 43,000 B.P., so comparable in age to other sites, such as Boker Tachtit in the Negev, where the transition occurs *in situ*.

At a later date along the Yenisei are the multilevel sites of Kokorevo and Golubaya recently excavated by Sergei Astakhov.[74] Farther north is the site of Afontova Gora with a radiocarbon date of 20,900 B.P. from the second site. Golubaya I is dated at 13,050 B.P. and the various sites and levels at Kokorevo between 15,460 B.P. and 12,690 B.P. The artifact collections have always provoked comment ever since Afontova Gora was discovered because it combines finely struck blades and advanced Paleolithic tools with pebble tools and Mousterian-like side scrapers. This makes it particularly important to have well stratified sites, such as Kara-Bom, whose ages are established by a radiocarbon chronology. Surface fines judged by their typology could be of any age.

The hearths at Golubaya contained bones of wild cattle, bison, steppe ass, red deer, hare, and wild goat. The charcoal came from spruce and pine although 13,000 years ago these would have been small in size and few in number.

To the east on the Angara is the well known site of Mal'ta where excavations began in 1928. Over 600 m^2 have been examined and several dwellings and windbreaks have been found.[75] Stone slabs, tusks, and bones were used to build walls around small circular depressions which had been dug to a depth of about 20 cm into the sandy loam. The houses were small as are recent dwellings in the Arctic. For example, the diameter of the fifth dwelling was 3 m and contained a central slab built hearth. The settlement contained carved figures of birds and waterfowl as well as female figurines. A child burial was found with an intricate bead necklace. There is a single radiocarbon date of 14,750 B.P. and the animal fauna consists of mammoth, woolly rhino, cattle, and bison. It is dominated by the bones of reindeer while fur bearing animals are barely represented.

The evidence from Siberia points to the major use of the region under the tundra conditions of the last glacial period. There are sites between the Yenisei and the Lena with absolute dates that may well extend this use back to 35,000 B.P. as at Malaya Siya and Varvaryna Gora. While the artifacts and artistic traditions are very different the timing is comparable to that found on the plains to the west of the Urals. As far as the coniferous forests were concerned, it may well be that the ebb of settlement in Tuva between the Middle Paleolithic and developed Late Paleolithic correspond to an increase in forest cover. The evidence for earlier occupation that might be associated with coniferous forests in the last interglacial is as difficult to assess as that for the use of tropical rain forests before agriculture. However, there is one other piece of evidence that points very strongly to either the tundra or the taiga forming a barrier to colonization. That is, of course, the absence of humans in North America until late in the last ice age.

10
Humans almost everywhere

In the summer of 1988 a paddle steamer left Yakutsk in northeastern Siberia to sail 200 km upstream on the River Lena. On board was an invited party of archeologists and geologists. The purpose of the journey was to visit the site of Diring-Ur'akh where, for the past six years, Yuri Mochanov had been digging what has been described as "the pearl of Yakut archaeology."[1] His finds of pebble tools, apparently similar to those from the lowest levels of Olduvai Gorge, had led to claims that Siberia was entered sometime between 1.5 and 2 million years ago, suggesting to some that Quatrefages' theory of 1879, where humans evolved in the northern circumpolar zone, should be dusted off once more.

Needless to say there was little agreement on board between supporters and detractors of these claims. The dating is tenuous as the river deposits in which the material lies have been redeposited and reworked so that in the opinion of S.M. Tseitlin, the leading geologist of the Quaternary in Siberia, the age of the artifacts has been unnecessarily exaggerated.[2] Others have criticized the approach to dating which assumes that all pebble tools are old either because in East Africa similar artifacts have been absolutely dated or because they look primitive.

Claims for the most ancient artifacts in the most unlikely places will, I am quite sure, be with us so long as there is Paleolithic archeology. Descriptions of eoliths, the "dawn stones" of East Anglia, clogged up many pages of archeological journals in the first half of the twentieth century with their supporters claiming them as the stone tools of Pliocene man.[3] They are all now regarded as the by-products of natural fracturing in beach and river gravels. The claims to common sense—that these and other geofacts are tools because they "fit the hand perfectly" and "look like" spokeshaves, points, picks, and even, as the Rev. Frederick Smith thought, a portrait of a Gibraltar monkey[4]—will no doubt continue to be made wherever gravel is dug.

Peopling the New World

Such "commonsense" claims have been made for finds in North America, from the "paleoliths" dug in the nineteenth century from the Trenton Gravels in New Jersey to Calico Hills, still under investigation, in California.[5] Currently there is no serious support for claims of Pliocene occupation or even for a human presence in the New World during the Middle Pleistocene, which ended 130,000 years ago. With this perspective on global colonization the claims made for Diring-Ur'akh seem strange if, as Larichev, Khol'ushkin, and Laricheva claim, the timewalkers who supposedly made these tools "were led into northern Asia by human minds, not by the instincts of unthinking animals."[6] Why then did they stay put and not make the trip to similar climates and lands in the New World for nigh on 2 million years?

There is, however, much less agreement about the arrival of humans in the New World in the past 50,000 years, although no one is in any doubt that it was extensively colonized after 12,000 B.P. The disagreements surface every few years as sites and artifacts are produced to claim occupation before this baseline. These are then examined and usually found wanting. For example, a worked bone, described as a flesher, from the Old Crow River in the northern Yukon was originally dated after its discovery in 1966 to 27,000 years old. However, some thought its serrated end was the result of gnawing by a wolf. Eventually a small piece was dated in the radiocarbon accelerator. It produced a date of less than 3000 B.P. thereby consigning the flesher to obscurity.[7] Whether it was made by an animal or a human suddenly did not matter quite so much.

The colonization of the New World has to be seen as at least three separate geographical problems:

1. colonization of the Beringian paleocontinent;
2. penetration south of the ice and peopling the New World;
3. colonization of the Arctic as the ice sheets retreated at the end of the Wisconsin glaciation 10,000 years ago (fig. 10.1).

Figure 10.1 Location of New World and Atlantic Moderns.

Beringia

With lower sea levels northeastern Siberia and Alaska were joined as a steppe tundra, dominated by mammoth and other large herd animals.[8] At the maximum drop of sea level 18,000 years ago 1 million sq. km of exposed shelf would have formed the central part of the Beringian paleocontinent. In the west it was broken by high mountains with local ice caps surrounding the Kolyma plain. Most of central and southwestern Alaska was unglaciated although east of the Canadian border the major Wisconsin ice sheet began. This extended south across the Alaskan peninsula and some of the Aleutian Islands. At such moments Beringia was at the end of the world, equivalent to northern Europe close to the Scandinavian ice sheets.

On the Siberian side the oldest evidence comes from very few sites. Of these the Diuktai cave on the Aldan river lies some 2,500 km from the Bering Straits and to the west of the glaciated Verkhoyansk mountains. There are several levels at this site excavated by Yuri Mochanov most of which contain small collections of stone tools but interestingly do include a few broken bifacial projectile points.[9] The small quantities of tools from these excavations are very reminiscent of the earliest phase of the European Upper Paleolithic, discussed in chapter 9. Level VII in the cave has been dated by radiocarbon to between 12,100 B.P. and 14,000 B.P. The lowest levels may extend back to 17,000 B.P. At the open site of Verkhne Troitskaya artifacts were recovered below a level dated to 18,300 B.P. Finally, at Ust-Mil, 60 km from Diuktai, radiocarbon dates extend occupation back to 35,000 B.P.[10] These Aldan river sites are almost 2,000 km northeast of settlements such as Mal'ta in the Angara basin.

Two sites in this vast region are, at between 1,700 km and 2,000 km, slightly closer to central Beringia. At latitude 71°N and with a radiocarbon age of between 11,830 B.P. and 12,240 B.P. the site of Berelekh in the Indigirka basin is the most northerly Pleistocene site in the world.[11] The artifacts include bifacial spear points as well as perforated pebbles which are thought to be pendants. The second site, Ushki Lake, was discovered in central Kamchatka in 1962 by Nikolai Dikov. A number of localities have been excavated around this small lake and altogether eight cultural levels recognized. Level 7 has radiocarbon dates of 13,600 B.P. and 14,300 B.P. while level 6 has been dated to 10,360 B.P. and 10,760 B.P.[12] Hearths were found in the older level and a "grave" pit which, although it contained no burial, did produce over eight hundred beads made from amber and four stone pendants.

Excavations in the upper levels produced plans of small houses the best preserved of which has a rectangular outline, a slightly sunken floor, and covers about 12 m^2. It also has a pronounced entrance passage on the west side faced internally by a stone lined hearth. Among the rich artifact collections in the lower levels are stone points which have stems for hafting while in the upper layers there are bladelet cores and finely made bifacial spear points. Large stone scrapers are also present.

In eastern Beringia there are two areas that have produced material of comparable age. The Nenana valley southwest of Fairbanks in Alaska has been

Plate 10.1 The modern Beringian landscape in southern Alaska. Herds of mammoths and bison once grazed in front of these glaciated peaks and along the braided river channels.

subject to intensive survey. Roger Powers and John Hoffecker have recovered stone artifacts dated between 12,000 B.P. and 10,000 B.P.[13] At the Broken Mammoth site on the Tanana river in central Alaska, David Yesner has excavated stone tools from an old land surface dated between 11,040 B.P. and 11,770 B.P.[14]

So few sites, so widely scattered from such a huge area. As the pace of research increases there may well be a pattern of Beringian settlement as complex as that found in the European Upper Paleolithic either side of the last glacial maximum 18,000 years ago. However, a recent comprehensive review of the present Siberian and Beringian evidence by John Hoffecker, Roger Powers, and Ted Goebel concludes that occupation first took place between 12,000 to 11,000 years ago.[15] An important element supporting this conclusion is the rate at which the exposed shelf was flooded and the date when the Bering Straits appeared. The current evidence suggests that most of the sites currently known from Beringia in fact postdate this event. Siberia and Alaska may have been sundered as early as 14,000 B.P. although sea level was still more than 30 m below its present level.[16] This timing corresponds well with the reconstruction of the late glacial environment in interior Alaska. At 14,000 B.P. there is a rapid shift to shrub tundra, as shown by the pollen profiles, and this marks a deterioration in the carrying capacity of the mammoth steppe as the forests gradually returned with spruce and birch, both widespread by 9000 B.P. The Nenana valley provided a temporary refuge for the animals of the steppe tundra and the remains of Dall sheep, wapiti deer, and steppe bison have been found; their presence acting as a focus for settlement. Mammoth is unknown and elsewhere on the plains of western Asia we saw in chapter 9 that their extinction was well underway between 14,000 and 12,000 years ago.

The New World

How and when humans reached south of the ice sheets is still very much an open question. Whether this was the result of a single pioneer population or multiple colonization is also keenly debated.

The question of how is reasonably straightforward although finding evidence is difficult. The choice is between either coastal colonization, as suggested by Fladmark,[17] or else an overland route, which the majority favor. David Hopkins has argued against the former on the grounds that between 25,000 B.P. and 14,000 B.P. sea ice was close inshore and marine productivity low.[18] The height of the Sartan/Wisconsin glaciation was instead the optimum time for inland resources in Beringia as Vereschagin and Baryshnikov have shown in their detailed study of the animals.[19]

However, it has been found that ice-free islands existed down the coast of British Columbia at this same time. Parts of the rugged Queen Charlotte Islands were unglaciated and favored by an oceanic climate.[20] This raises the possibility of a productive fringe to the edge of the ice sheets that could only be exploited by sea travel. No archeology has yet been found to back up the environmental evidence.

As a result the overland route commands greater support. Here, of course, the difficulty lies in the knowledge that the way south was blocked by the late Wisconsin ice sheet. This consisted of two separate sheets: the Cordilleran, centered to the west in Alaska and the Aleutian Islands; and the Laurentide in the east. These came together to form the barrier. The timing of this closure and the dates at which an ice-free corridor might have existed to allow a southern passage have been a major focus of research. When open this corridor stretched down to the southern border of Alberta south of Calgary and varied in width between 25 km to 100 km. Reeves has argued that it was only closed—if at all—for a short period of time around 18,000 B.P. and then probably by swollen lakes and rivers rather than by ice.[21] Others argue that ice closure happened and it was not until 13,000 B.P. that a narrow ice-free corridor dotted with glacial lakes existed.[22] Estimates for the length of the corridor vary from 2,300 km to 1,600 km.

While the details of this parting of a glacial "Red Sea" still have to be finally agreed upon it is obvious how critical the timing will be for the acceptance of early occupation in the New World. A solution, dispensing with a geological Moses, would be to find much older sites of the order of 30,000 B.P. to 45,000 B.P. when the ice sheets were far apart. This coincided with a productive marine environment so that the alternative coastal route could also be considered.

This, however, is where the evidence becomes very thin. As I mentioned earlier, it is not for lack of research or claims of great antiquity. Every year material from South and North America is presented as evidence of initial settlement within this age bracket and every year the data are demolished except in the eyes of those who have staked their reputations on them.[23] But as James Adavasio has remarked, for all these sites that enjoy their fifteen minutes of Andy Warhol like fame there are some which refuse to go away.[24] The two current survivors, separated by 12,000 km, are Meadowcroft rock shelter near Pittsburgh and Monte Verde in south central Chile. What is significant about these two sites is the evidence they contain for occupation prior to 12,000 years ago, but which bears no obvious relationship to the widespread, well dated, and well documented Paleoindian adaptations found over much of North America after this date.

To understand the importance of Meadowcroft and Monte Verde we have to appreciate what expectations these later, widespread Paleoindian traditions engender about the colonizers. After false starts with the paleoliths from Table Mountain, Tuolomne County, California, the question of human antiquity in North America was finally laid to rest with the discovery in 1927 of an unquestioned projectile point embedded in the ribs of an extinct species of bison.[25] This took place at Folsom, New Mexico. The site has since lent its name to a common Paleoindian tradition marked by distinctive hafted stone knives and points with beveled bases. These are often finely worked on both faces into a lethal lance shape with a channel flake, or flute, struck from the base and running up the piece to assist hafting. This requires great knapping skill.

The Folsom fluted points are not the oldest. Equally widespread on the Great Plains and eastern North America are the earlier Clovis points. These appear from the Canadian plains to central Mexico and from coast to coast after 12,000 B.P. In a few instances they are associated with mammoth skeletons.[26] Stratigraphic information shows very clearly that there is no chronological overlap between Clovis and the later Folsom. Radiocarbon dating pins the transition down to some time between 10,700 B.P. and 10,500 B.P. at sites such as Murray Springs in Arizona. The change is quite dramatic. While the spear points are still fluted and bifacially worked they now follow very different designs. Moreover, mammoth is never found on Folsom sites. Extinction of this and other megafauna had occurred. One animal that survived was the bison which became a mainstay of the plains Paleoindians at jump sites such as Head-Smashed-In, Gull Lake, Garnsey, and Casper located between Saskatchewan and New Mexico.[27] Herds were either driven over cliffs, up steep gullies, or into specially built wooden pounds and corrals where they were slaughtered and processed for hides and stores of meat and fat.

The large geographical range and the comparatively short timescale covered by Clovis is very striking. The fluted points and knives vary in size and shape. They do not relate to the different habitats where they are found. David Meltzer has shown in eastern North America that adaptations to local conditions are reflected instead in the number of finds of isolated fluted points.[28] The Pleistocene tundras to the north produced very different archeological signatures in terms of density of finds and size of campsites to the southern forested environments. Specialist caribou hunters, with considerable movement of individuals, were found in the north. In the forests a wider range of resources was used and mobility was much less. Big game hunting was very rare throughout the region. This quieter side of Clovis life is, however, swamped by questions of colonization, big game hunting, and extinction.

In a justly famous paper published in 1967 Paul Martin put forward the view that once through the ice-free corridor they encountered a rich environment with few carnivore competitors.[29] Martin argued that meeting such bounty unleashed big game hunting which fueled a phenomenal rate of colonization. Pleistocene overkill provided the energy while human reproductive patterns created the motor for incessant and rapid colonization. Generating estimates from the model depends primarily on establishing a population growth rate. Records show that humans have achieved 3–4 percent per annum while among agricultural pioneers a doubling of population every fifty years is not uncommon.[30] A staggering rate of increase is recorded historically among the mutineers on Pitcairn Island who numbered thirty-one when found in 1808. Three generations later in 1856 the population was 193 and had to be removed to a larger island.[31] While these cases are exceptional, the ability of human populations, as Malthus first pointed out, to increase at a geometric rate should not be forgotten.

Martin argues for a front, 160 km deep, with a population density of 0.4 persons per km^2. Behind this population density dropped dramatically to 0.04 per km^2. He assumed a constant rate of advance for the bow wave through the plains of North America and down the mountain spine and plateaux of South America. Every year a further 16 km was advanced to find room and resources for the growing population. These figures for population growth fueled by cheap energy from killing megafauna predict that about a thousand years after the arrival of hunters through the ice-free corridor, Tierra del Fuego would be in view. Some confirmation for his simulation comes from the date of 10,700 B.P. for large stemmed points from Fells cave at the tip of Patagonia; 14,000 km and a wide array of habitats covered in about a thousand radiocarbon years.[32]

Martin's blitzkrieg version of the overkill model literally sweeps all before it. It represents maximum human impact on a new environment but because of the speed yields minimum visibility. Negative evidence is thus conveniently turned to his cause. These super-predators "who preferred killing and persisted in killing animals as long as they were available,"[33] had no need, while prey was abundant, to make traps and drives. The animals were also wiped out before they could be carved, painted, or engraved. The trail of extinctions south of central Mexico where the fluted points peter out is just the inexorable roll of the wave.

This is the background for investigating a pre-Clovis tradition more than 12,000 years old and which, if found, would contradict Martin's model of rapid colonization. At the richly layered site of Meadowcroft there is a full suite of radiocarbon dates in excellent stratigraphic order.[34] The lowest levels with cultural material extend back to 15,950 B.P. while less certain associations could extend occupation back to 21,000 B.P. Those are less important than the unambiguous demonstration of over 3,000 years of pre-Clovis occupation in this corner of Pennsylvania.

But what do the artifacts look like? Is there a hint of the Clovis to come? At

Meadowcroft there is a single lance-shaped projectile point that is bracketed by dates of 11,300 B.P. and 12,800 B.P., which puts it before the Clovis efflorescence at 11,500 B.P. The single point is well made but lacks the fluting which is the hallmark of the later Paleoindian types. Otherwise the early stone tools at Meadowcroft are not giving much away—small blades and a range of bone and organic artifacts.

For archeologists working in the "lower 48 states" the northern stone-tool "ancestors" of the widespread Clovis complex are not readily apparent. However, archeologists working in Beringia point to the artifacts from the Nenana valley sites which include well made, but small, bifacial projectile points. They conclude from this evidence that the Beringian sites share many common artifacts with the Clovis sites to the south. Together the evidence from these projectile points and their associated technology probably represents the same wave of migration that swept into and through Beringia 12,000 years ago.[35] Another cultural similarity that links the archeology of the Paleoindians over such a huge geographical area of North America is the absence of art, even though organic artifacts are well preserved.

The rarity of these earlier sites does not invalidate the evidence they contain. Low visibility, championed by Martin for the Paleoindians, works just as well as an explanation for much earlier occupations. A single site can force another look at the overkill hypothesis. A 3,000 year wait suggests much less concern for continuous oversize helpings of red meat and an explanation of colonization and extinction that depends less on a single cause. Vance Haynes, for example, has recently suggested that the dry interval which occurred between 11,000 B.P. and 10,000 B.P. put the megafauna under extreme stress, particularly in the southwest United States, and increased the effect of human hunting.[36] This led to extinction. If that argument could be extended to others parts of the continent then the Paleoindian tradition might, as Robert Kelly and Lawrence Todd have argued, be a reaction to stress in a declining environment rather than the response to an unexpected bonanza resulting from initial colonization.[37]

All of these problems are drawn into sharp focus by the site of Monte Verde in southern Chile. This site commands considerable respect and, it is true to say, any disbelief comes from the unexpected character of the finds made by Tom Dillehay and his team.[38] The site has two well dated but separate components. The youngest settlement is found on both sides of a small stream and is stratified beneath peat which has led to good preservation and excellent conditions for dating. In this settlement, dated to 13,000 B.P., there are two zones. In one there are the preserved remains of what Dillehay interprets as huts formed by logs and branches. The other area contains the remains of upright posts, mastodon bone and skin. It appears to be a butchery and processing area. The stone tools are nondescript and expedient in terms of finishing, selection of local raw material, and rudimentary flaking. There are no links whatsoever to either Meadowcroft or any of the later Paleoindian lithic traditions.

Well away from this 13,000 year old occupation is a much smaller locality where eight stone tools have been recovered along with what are described as hearth like features. Charcoal from these has given dates of 33,000 B.P. The stone tools are lumps of local stream pebbles minimally flaked. At the moment the jury is still "out" on these finds as the full details await publication and discussion but undoubtedly such an early claim has improved the chances of general acceptance for the 13,000 B.P. material.

Vast distances and often ambiguous radiocarbon dates have produced a variety of explanations for the archeological patterns. While Hoffecker, Powers, and Goebel argue strongly, and in my view correctly, for significant Beringian and hence North American colonization in the period 12,000–11,000 years ago other solutions have been found for the earlier, fainter traces. One common solution to the pre-Clovis "wait" in the lower forty-eight states has been to fall back on the thesis of multiple diffusion with several waves of people, each with a distinctive toolkit, coming to America. Writing before the discovery of Monte Verde and citing excavations from some highly contentious sites in Mexico, Venezuela, and Peru, Janusz Kozlowski and Hans-Georg Bandi proposed four such arrivals.[39] These occurred at approximately 22,000 B.P., 16,000 B.P. when the first foliates arrived perhaps from the area of Diuktai cave, 13,000 B.P. for another population movement into Alaska, and lastly coinciding with the appearance of Folsom about 10,000 B.P. I believe the data are too sketchy to support their scheme in detail. It depends on fitting the facts to a model of lithic homelands in the Old World, even though the fluted points are an American original. Dikov also traces four movements but sees the origins of the last two in his Kamchatkan sites, such as Ushki, where other types of leaf points have been found. One wave or several waves, from this direction or that, research into the problem continues.[40] But what *is* widely agreed is that it all occurs after 40,000 B.P. and, in my opinion, after 12,000 years ago.

The Arctic

The springboard for archeological research in the eastern Arctic began, appropriately, as a consequence of the Cold War. This led to the construction of early warning stations in Canada and arctic Greenland in 1954 and provided both opportunities and facilities for archeological research. As a result we now know that the major expansion into the Arctic, as far east as Greenland, took place between 4,500 and 3,200 years ago. The headland site of Qeqertasussuk, dating to 4,500 B.P., has recently been excavated by Bjarne Grønnow north of the polar circle in west Greenland.[41] Both Don Dumond and Moreau Maxwell associate this expansion with the arrival of Paleo-Eskimo populations. Survival in this environment is hazardous, to put it mildly, and as might be expected coping with such high risk is abundantly reflected in the sophisticated stone and bone weapons and artifacts, burials, and art.[42]

Inuit origins have been tackled linguistically, biologically, and archeologically with the usual results of non-concordance. All too frequently simplistic interpretations are made of novel material culture as waves of new settlers.

According to Hanihara Kazuro and W.W. Howells the evidence from physical anthropology shows that North America experienced at least two major waves of Mongoloid immigration from somewhere in eastern Asia.[43] The earlier

Plates 10.2, 3 Extreme seasonality at the Qeqertasussuk site in west Greenland. The resources of the sea are rich and crucial to survival, but only available for limited times of the year. The summer picture shows the excavation, some of it in the permafrost, in progress.

penetrated south and all Native Americans are descended from them. A later intensified Mongoloid form came from a small population trapped in northeast Siberia. They expanded, for reasons unknown, in dramatic fashion between 10,000 B.P. and 3000 B.P. From this population, assisted by local gene flow, came the modern Koreans, Japanese, and Chinese as well as the Inuit.

From the archeological evidence Don Dumond prefers to see western parts of the Arctic peopled by southern hunters moving north as a response to deglaciation and climatic warming sometime after 6000 B.P.[44] This assessment is largely based on cultural affinities between finds in Alaska, eastern North America, and the Great Basin of the southwest United States.

Wherever they came from, the key to Arctic colonization was the orientation of peoples to the sea and its resources. Maxwell believes that colonization after 4400 B.P. resulted from population pressure and declining resources in Alaska as sea temperatures rose.[45] While there are many details still to discover, the rapid expansion through the Arctic from Alaska to Greenland can be traced through a series of related stone toolkits known as the arctic small tool tradition. Between 4150 B.P. and 3850 B.P. all those parts of the Arctic which were ever going to be occupied were first colonized. Among the stone tools, pressure flaked projectile points are common as well as complex sea mammal hunting equipment such as toggle headed harpoons. When these all-important maritime adaptations first appeared is difficult to say as earlier archeological traces have probably been lost when the Beringia Shelf was flooded and present shorelines reached between 7000–6000 B.P. Robert Ackerman thinks the date of 3360 B.P. for a campsite with marine hunting equipment on Wrangel Island, in the Arctic Ocean off northern Siberia, marks the widespread appearance of economies based on hunting sea mammals such as whales, seals, and walrus.[46] Previously the Beringia Shelf provided good coastal resource locations and lake and/or river fishing as well as caribou hunting in the interior.

Subsequent Arctic prehistory is far from static. There was, for instance, the rise of whaling societies around the Bering Straits and the appearance of reindeer herders on the Siberian side. Some of these developments are put down to the arrival of new peoples. Ackerman, on the other hand, places them in the much wider context of state formation in China and Mongolia. The effects of civilization rippled north along the Ordos and Amur rivers and finally reached Beringia and the Arctic during the past 2,000 years.[47] Art styles and burial customs indicate contact with the Amur 2,500 km to the south. Warfare was very common in the Alaskan Eskimo whaling societies as shown by suits of whalebone armor found in elaborate burials, perhaps reflecting increased contact with the powerful societies to the south. Social, political, and economic developments in areas as remote as the Arctic may therefore have been due partly to the effect of processes far away. Perhaps the effects of earlier world systems precipitated the colonization of the Arctic, traced archeologically through the Arctic small tool tradition and maritime hunting.[48] This would have been a form of distant intensification within a wider competitive sphere. Colonization was the result, just as we saw with the settlement of tropical rain forests and deep deserts in chapter 9.

Colonization, as a consequence of participation in a much larger social and economic system, was not the case in the Arctic regions of the Old World. After 10,200 B.P. the ice retreated into the mountains of Scandinavia and a coastal fringe gradually appeared that was rapidly colonized. Sites above the Arctic Circle are dated to the earliest phases of the postglacial at 9200 B.P. while the Fosna–Hensbacka sites in southern Norway and Sweden are widespread by the tenth millennium B.P.[49] The Finnmark province of northern Norway has Komsa sites dated to 7000 B.P. while in the Varanger fjord there is abundant occupation at 70°N by 5600 B.P. The retreat of the inland ice saw the development of hunting based on reindeer and elk. Signe Nygaard notes that the northern marine based economies had much more art, for example the Alta fjord engravings the oldest of which are dated to 5500 B.P., as well as more formal networks of trade and alliance as shown by raw materials and the distribution of common items. These later developments coincide with the appearance of agricultural economies in southern Scandinavia.

Farther east there is evidence that the Taimyr peninsula was occupied at least 3,000 years ago and probably earlier. This means that the continental shelf islands of Novaya Zemlya and Severnaya Zemlya were probably also occupied by this time. New evidence points to much earlier dates than previously believed for the colonization of parts of the Siberian Arctic. In 1989 an expedition, by archeologists from St. Petersburg to Zhokhov Island in the New Siberian archipelago (latitude 76°N), found a prehistoric site with thirteen dwellings built from driftwood and preserved in the cold climate. Tools were made from ancient mammoth ivory while obsidian and flint had been brought to the site. These reindeer and polar bear hunters reached Zhokhov Island between 7,000 and 8,000 years ago.[50]

However, there is no evidence at present that the Svalbard-Spitzbergen archipelago which lies between 76°N and 80°N was ever occupied before the islands became an important whaling base in the medieval period. A voyage by Martens in 1671 describes the fleets which were there in force, as well as abundant bear, fox, and reindeer on the islands. In such a productive environment evidence for earlier populations might still be forthcoming, although at first western contact they were vacant.[51]

Australia, Papua New Guinea, and Sahul

How different from the Americas has been the history of research into Australia the continent of hunters and gatherers. Written off for many years as having little time depth to its prehistory and marked by stone tools described only twenty years ago as "crude and rather colourless,"[52] there was a feeling that its prehistory matched the attitude of the British Empire, which once used the continent as somewhere handy to dump the empties.

This has now all changed. In 1962 Australian prehistory broke the Pleistocene barrier with John Mulvaney's excavation at Kenniff cave in central Queensland and a date of 16,130 B.P. for stone tools from this multilayered site. Other dates earlier than 10,000 B.P. soon followed. Sandra Bowdler extended the sequence of

Pleistocene occupation on Tasmania with a date of 22,750 B.P. from Cave Bay cave on Hunter Island. Jim Bowler's study of the fossil lake and dune system at the Willandra lakes in western New South Wales not only provided a rich sequence of landscape change but also discovered, in the lunettes around Lake Mungo, cremations and burials, and campsites dated as early as 32,000 B.P.[53]

Australian archeologists made full use of the advent of radiocarbon to unravel its prehistory and so largely escaped from the strait jacket imposed by flint typologies and their arguments of lithic ancestry. The prehistorians were also fortunate in that once the early dates started to appear they were not rejected or treated suspiciously on the ground that nothing of similar age had yet come from a potential homeland. The Java finds of early and late *Homo erectus* provided the data.

At the moment the earliest dates for occupation in Papua New Guinea come from the coastal terraces of the Huon peninsula where well stratified artifacts have been dated to at least 40,000 B.P. and the excavators, led by Les Groube and John Muke, believe they may be older.[54] The most interesting artifacts are the large axe-like forms made of andesite and weighing as much as 2.5 kg. They have been flaked on one face only and their sides have been grooved to form a "waist." This may have aided hafting. Their most likely use was for ring barking trees and clearing ground. Whether this implies a forest management technology, already beginning to select food plants such as yams and tree fruits, is now up for discussion.

In Australia the earliest radiocarbon dates come from the Swan River near Perth. A date of 39,500 B.P., almost at the limit of radiocarbon techniques, was obtained from carbonized wood and associated with almost a thousand small chert artifacts. Many believe that earlier finds are still to be made. Thermoluminescence (TL) dates from the Malakunanja rock shelter in northern Arnhem Land suggest initial occupation by 60,000 B.P.[55] Sterile sands continue beneath this date which is associated with the lowest artifacts in the sequence. While some technical questions still surround the TL dating of sediments none of the current sceptics finds it difficult to accept that such an age for the initial colonization of greater Australia is possible.[56]

Australia was connected during periods of low sea level with Tasmania and Papua New Guinea to form the paleocontinent of Sahul. Together, Sahul presents the full range of environments encounted in chapter 9. These range from the equatorial rain forests and high mountains of New Guinea to the once glaciated peaks of Tasmania, now covered in the southwest of the island by temperate rain forest. Plains and highly varied arid regions make up most of the Australian part of the paleocontinent fringed by the eastern highlands and mountains of the Great Divide. All these environments have been sampled, so that Sahul provides the best opportunity to test and confirm the pattern of colonization that we have seen before. It is also an extreme test since once the Wallace line separating the fauna and flora of Sunda and Sahul had been crossed modern populations were entering a completely novel land where the trees shed their bark rather than their leaves. While the indigenous North American fauna held many surprises like short-faced bears the size of horses, giant sloths and beavers, not to mention the armadillos of South America, there were at least familiar species such as mammoth, bison, horse, and deer. In Sahul there were instead mammals that laid eggs, wolves that carried their young in pouches, and marsupials such as *Diprotodon*, the size and roughly the shape of a rhino.

There was never a dry land crossing to Sahul. At the height of the last glacial maximum 18,000 years ago the sea was 130 m lower, which still left an open sea crossing of 65 km.[57] At most other times the distance to be crossed ranged between 80 km and 100 km. Leaving aside the improbability of a man and woman floating across on a log, human presence in Sahul is the earliest evidence for the use of sea-going craft. What is of course significant, as we saw in chapter 9 for the Tasmanians, is that Australian watercraft were extremely rudimentary. Not all of them became waterlogged after half an hour or at most a few hours in the water, but nonetheless the simple wooden and bark punts that were used for island hopping and fishing would not have been up to the task. This of course ignores the fact that the other major part of Sahul, Papua New Guinea, has advanced watercraft as shown, for example, by the islanders of Mailu who maintain their place in a wider regional economy by producing pots which are then traded in sail driven canoes to mainland New Guinea and through large areas of Melanesia.[58] Indeed, outrigger canoes are described from the coasts of Queensland.[59]

Arrival by sea has led to some intriguing models for colonization which, as Peter White and Jim O'Connell have shown, provides an informative contrast to the American debate.[60] Basically the camps are drawn up between overlanders and coastal huggers. For the first, Joseph Birdsell proposed that wherever they landed (and he favored the north), population would then spread in all directions.[61] He started with a colonizing unit of twenty-five persons and an interval between generations of sixteen years. Population budded off and colonized new land when numbers reached 60 percent of the carrying capacity. He showed, using an estimate of 300,000 as the size of Australia's Aboriginal populations in 1778, that this could be reached 2,204 years after their arrival. Even allowing that this figure is a gross underestimate of Aboriginal numbers, the prehistoric time depth now available more than covers such revisions.

Variations on the overland model plot different routes to skirt the central deserts and other areas which became more arid during the course of the last ice age. One element that has received a good deal of attention has been testing the blitzkrieg model of colonization by tracing the extinction of the Australian megafauna. The results show conclusively that the megafauna was not wiped out across Australia at the same time but persisted in some regions for at least 20,000 years after humans had arrived.[62]

Human predation of the megafauna has also been difficult to prove. But whether the first Australians hunted the giant, 3 m tall, short-faced kangaroo to extinction or not, they did it without fluted points. The earliest stone technologies contain few retouched tools. Bone tools are not common and when they occur are whittled points with no decoration. It is easy to write off such toolkits as technologically simple, but the missing wooden component has to be especially remembered in an Australian context. One of the main uses of stone tools in traditional technologies is to make wooden artifacts such as bowls, boats, spears, ceremonial paraphernalia, huts, shields, and spear throwers. The last item is of great importance since in desert technologies it serves as a multipurpose tool. At one end is the hook for the spear which is laid along the flat shaft of the thrower. The other end is then gripped, aim taken, and the spear thrower flipped forward. The extra propulsion this gives to the spear carries it farther and faster and makes it an effective means of impaling animals such as kangaroos and wallabies to the ground. This is a very different approach to the Kalahari bushmen where the bow and arrow is used as a delivery system for injecting poison into large animals which are then tracked until they fall. The flat section of the spear thrower very often doubles as a board or tray while at the handle end a stone adze flake is mounted in a ball of spinifex resin. This forms an efficient woodworking tool. The earliest preserved clubs and boomerangs come from Wyrie Swamp in southeastern Australia and are 10,000 years old. The true antiquity of these devices is unknown.

The second colonization hypothesis has Moderns sticking to what they know. Sandra Bowdler proposed a two phase process with coasts and rivers being colonized first while deserts and mountains were settled later.[63] This is a very conservative view of colonizing populations, unwilling to leave the security of the beach for the uncertainties of the Outback. Not surprisingly her argument fitted the data nicely since the majority of prehistoric research follows the density of modern population and is concentrated in the eastern coastal fringe. A site such as Kenniff cave which is 300 km from the coast and a long way from any major river was explained away as an "early, tentative attempt at non-riverine/coastal exploitation."[64] The model, it seems, cannot lose.

The direction of migration into Sahul is as difficult and elusive to reconstruct as much earlier migrations of Ancients into Europe. Even with shorter and better dated timescales, present evidence is too sparse and the resolution of dating too coarse to draw the arrows accurately. Moreover this misses the point about the process by trying to pin it down like an historical trek or voyage of discovery when in fact it can only be seen by us in broad terms, against the background provided by different types of environment and their varied conditions for colonization.

Arid zone

The arid and semi-arid zones of Australia make up most of the continent. Population densities were always low. The detailed work around the now dry Willandra lakes by Harry Allen has shown that between 50,000 B.P. and 25,000 B.P. they were full with a rich freshwater fauna.[65] This is shown by the prehistoric shell middens and food dumps around them. These camps also contain the bones of land animals. After 25,000 B.P. and until 15,000 B.P. the lakes became closed and highly saline while after this date they dried up altogether and the scrub mallee vegetation that characterizes the area today took over. The area was used by small populations practicing a generalized hunting and gathering subsistence from 30,000 B.P. to 15,000 B.P. The cremations and burials date to this phase. Then the Willandra lakes were abandoned until 5000 B.P. when many artifacts are once again found. This does not coincide with any upturn in climate, as I shall discuss later.

To the west, in the arid lands of northern South Australia a new technique of absolute dating has been applied by Dorn and his coworkers to the silicates, or desert varnish, that cover pecked and engraved rock surfaces. At Karolta one of these dated "skins" is as old as 31,000 B.P. and eight exceed 20,000 B.P. Animal tracks form one recognizable motif.

Confirmation for Pleistocene age art in the arid zone comes from the Koonalda cave located in the dry featureless Nullarbor Plain facing the Southern Ocean. A sequence of hearths and stone tools has been dated between 23,000 B.P. and 15,000 B.P. while the soft limestone walls along deep passages have been scratched with fingers and sticks leaving various patterns. A burnt torch stick near one of these panels was radiocarbon dated to 19,900 B. P.[66]

Work by Mike Smith in the Red Centre area of Australia has demolished the idea that the arid zone was avoided until a late date. His excavations in the Puritjarra rock shelter in the Macdonnell Ranges west of Alice Springs have shown that by 22,000 B.P. the arid center was inhabited. Occupation may well extend back farther. In this context it is very significant that from 25,000 B.P. to 16,000 B.P. the arid zone was expanding as a result of the effect of the last glacial maximum on climate and in the area lakes were drying up and dunes were being built by the action of the wind.[67]

Smith's work also points to a more complex picture concerning the longterm use of the arid regions. For some time the view expressed by Richard Gould of an unchanging desert adaptation from prehistory to the present has held sway.[68] However, Smith has shown that one key feature of the contemporary economy, the use of grass seeds such as *Panicum*, first occurred about 5,000 years ago. Only then do slabs of sandstone used for seed grinding appear in the sites. Previously more use may have been made of desert tubers such as *Ipomoea*.

Coasts and islands

The early coastal sites have of course been lost as the seas rose and drowned the continental shelf. Other rich coastal environments emerged only recently, one example of which are the tropical savannahs of the Northern Territory. Until 9000 B.P. the coast was over 300 km from the massive sandstone escarpment of Arnhem Land. Then the Arafura Plain joining Australia and Papua New Guinea was drowned with present shorelines reached about 6,000 years ago.[69] The seasonal wetlands that today support a remarkable diversity of fish, birds, and reptiles only appeared a thousand years ago as the Alligator rivers produced embankments that held back the fresh water and so reversed the previously saline conditions.[70] Not surprisingly these rich wetlands supported some of the highest population densities in Aboriginal Australia, forty times greater than the arid sandstone plateau to the south.

Many of the sandstone cliffs and rock shelters in the escarpment and its massive outliers are painted in different styles and subjects. George Chaloupka

Plate 10.4 The massive Puritjarra rock shelter in arid central Australia, occupied during the arid maximum of the last glaciation.

has interpreted these as a record of landscape change, perhaps over the past 35,000 years.[71] He argues the art changed as the sea came closer with the first pictures of saltwater crocodiles, barramundi fish, and turtles, often painted in a large X-ray style which shows their insides and backbones. This tradition still exists in modern bark paintings and shelters known to have been painted as recently as 1964.

His chronology for the art is not without its critics. What is undisputed, however, is that between 24,000 B.P. and 16,000 B.P. in the rock shelters of Malangangerr and Nawamoyn at Oenpelli stone tools including thick flakes and cores looking like a horse's hoof were excavated. These are very similar to the oldest material from the Willandra lakes. Associated with these, however, were some small stone hatchets which had been ground along their edges to produce the finished shape. They are made on igneous rocks and may have been hafted. While nowadays located in the coastal savannahs, the sites indicate the lower population densities that once existed on these former plains of the last glacial maximum.

Huge open sites appear on the wetlands as soon as they are available after 1000 B.P. One of these contains 1.5 million stone artifacts all of which were brought into site from the escarpment 30 km away since no stone occurs on the wetlands.

The coasts of Arnhem Land are very rich in shell fish and when the sea reached present levels extensive middens are found. The use of such hard earned but prolific resources is common on all the coasts of Australia and Tasmania and was traditionally a mainstay of the female economy much as seed processing was in the arid zones. Fish traps built of stones and utilizing the fall of the tide to pen fish are also common along the coasts but how old they are is unknown. In the southerly latitudes elephant seals were taken in great numbers on their breeding grounds as at the massive middens of shell and bone at West Point in Tasmania, two and a half thousand years ago.[72]

Australia also provides an opportunity to look at island occupation which I will deal with in more detail below. Rhys Jones has shown in the Tasmanian archipelago how island size and distance from the mainland accurately predict whether they were vacant at western contact.[73] Residential islands were all above 100 km^2 and less than 5 km from the mainland. Seasonally visited islands were less than 70 km^2, and usually much smaller, but could be up to 8 km away. Irrespective of size any island farther than 13 km was never used. Mike Rowland's study of the tropical Percy islands off the Queensland coast shows how these distances are less limiting in more productive habitats.[74] These are very small islands, the largest is only 1800 ha, and lie between 50 km and 70 km from the mainland. Yet archeological evidence has been found on all those so far studied. They may have been used seasonally. His experiment with local Aboriginal people involving building and paddling a bark canoe from the mainland to the Keppel Islands, 60 km from the mainland, showed what could be done. The canoes were simple but robust enough. No attempt was made at a return journey.

Mountains and forests

These last two habitats have to be taken together. Pleistocene climates turned much of highland New Guinea above 2,000 m into open grasslands. Moreover,

Plate 10.5 What every Australian hunter needs — spearthrower, spears, dilly-bags. An example of Kakadu rock art.

throughout Sahul the forests are human artifacts. Horticultural gardens in Papua New Guinea are cleared by fire and axe. The widespread practice of burning by Australian Aborigines, either to clean the country of snakes or provide hunting opportunities as animals come in to feed on new growth, has produced the great artificial "wilderness" of the Outback. Australian hunters and gatherers were not alone in this practice. Native North Americans fired the landscape while pollen evidence from the postglacial of Europe indicates similar practices.

The eucalypts are especially adapted to fire. Periodic burning by Aborigines created a mosaic of vegetation at different stages of succession. This served to limit the risk of major bush fires as well as increase productivity in plant foods in what Rhys Jones has termed "firestick farming."[75]

At the moment the highest site in Papua New Guinea is Kosipe at an altitude of 2,000 m with a few stone tools dating back to 26,000 B.P.[76] It would have been located within the upper montane forest and below the Pleistocene meadows. The same holds for the limestone rock shelter at Nombe, 1,720 m above sea level, and backed by a mountain range whose peaks rise to 2,500 m. The diversity of forest game, albeit small sized, and plant foods was evidently preferred to foraging on the grasslands.[77] As the forests returned at the end of the Pleistocene evidence for swamp management and clearance is available as early as 9000 B.P. in the highland valleys. At Kuk in the Wahgi valley, 1,650 m a.s.l., Jack Golson also has good evidence at 6000 B.P. for taro cultivation while at the same time pig, which must have been brought from South East Asia, makes its appearance in highland rock shelters.[78] There is evidence from charcoal in the pollen cores at Kosipe that people were burning the forests 30,000 years ago. Would such practices have kept the regenerating forests of the postglacial at the required successional rate? It is tempting to ask if the agricultural management, seen at Kuk, was necessary for the continued use of the highland forests and whether without it settlement would have vanished.

As we move south into Australia the altitudes drop markedly and management by fire is very apparent in the pollen cores. Kenniff cave at 900 m in the central Queensland highlands was occupied from 1600 B.P. to an estimated 22,000 B.P. Just before 4000 B.P. the stone tools change from large, hand held scrapers to small hafted pieces. At the same time new foods such as macrozamia nuts appear for the first time, indicating that the knowledge of how to detoxify the lethal carcinogens, through leaching and roasting, from a potentially rich food resource had now been developed.[79]

Early art is known from Laura in Cape York in an area of tropical rain forest. Tracks of emus and wallabies were engraved on to the back wall of a rock shelter and covered by deposits dated to 13,000 B.P.[80]

There is still scant evidence for the exploitation of the southeastern highlands. Cloggs cave has early occupation at 17,000 B.P. but is at low altitude on the fringes of the Victorian uplands. A highly seasonal resource was to be found in a small area of the Snowy Mountains where many millions of Bogong moths, a rich form of fat and protein, estivated to escape the intense summer heat in the lowlands. Josephine Flood has documented the recent use of this resource and shown how it formed a key food in bringing scattered groups together for feasts and corroborees where marriages were transacted and alliances forged.[81] Flood believes that these highlands were not used until sometime after 4000 B.P.

The best evidence for the use of highlands comes from an area today of dense forest which was vacant at the time of first European contact. The temperate rain forests of southwest Tasmania centered on the Gordon, Franklin, Weld, and Florentine rivers were a true human jungle. Today the region supports huge *Nothofagus* (southern beech), *Eucalyptus regnans* (mountain ash) and some of the tallest flowering plants in the world. At ground level there is an impenetrable mesh of ferns, bushes, creepers, and fallen trees. This seems to have been one area where "firestick farming" made little impression on the forests as they expanded back from their glacial refuges. It is worth asking why. The answer, as Richard Cosgrove discovered, is that although fire had an impact by turning forests into sedgelands their productivity was only marginally increased.[82] The only tangible benefit would be easier movement. It is for these reasons that the areas were gradually abandoned as climate improved.

Recent research has demonstrated that between 31,000 B.P. and 14,000 B.P. the caves of the area were intensively used.[83] They have been found by searching along the rivers for limestone outcrops and then hacking through the bush in search of caves and rock shelters, a strategy that has proved immensely successful. Since 1981, when the first trench was dug into Kutikina cave and dates of 15,567 B.P. to 19,770 B.P. obtained for a rich stone industry, many other sites have been located and sampled. Dates of 30,000 B.P. now come from the lower part of layer 3 in Nunamira cave on the Florentine river, excavated by Richard Cosgrove. He makes the point that at 400 m above sea level the occupation of this cave pushes back by more than 10,000 years the occupation of upland environments, which suggests that on mainland Australia there are still older sites to be found in these settings. Three points emerge. Firstly, there is no hint of extinct megafauna among the animal bones from his excavations. Secondly, human populations lived in an open landscape close to the glaciated uplands of central Tasmania for some fifteen thousand years. Emu eggs, birds as well as kangaroos were all exploited on a seasonal basis. Finally, dates of 30,000 B.P. raise the possibility that Tasmania was not reached by dry land but was also colonized across open water.

The shorter sequence at Kutikina dates to the last glacial maximum 18,000 years ago. A quarter of a million bone fragments came from the small excavation trench. Many showed signs of burning and 75 percent of those which could be identified to species came from red-necked wallaby which weighs about 15 kg and today lives in open, wooded, or scrub country. These small animals do not form large herds. They are, as might be expected from their size, prolific seasonal breeders under "r" selection (chap. 3), occupying small home ranges. Between 1981 and 1986, 2 million were harvested by commercial and non-commercial shooters, representing a staggering 6 million kg of carcasses every year. Prehistoric hunters would have been culling them at much lower rates, but the abundance of these wallabies in the open grassy woodland conditions of the

Plate 10.6 The impenetrable temperate rain forests of southwest Tasmania were vacant at the time of the first Western contact. The limestone bluffs on the Franklin river have caves with rich archeological deposits dating to times of more open conditions.

ice age must go a long way to explaining the remarkable number of occupied caves in this modern wilderness.

Another feature of these sites in southwest Tasmania is the quantity of finds. The density of chipped stone ranges from 30,000 pieces per m^3 at Nunamira cave to 70,000 per m^3 at Kutikina, which compares very favorably with the open sites in Wadi Jilat, Jordan (chap. 8). At Kutikina the excavators found a few flakes of Darwin glass which was formed when a meteorite hit the earth's surface.[84] It formed a fine grain, high quality raw material. Darwin Crater, where this happened, lies 26 km south-southeast of Kutikina and even under open conditions is separated by very broken country. Some Darwin glass was found at Warreen cave on the Maxwell river 40 km away, while at Nunamira 70 km from the crater only five pieces have been recovered. Distance continues to exert its influence on the distribution since only a single flake of Darwin glass has been found in Bone cave on the Weld river over 100 km southeast of the meteorite crater.

Finally the caves of southwest Tasmania have produced traces of wall art. At Ballawinne on a tributary of the Gordon a series of hand stencils, which are very common in mainland Australia and indeed as a motif in most parts of the prehistoric world, were found 20 m from the entrance. There were sixteen prints from at least five people. The coloring material was red ocher (hematite). Two more sites in the Weld catchment contain similar stencils and in Judd's Cavern the ocher was mixed with blood. This has now been dated by the radiocarbon accelerator, which can cope with very small organic samples, to between 10,730 B.P. and 9240 B.P.[85]

The key to occupation in this region was the red-necked or Bennet's wallaby. As Richard Cosgrove, Jim Allen, and Brendan Marshall have pointed out, the wallabies were susceptible to changes in rainfall and vegetation as climate improved.[86] As their predictability as prey declined the role of caves as hunting camps for this tethered resource changed dramatically and the caves fell out of use. Burning could not compensate by increasing productivity and alternative systems of intensification like forest management which were available, for example, in the highlands of Papua New Guinea, did not exist.

This begs the question why crops such as taro and animals such as pig never reached Australia, particularly as Papua New Guinea was still joined when they were introduced there. Only one domestic animal reached Australia, the dingo, sometime after 4000 B.P. It may be that pigs and garden crops were a means of staying in areas under active forest regeneration at the end of the Pleistocene. They were not, as elsewhere in the Old World, part of the domestic plant and animal package associated with population migration.

Although early occupation may be sparse in the uplands and forests, it is there. What is noticeable is that the number of sites increases dramatically after 4000 B.P. when over a period of time we find new types of tools as well as small size, often expensive to process, but highly productive foods, such as seeds and macrozamia, being exploited for the first time. Sandra Bowdler and Harry Lourandos have proposed that the move to these resources marks an intensification in social life.[87] The ceremonial cycles supported by alliances and enacted through ritual provided the context for intensifying subsistence. As a result down came the cost barriers to exploiting these classic small "r" selected foods.

This picture of late intensification ignores the southwest Tasmanian evidence. Site densities are high in the landscape at a much earlier period. Art, and its association with ritual and ceremony, is also Pleistocene in age. The presence of wallabies may have been predictable but they were still costly to chase and process since they were small in size. The same case could also be made for the archeological evidence from the Willandra lakes during the Pleistocene with intensive use of fish and shellfish. The conclusion must be that intensification has been a recurrent feature in Australian prehistory, affecting different regions at different times, involving a wide range of resources, and leading to a variety of expressions in surviving material culture. It did not occur just once, or simultaneously across much of the continent during the middle of the postglacial warm period.

Lourandos's model of intensification after 5000 B.P. also recognizes a significant rise in population. Social evolution, he argues, unlocked the expensive but abundant resources by providing a reason to work that bit harder. Population growth followed. Instead of rapid saturation of the continent, as Birdsell proposed, population stayed low for many thousands of years then rapidly expanded. Steve Webb has shown how, using an annual growth rate of only 0.067 percent, a local population could expand from 3,000 to 70,000 persons in about 4,500 years.[88] He favors overcrowding as one explanation for the unexpectedly high levels of pathological skeletal conditions among Aboriginal populations living along the central Murray river. Here, indications of chronic food stress revealed by growth deficiencies were much higher than for groups living in the deserts and other parts of Australia. But once again, as the early Tasmanian data show, this may have been a recurrent pattern of rise and fall in local populations partly determined by climate and environment as well as by the shifting alliances within a continent of hunters and gatherers. The presence of Pleistocene art from Tasmania to Cape York and into the arid zone strongly suggests that the central feature of the Dreamtime, which identifies paintings with sacred places and formed the foci for landowning local groups throughout the continent, is of great antiquity and even essential to the process of colonization by providing the Ancestors with their ceremonial tracks across a social vacuum.

Colin Pardoe has added another dimension by pointing to the similarities between the Murray and the Nile.[89] Both are rich valleys bordered by desert. These facts of existence impose a linearity on settlement and contact. Both rivers have graveyards, as we saw for the Nile in chapter 9. Along the Murray three of the best known are at Kow swamp, Roonka, and Coobool Creek. These graveyards stood as symbols in both river corridors. They marked territories and symbolized corporate ownership of territory. Variation between the skeletal remains from each graveyard points to the imposition after 13,000 B.P. of social boundaries and exclusivity on the exchange of genes. Low vagility as opposed to very high values for the surrounding desert groups now marked the Murray populations. Here is an example of modern social behavior leading to biological variation—another example of exaptive outcomes rather than historical adaptations.

For the inclusive systems of the arid zones, John Mulvaney and Isabel McBryde have both shown how the continent was crisscrossed by "chains of connection."[90] These linked individuals and local groups through networks of kin and alliance and crossed language and cultural boundaries. In many areas of central Australia they still do. Movement of raw materials from the coast into the interior and vice versa, as well as shared song cycles, myths, and initiation rites are tangible evidence of a pancontinental system. But at a local scale the country is divided into discrete territories based on closely guarded ritual knowledge and the initiation of men into that knowledge. Paintings and the mythical associations of places are very important.

One result of these systems of inclusion and exclusion is to match the rhythms of feast and famine that such seasonal environments undergo. The risk of dietary failure is spread outside the local area so that failure of water or resources can be countered by moving to neighboring kin and allies in times of need. Without such social means to cope with unpredictable rainfall, low population densities as well as the need for an encyclopedic knowledge of the landscape and its resources, it is inconceivable that all the environments of Australia would have been so rapidly colonized. The journeys of the Ancestral Beings across Australia creating and naming the landscape for the human custodians who followed and continued to steward the tracks should not be read as myths in hindsight but rather as the elemental threads of purposive exploration and colonization.

The Pacific

The Pacific is 20,000 km across and encompasses a great many islands. These vary enormously in size from New Zealand at 268,676 km^2 to Anuta, a high island of only 0.4 km^2 but inhabited by some 150 people (fig. 10.2).[91] The ocean is traditionally divided into three parts: Micronesia, Melanesia, and Polynesia—the small, black, and many islands. The latter is often described in terms of a

Figure 10.2 Location of Pacific and Indian Ocean Moderns.

triangle with Hawai'i, Easter Island, and New Zealand forming the points. Within this triangle the distances are immense. Hawai'i is 3,862 km from the Marquesas and 4,410 km from Tahiti in the "center" of Polynesia. Easter Island lies 3,703 km west of South America and 1,819 km east of Pitcairn, its nearest neighbor.

It has always been thought that this great ocean was the last to be colonized with the process signaled by the appearance of distinctive Lapita pottery throughout western Polynesia between 3400 B.P. and 2000 B.P. Recent work by a large team led by Jim Allen and Chris Gosden has shown that at least around the edges of the Near Pacific the process had begun very early indeed.[92] While trying to tie down the Lapita homeland their excavations on New Britain and New Ireland, some of the closest islands to Sahul in the Bismark Archipelago but never joined by lower sea levels, produced material of great antiquity.

At the Matenkupkum cave on New Ireland, Chris Gosden's excavations have shown that occupation begins at 33,000 B.P. while Panakiwuk and Balof caves on the same island have dates of 15,000 B.P. and 14,000 B.P. respectively.[93] New Ireland lies some 600 km to the northeast of the Huon peninsula, which yielded dates for its waisted axes of 40,000 B.P. It would have been possible to reach New Britain and New Ireland by hopping between visible islands. However, in order to reach Buka in the Solomons, the next chain of islands, an open-sea crossing of 170 km is needed and the destination is not visible from New Ireland. A date of 28,740 B.P. from the Kilu rock shelter on Buka Island, recently obtained by Stephen Wickler and Matthew Spriggs, shows that occupation was achieved 25,000 years earlier than previously thought.[94]

Such early colonization was not, however, confined to Melanesia. Farther north the Ryukyu chain of islands was likewise never joined at times of low sea level to either the Japanese archipelago or the continental shelf in the East China Sea. Consequently, the find of a human skeleton at Yamashita-Cho cave on Okinawa dated to 32,100 B.P. must be added to the evidence from Sahul and the Bismarks as the earliest evidence for sea crossings by boat.[95]

It is also possible that the first settlement of Japan involved a sea crossing. The geological and archeological evidence is hotly disputed. Serizawa has claimed an early industry at the Sozudai site of apparently 100,000 B.P. but this seems largely based on the crude/fine principle applied to the stone tools and is strongly disputed by almost everyone else.[96] The majority currently accept several sites in Miyagi Prefecture. The stone tools are made on coarse grained rocks with little retouch and are only broadly dated between 70,000 B.P. and 30,000 B.P. Although a date of 50,000 B.P. is often put forward, a variety of absolute dating methods in fact favor 40,000 B.P. as the age for this material.[97] If this is the case then colonization occurred when Hokkaido, Honshu, and Kyushu were fused and joined to Sakhalin Island and the Asian mainland to the north. However, a deep trench of 200 m still separated the southern end of this peninsula from Korea.

After this Japanese version of the Clovis and pre-Clovis debate in North America we find, 30,000 years ago, well dated, uncontested flake tools, some

Plate 10.7 Tropical island, tropical forest. Matenkupkum cave on New Ireland contains evidence which points to the start, 33,000 years ago, of a maritime economy.

edge ground pieces and large pebble choppers.[98] Then comes the big shift not duplicated elsewhere in South East Asia.[99] After 20,000 B.P. a prepared core technology appears that produces large blades which are retouched into knives. These vary greatly between regions within Japan. Miniaturization begins about 15,000 years ago with leaf-shaped projectile points appearing by 14,000 years ago and microliths a thousand years later.

Later Pleistocene sites are also known from the islands. At Minatogawa cave on Okinawa more Pleistocene age skeletons have been dated to 18,250 B.P. and 16,600 B.P., but there are no artifacts.[100] Obsidian used for making stone tools on the Kanto plain of Honshu between 15,000 B.P. and 20,000 B.P. can be sourced to the small island of Kozushima.[101]

Nothing is currently known about the boats that made these first voyages into the Pacific. Adequate marine technology was obviously necessary but hardly sufficient reason. Given the evidence we have now looked at in chapters 8 and 9 for the burst of Modern colonization, I regard the Pacific Pleistocene dates of *c.* 30,000 B.P. as very significant. Rather than a technological breakthrough the colonization of the Near Pacific was associated with comparable organizational skills that overcome former hurdles to settlement and permanent occupation. Furthermore, as Geoff Irwin has convincingly argued, this process began as exploration and led to colonization.[102]

But the colonization of the Pacific also reminds us that the process was far from instantaneous. Peter White regards northern Sahul and the large islands of Melanesia as the area where the cultures "marinaded" for 50,000 years before colonizing western Polynesia.[103] Based on the appearance of sharks teeth and obsidian from New Britain in the three New Ireland sites a full maritime adaptation did not appear until much later between 8000 B.P. and 6000 B.P.

The appearance of Lapita pottery outside Melanesia marks the next push into the Remote Pacific 3,500 years ago.[104] After this time obsidian from the volcanoes of New Britain has been found in a Lapita settlement on New Caledonia, 2,500 km to the southeast.[105] Lapita settlements are then found on the Polynesian islands of Fiji, Samoa, and Tonga.

The origins of the Polynesians have been the focus for much discussion among archeologists, anthropologists, and linguists. The latter place the homeland of the proto-Austronesian languages on Taiwan at around 5500 B.P. This breaks up after 4000 B.P. as migrants move out into the Pacific, although John Terrell reminds us that linguistic geography is not historical geography.[106] In Melanesia there are 400 languages representing thirty local subgroups. The rest of Polynesia and Micronesia has only a further fifty languages. It will be interesting to see if linguists can match this disparity with the different archeological timescales now emerging between Melanesia and Polynesia, the former showing colonization by 33,000 B.P. and great language diversity, the latter nothing, until the expansion of Lapita pottery after 3500 B.P. As a rough guide both regions have produced twelve languages per thousand years of regional occupation to reach their modern totals. Whether this implies a constant rate of expansion remains to be seen.

The physical anthropology of the Pacific Islands is extremely complex and has been open to racist distortion. As a reaction, current thinking is that the Polynesians did not come from anywhere. Instead the islands of Fiji, Samoa, and Tonga were, as Les Groube pointed out in 1971, the place where the Polynesians "became," and the center of their becoming was this triangle of islands.[107]

The Polynesian wave then set out from this center 2,000 years ago and completed the process of colonization by A.D. 1000. The points of the larger Polynesian triangle were colonized as follows: A.D. 300 Easter Island; between A.D. 300 and A.D. 500 Hawai'i; and A.D. 800 New Zealand. The speed is comparable to the peopling of the Americas and Sahul and is not entirely accounted for by the means of transport. The appearance of the South American sweet potato in Polynesia also shows that this wave hit the limits of Pacific colonization and bounced back, although the timing and mechanisms still need to be investigated.

But what do we learn about humans as colonizers from this remarkable wave of colonization? In the first place, the speed is so fast that models from island biogeography offer little help in plotting the pattern. In a much quoted passage Crawford summed up the situation in 1852:

> The Pacific sea . . . encouraged maritime enterprises. Instead of being like the forest, marshes and mountains of a continent or great island, a barrier to communication, it was a highway which favoured intercourse and migration.[108]

Secondly, the idea that it was an aimless process has to be rejected. This can be done by using a computer simulation of founding populations in the Pacific developed by Michael Levison, Gerald Ward, and John Webb.[109] They show that within the thousand years involved it would be extremely unlikely that, using a chance and drift model, the three outliers would be discovered. They also tested for drift voyages from the east. Starting from the Galapagos they found that only one, from a run of 732 computer voyages, managed to get close to Polynesia, but that after 182 days at sea and only reaching 127°W the imaginary crew expired. Geoff Irwin is quick to point out that this finding adds support to purposive exploration rather than human drift. His computer simulation with Simon Bickler and Philip Quirke confirms the navigational sense, given that the trajectory of Remote Pacific colonization was from the west, that first they went into the predominant winds and only then across and down them.[110] Since these voyages were undertaken at a time when populations were very small few risks would have been taken that led to loss at sea. Surplus population was not simply thrown at the ocean in the hopes of finding land.

A real life simulation bears this out. Thor Heyerdhal's all male crew did rather better than the computer. Their voyage from Peru lasted ninety-seven days and reached the Tuamotus east of Tahiti. But while he was quite right to point out the uniformitarian principle, where "the trade winds and the Equatorial currents are turned westwards by the rotation of the earth, and this rotation has never changed, in the history of mankind,"[111] and courageously showed that his eastern migration theory was possible, the weight of archeological, linguistic, and anthropological evidence is against such seafaring common sense. He did show, however, that if your exploration has a purpose your chances of finding land greatly increase. Irwin provides this lifebelt for the prehistoric Pacific colonists coming from the west, "A means of safe return is latitude sailing, a traditional skill which simply involves returning to the latitude of one's origin island, while still upwind of it, and then running with the wind along the latitude."[112]

The likelihood that the colonization was purposive, against the currents, with islands being actively sought raises a third point about human motives. Were they impelled out into the ocean by overcrowding or starvation, reminiscent of much earlier explanations (chaps 2 and 3), or was it a heroic tradition inspired by a worldview of discovery in which prestige went to those who founded colonies and made return voyages? I believe we can discount the former set of motives since the process took 50,000 years to come to fruition and only fifteen hundred years to cook. If land hunger was the reason then surely this would have occurred much earlier in Melanesia and continued to be a problem. Purposive voyaging is strongly supported by the fact that they took both sexes and carried domestic plants and animals with them. The pattern is variable across the Pacific with pig, dog, and fowl reaching some but not all islands. The same is the case for staples such as breadfruit, taro, banana, sago, and manioc.[113] One feature is universal. Mass extinctions of wildlife, particularly the birds, accompanied colonization. On Hawai'i archeological excavations have shown that after the arrival of humans ninety species of bird became extinct.[114] New Zealand saw the hunting to extinction of the giant moas, flightless birds the largest up to three and a half meters tall with neck extended. The many moa hunting sites, with the charred remains of these birds in hearths, leave us in no doubt as to the cause of their demise.[115]

Following the colonization of the Polynesia outliers there was a change, after A.D. 1000, in canoe design. At contact the canoes of Hawai'i, New Zealand, and the Chatham Islands were inshore or constructed for short hops between islands, and the double canoes best suited to ocean voyaging were only found in central and western Polynesia. No canoes were present in Easter Island where the inhabitants had long since chopped down and burnt the last tree. Furthermore, at western contact some islands were vacant but prehistoric evidence has been found on them. Patrick Kirch provides a full list and it is noticeable that the nearest islands to the three outliers who had given up on ocean going craft, Fanning and Christmas in the Equatorials near Hawai'i, Henderson and Pitcairn closest to Easter Island, and Raoul in the Kermadecs next to New Zealand, were all vacant (see fig. 10.2).[116] They vary greatly in size (Christmas Island 645 km^2; Pitcairn 5.2 km^2) and shape (Raoul high volcanic; Fanning atoll).

The abandonment of these islands and the changes in canoe design mark the cessation of voyaging, a retreat in occupied range, and the severance of many inter-island links. Geoff Irwin has argued that return voyages from the outliers did take place and contributed to the end of voyaging by providing information on the ocean and its inhabited islands. Two interpretations are suggested. Either the frequency of finding further desert islands had decreased to a point where new voyages produced poor returns or the worldview supporting the ideology of voyaging changed dramatically.

Choosing between the two is not easy, but a pointer comes from considering what are for us the remarkable navigational skills involved. Medieval seafarers were just as proficient in navigation as the Polynesians and with only a few more technical aids. The human mind is very capable of producing cognitive maps to solve such practical problems so long as the task is sufficiently necessary, challenging and clear in outcome. Maps, written logs, sextants, and compasses undoubtedly reduce the risks of failure but do not make us more intelligent mariners. As Paul Graves has pointed out, navigation and negotiation have many points in common in terms of memorizing complex, changeable data while at the same time being able to turn it to practical advantage either in constructing alliances on dry land or charting a course through oceans.[117] These are the twin hallmarks of modern humans.

The implication of these comments is that ideology provides the only sufficient reason for voyaging and its cessation. Could it be that the shift in worldview was one from a belief in desert islands still to be found to one where humans, or their mythical counterparts, were expected everywhere?

Support for this last interpretation comes from *our* knowledge that there was still something for these navigators to find. The Galapagos, San Ambrosio, and Juan Fernandez were desert islands. Furthermore, why did the Polynesians not settle Australia and the Americas? They certainly reached the latter as the subsequent presence of sweet potato in Polynesia shows. Peopled lands rather than desert islands had been encountered.

Diminishing returns in the western Pacific also played an important role in changing a worldview. The Polynesians were not discovering the world, as Captain Cook had done, or the maritime empires. They were only following the particular ideology of their world. Irwin suggests that at the margins of Polynesia, where distances are vast and islands few, most voyages became two-way passages of non-discovery.[118] Over time a distinction grew up between voyages equipped for exploration and those designed for colonization. The latter carried extra men, women, and domestic plants and animals. Taking these points together I can only agree with Irwin's opinion that we do not need either to romanticize Pacific colonization or reduce it to mechanistic or chance events. Motives may remain as tantalizing as ever but the conclusion is indisputable. The colonizers were deliberate in their actions, concerned for their safety, and rational within their own worldview.

Islands and oceans

The appeal of mechanistic explanations for island colonization is, however, strong. The theory of island biogeography, developed by MacArthur and Wilson and formulated on the basis of extinction and replacement rates, states that distance between and size of islands will influence the pattern of colonization.[119] Their work mostly concerned birds but has been more widely applied. It is ironic that the Pacific islands witnessed catastrophic extinction rates among their local fauna due to human colonization. Yet prehistory is not part of their ecological theory. A rethink of their "natural" process is urgently needed.[120]

William Keegan and Jared Diamond have extended these unmodified ideas in a worldwide summary of prehistoric island colonization.[121] In their view the motives for voyaging came from a combination of trade, searching for prized resources or unoccupied lands to relieve overpopulation, and just plain curiosity. The concept of human drift (chap. 2) is yet again close at hand. They believe the most important ingredient is the pattern of islands. Against such fixed geography the rewards for voyaging are measured and the direction explained: "the configuration of islands in the Pacific . . . rewarded Pacific peoples, more than peoples of other oceans, for developing maritime skills."[122] Moreover, the discovery of some islands led to the expectation of more islands to be discovered. They call this an autocatalytic process. Colonization is like playing a one-armed bandit with a regular jackpot; a reduction in payouts leads to a cessation of colonization.

The problem with this adaptive model is that it tells us how the world should be for good functional reasons rather than how it is, or rather was, used. Just as Heyerdhal based his model of Pacific colonization on the pattern of currents, we can see with the pattern of island colonization that environmental common sense is often contradicted by human prehistory. Reducing the global process to a mechanistic pattern tells us little more than the obvious.

Size and distance

Islands pose their own problems to colonization. It is not so much their remoteness due to isolation—nowhere on earth is as geographically isolated as Easter Island, which the islanders knew as Te Pito o te Henua or "land's end"—but instead maintaining population on often limited, and circumscribed resources. The link between Polynesian expansion and the availability of domesticates is in that sense analogous to colonizing the deep deserts and tropical rain forests (chap. 9). Hunters and gatherers may have solved the problems of pancontinental contacts and social networks. Domestic food resources, which are both more dependable and easier to reproduce, were neither part of their technology nor required to sustain their social life.

The pay-off from domestication came in spreading the risk of failure in food resources that is accentuated on small islands where alternative supplies may not be available at moments of shortage. The desertion of islands in the Tasmanian archipelago as well as Kangaroo Island off South Australia are witness not so much to the lack of "natural" resources they support, but to their predictability. The alternative for fisher-gatherer-hunters would be an even greater effort in inter-island communication to service alliances with the mainland. Unpredictability and risk of starvation could then be buffered by social rather than technological means. It is at this level that distance becomes a critical factor. Traditional boat technology had little to do with solving such problems. In the absence of domestic resources fixed, owned, and husbanded in small areas, the mid-latitude islands were abandoned.

By contrast, on Easter Island a population estimated at 7,000 lived on 160 km^2 for almost one thousand years until discovered on Easter Day 1722 by the Dutch navigator Roggeveen. The islanders achieved this self-sufficiency with just rats, chickens, and humans as sources of protein, only a few of the Polynesian domestic plants, and poor fishing.[123] But existence was not easy. The investment first in statuary then later in bird man cults, rock engravings, and endemic warfare is strongly reminiscent of the Arctic "hysteria" in the Upper Paleolithic of the Old World (chap. 9). The social order based on the annual selection of the warrior-chief on the cliffs at Orongo was, as McCall reminds us, "as fragile as the shells of the eggs that were the focus of the annual quest" and whose discovery led to his election.[124]

Small populations and limited geographical universes may not be an ultimate barrier to colonization. But there are still instances that either break the rule that humans were everywhere or confound the common-sense view of proximity and ocean currents as spurs to early colonization.

The Galapagos overturn our expectations on both counts. They lie on the Equator 1,000 km off the coast of Ecuador. Together they comprise 7,850 km^2 including six major islands several of which are still volcanically active. Their first mention follows the shipwreck of Bishop Tomas de Berlanga in 1535. In his account he writes of turtles, iguanas, giant tortoises, and seals—but no natives. Volcanic activity may be good for allopatric speciation but when combined with the scarcity of fresh water was not conducive to colonization. The first permanent settlement was in 1832. Archeological evidence in the form of a small number of South American pot sherds was found by Thor Heyerdhal and Arne Skjølsvold during their excavations in 1953. But Suggs has dismissed these as evidence for earlier occupation. He explains them as ballast dumped by Conquistadors and buccaneers. As John Terrell concludes, an ebb and flow model of occupation on these harsh islands is not impossible.[125]

What is abundantly clear, however, is that no voyaging tradition left the coast of the Americas, even though the current was in their favor, to explore Polynesia from the east. The 13,000 B.P. dates from Monte Verde in Chile provided plenty of time but the San Ambrosio and Juan Fernandez Islands indicate, as do the Galapagos, the complete absence of such a tradition.

Lack of fresh water seems to be the explanation in these instances since many Pacific islands are volcanic and yet densely peopled. Elsewhere in the Pacific colonization did occur, although hardly as a result of ocean voyaging. The nine California Channel islands ranging in size from 2.6 km^2 to 506 km^2 lie between San Diego and Santa Barbara.[126] The closest is 30 km from the mainland and the farthest 110 km. All of them contain prehistoric remains with the oldest middens dated by radiocarbon to 7500 B.P. and 8000 B.P. One interesting feature is the presence of dwarf mammoths on at least three of the islands.[127] These are believed to have swum there and then dwarfed to meet island conditions. Some of the bones are burnt which has encouraged interpretations that they were barbequed, but the dates for the bones are much older at 15,000 B.P. to 29,000 B.P.[128] They were probably burnt as fuel. However, here is a case of hunters and foragers using small parcels of land farther from the mainland than was the case with the offshore Tasmania islands or Kangaroo Island near Adelaide. The latter lies at a comparable latitude to the California Channel islands. Weather patterns may well have more of a role to play in determining occupation as either seasonal, permanent, or absent.

Indian Ocean

When we look at the Indian Ocean we find plenty of exceptions to common-sense colonization determined from currents and proximity to the mainland. None is more "surprising" in its pattern of colonization than Madagascar of the Malagasi Republic. This huge island 586,486 km^2 lies 400 km from the coast of Mozambique with the small Comoro Islands acting as stepping stones in the north. The fauna and flora of Madagascar have many unique elements, particularly the lemurs, which reflect allopatric speciation along this segment of the continental plates. Some mainland species made it to the island and among these the extinct pigmy hippo is intriguing since no landbridges ever existed to facilitate its passage.[129] The voyages of the ancient explorers of Greece and Rome were recorded as far south as Dar es Salaam. A circumnavigation of Africa is even hinted at in a Phoenician story.[130] But Madagascar never figures in these reports. Humans were absent until 500 B.C. when pioneers arrived from Indonesia.[131] Colonization from Africa occurred in the ninth and thirteenth centuries A.D. while in 1801 the Comoros were described as inhabited by "savage moors." Keegan and Diamond put this late colonization down to the "different overwater colonizing abilities of the peoples involved." This they believe was stimulated by differences in the pattern of islands they had in front of them for colonization. Such prescience seems strange given the size of Madagascar and the thousands of islands that lie beyond it throughout the Indian Ocean. As in the Pacific, these lay against the current.

Arab traders certainly knew of the Seychelles, a chain of forty low islands to the north of Madagascar, but no permanent settlement was ever established. John Jourdain landed on Mahé, the largest at 145 km^2 of the granitic islands, in 1609 and wrote, "It is a very good refreshing place for wood, water, coker nutts, fish and fowle, without any feare or danger, except the allargartes; for you cannot discern that ever any people had been there before us."

The islands of Mauritius, Réunion, and Rodrigues in the southern Indian Ocean were likewise valued as revictualing stops by Malay and Arab sailors on the route from Madagascar to Malaya. Mauritius appears on Arab maps and was rediscovered by the Portuguese in 1511. By 1680 its dodos were extinct. Réunion was the home of the white dodo and altogether some forty-two species of birds, most flightless, were clubbed into extinction on the Mascarenes. Closer to the trade routes that hugged the coasts of India and Arabia the island chains formed by the Laccadives and Maldives were, according to the historian Baladhuri, first colonized by Arab expansion in the ninth century A.D.

Farther east in the Bay of Bengal the spice islands, Andaman and Nicobar, have aboriginal populations. Archeological remains are also plentiful but undated.[132] The three main Andaman Islands are large, 359 km^2, 490 km^2 and 561 km^2, with high population densities encouraged by the rich marine resources rather than the plants and fruits of the interior tropical forests where dogs are used to hunt feral pigs. The Andamans are 300 km from Burma but some smaller stepping stone islands facilitate the journey.

The most outlying islands in the Indian Ocean lie southwest of Java. Christmas Island is 190 km away while the Cocos and Keeling group are 550 km distant. A visit to Christmas Island in 1887 showed it was uninhabited and "there was no indication that it ever had been occupied."[133] Scarcity of water seems to be the reason. The Cocos and Keeling group did support permanent settlement in 1900 and was a convenient stop for whalers. There is no indication that they were inhabited in prehistory.

The colonization of the Indian Ocean therefore stands in stark contrast to that of the Pacific. All colonization apart perhaps from Andaman and Nicobar is historic rather than prehistoric, dependent upon the maritime economies that emerged in the surrounding civilizations. The present population densities on many of the islands equal that for Pacific islands. Inhospitability factors provide few answers to their late settlement. The counterintuitive colonization of Madagascar, from the east, is analogous to the western colonization of the Pacific. But at least the Indonesians had the currents, if not the distance, in their favor. What is important to recognize is that in both cases colonization was highly focused and directed, firmly rooted in the ideology of the homeland society, and due neither to *chance drift* nor a simple matter of *enough time*, as is argued for the dispersal of animals and plants (chap. 3).

Let me state the obvious. Islands are necessary for island colonization. They are not by themselves sufficient reason for the timing of the prehistoric process. Ocean voyaging is therefore less of an adaptive outcome and more of a cooption of those intertwined skills of navigation and negotiation. These are revealed exaptively through the record of prehistoric colonization rather than selected adaptively to meet a current problem such as overpopulation, need for trade goods or to satisfy curiosity.

The Atlantic

Nowhere is this more apparent than in the next ocean to consider, the Atlantic. In the north there is the interesting case of Iceland, almost 103,000 km^2 in size, named after a visit in a hard winter in A.D. 865 but not settled until nine years later by the Norseman Ingólfur Arnarson.[134] The Vikings found Irish clergy in residence as they had on the Faroes in A.D. 800.[135] The Eskimos never reached these islands but were there to meet Erik the Red when he rediscovered Greenland in A.D. 980 during a warmer century. Hence its optimistic name was not just a ploy to encourage settlers.

The small islands around the shores of Britain such as the Scillies, Skomer, Lundy, the Channel Islands, the Isle of Man and the Scottish Islands all have abundant prehistoric remains, particularly Neolithic and Bronze Age as on Orkney, Shetland and the Hebrides. Ireland has never produced convincing evidence for Pleistocene occupation when much of the country was covered by an ice sheet, but the postglacial is well represented by hunting and gathering settlements 9,000 years old.[136]

Farther south, both the Azores and Bermuda were desert islands at European contact. Gonçalo Velho Cabral most probably found the Azores in 1432 and Columbus visited them on his way back from the West Indies in 1493. Bermuda, 900 km from Cape Hatteras, was found by the Spanish, perhaps as early as 1503. It was stocked with pigs which the English made good use of when the *Sea Adventure* was shipwrecked in 1609. Their experience is thought to have inspired Shakespeare's *Tempest* although no Caliban and Ariel walked the enchanted isle.

All the Canary Islands were colonized in prehistory. A recent study by Jorge Onrubia Pintado puts the earliest occupation on Tenerife at 2500 B.P., although both here and on Gran Canaria most occupation comes after 2000 B.P. with the smaller islands in the group, La Palma, La Gomera, and El Hierro, colonized slightly later still.[137] The tall peak on Tenerife, 3717 m high, can be seen from the African mainland. The Fortunate Islands, as they were known in antiquity, were settled from Africa (they take their present name from *canis*, the dog). However, the large island of Madeira 450 km to the north of the Canaries was vacant when discovered by Henry the Navigator in either 1418 or 1420. It was settled in 1425. No prehistoric remains have been found. One of Diodorus' stories raises the possibility that a Carthaginian vessel was blown off course to the island in about 500 B.C., while Plutarch mentions Roman sailors who found the empty "Happy Isles" in 80 B.C. Neither favorable report led to colonization. The Cape Verde isles, 700 km from the coast of Africa were also uninhabited and have no prehistory. They were found by the Portuguese in 1460.

The isolated islands in the South Atlantic of Ascension, Trindade, St. Helena, Tristan da Cunha, South Georgia, and in the southern Indian Ocean, Kerguelen, Crozets, Amsterdam, and St. Paul were all deserted. Extreme isolation, small size, and in some cases harsh southern latitudes explain very well why they were not inhabited.

Less obvious is the lack of earlier occupation on the Falklands (Malvinas), 500 km from the coast of South America and 15,800 km^2 in size. Indeed, it is still unclear which western explorer first saw and described them. A letter which may or may not have been written by Vespucci in 1504 claims that a new land was found following a storm in April 1502, "all barren coast and we saw in it neither harbour nor inhabitants." At the close of the sixteenth century Hawkins gave a completely contradictory account, "the land is a goodly champion country, and peopled: we saw many fires, but could not come to speake with the people." Some believe he was mistakenly describing part of the Patagonian coast while others claim that what he called the Maiden Land in February 1594 was an accurate description of the Falklands by one of England's most celebrated navigators.[138] Prehistoric remains have never been reported and when the islands were settled by Bougainville in 1764 there was no mention of natives and the largest animal was the Falklands fox. Dr Johnson was in no doubt about their value: "A bleak and gloomy solitude, an island thrown aside from human use, stormy in winter, and barren in summer: an island which not the southern savages have dignified with habitation."[139]

Smaller seas

The last two oceans are the Caribbean and the Mediterranean, both settled in prehistory. In the Caribbean the earliest occupation is radiocarbon dated to 4500 B.P. but settlement may extend back another two and a half thousand years.[140] The earliest dates from the Dominican Republic in the Greater Antilles and St. Kitts in the Lesser Antilles are for sites without ceramics but with rich collections of stone artifacts. Fishing formed the mainstay while some forest management was undoubtedly practiced. The earliest ceramics appear about 2000 years ago and include clay griddles for cooking manioc. It is thought that these spread from the Orinoco through the lesser Antilles. Several population movements have been suggested in the Caribbean along with that of the Arawaks and the Caribs whom Columbus met in the Bahamas, supplemented by a host of peoples named after successive styles of pottery whose historical claims are best ignored.[141]

There are a few hints in the Mediterranean that occupation may predate the arrival of domestic crops and animals in the region. Excavations in the Franchthi cave on mainland Greece have produced obsidian tools from levels dated between 12,830 B.P. and 9500 B.P. These came from the island of Melos, 150 km to the southeast in the Cyclades. Settlement traces from this period are very faint but, as John Cherry points out, from this time all the islands could have been, but were not, settled.[142] There are claims from Cyprus, Mallorca, and Sardinia that bones of dwarf species overlap with the earliest human occupation. Alan Simmons has found evidence on Cyprus for occupation by 10,000 B.P. which he believes is connected with hunting pygmy hippos the size of a large pig.[143] The ancestors of these dwarf elephants, hippos, and giant dormice may well have made the journey to the islands during the Messinian salinity crisis when dry land was available. Carnivores did not.

The permanent occupation of all the islands took several millennia after agriculture appeared in the region. Until 5000 B.P. the ephemeral archeological traces on most islands are interpreted by Cherry as an extension of risk buffering practices from the large islands such as Crete, Cyprus, Sicily, and Sardinia as well as the mainland. Here small Neolithic farming communities spread the risk of drought and soil exhaustion by splitting up plots and diversifying their crops and animals. By putting their eggs in several local baskets a degree of self-sufficiency and the retention of household autonomy over subsistence was retained. The smaller islands provided stop-overs for seasonal fishing parties and additional lands to forage in times of need. Cherry favors the model of island biogeography to explain the timing and pattern of colonization. Island size and remoteness were the key factors. Small islands would "fail" hunter-gatherers and subsistence farmers more regularly than large ones. In this case the radiocarbon dates bear him out. Moreover, colonization in the west Mediterranean was much faster than in the east since 99 percent of its island area consists of only six big islands. In the eastern basin the six largest only account for 74 percent of the area.

Plate 10.8 The end of prehistory. The defended Bronze Age port of Phylakopi on the Aegean island of Melos.

These geographical predictions can also be extended to consider the big push through the small islands of the eastern Mediterranean which coincided with the growth of regional powers in the Bronze Age. By 4000 B.P. most of the smaller islands have their first substantial settlement and the palace bureaucracies at Knossos on Crete and Mycenae on the mainland appeared by the middle of the millennium. Inter-island risk buffering was now embedded in a pattern of trade and exchange, organized from palace centers which set quotas on agricultural production, recorded it on clay tablets, and stored the surplus in granaries. Local autonomy was a thing of the past. Towns, such as Phylakopi on the small obsidian rich island of Melos, now sported impressive fortifications, to keep people either out or in. The social solution to the problem of island occupation was now written on the wall.

11
Why people were everywhere

Now that the timewalk is over I feel rather like Sir John Mandeville who wrote in his fourteenth-century travels, "Of Paradise I cannot speak properly, for I have not been there; and that I regret."

The past is also closed to travelers. We visit it in our imaginations. We reconstruct its process from the evidence that survives. We interpret the results according to our principles and purposes. But we can never experience the past, not even vicariously if we were to live in a cave, make fire by rubbing two sticks together, and hunt animals with flint tipped spears. Anyone who claims otherwise runs the risk of equaling Mandeville's fictitious boasts about human monsters and the mythical Prester John.

I have followed the prehistoric timewalkers not to share their experiences but instead to understand our past. This has meant looking at inherited ideas about human culture, race, and the rediscovery of prehistoric colonization, as well as providing an update on the evidence and theories surrounding human origins and evolution. The question remains, why were people everywhere? The answer points to a rationale for a world, rather than a regional or period, prehistory. As Grahame Clark explained in his seminal text,

> Prehistory is not merely something that human beings passed through a long time ago: it is something which properly apprehended allows us to view our contemporary situation in a perspective more valid than that encouraged by the study of our own parochial histories.[1]

The curtain of civilization

My timewalk has now come full circle and I have arrived back in the harbors of civilization. From here the creators of a prehistoric past—Locke, Smith, Dégerando, Darwin, and Lubbock to name only a few—set off either in fact or theory to discover living prehistory in their present day.

The islands which I examined in chapter 10 were often those upon whose shores stood "savages," once thought to epitomize Europe's past. These islands, it now appears, were settled because of the impetus to remote colonization either on land or across oceans facilitated by an increasing knowledge of cultivation and the ability to domesticate animals. These gifts of civilization may have been delayed by distance in the global process of colonization. But the purpose of varied maritime societies, only some of which were nations, eventually brought all remote places within their influence and then exploited their usefulness.

This all adds up to the following conclusion. Humans went everywhere in prehistory because humans have purpose. The timewalk has brought this out time and time again. We no longer have to accept a world prehistory which denies the significance or denigrates the process of prehistoric global colonization. Instead we can see it for what it was—another example of humanity unfolding—variable and different but always recognizable. No longer is it sufficient to explain the past away by human drift or the diffusion of ideas or people from one center to the rest of the world in a rerun of recent imperial history. We can see clearly, as did the Neapolitan philosopher Giambattista Vico as long ago as 1725, that common ideas are the independent discovery by every society of what it needs at any given stage in its own development. To admit a purpose in the process of prehistoric global colonization is to regain this historical track where, as R.G. Collingwood maintained, "the plan of history is a wholly human plan."[2]

Optimists by nature

The notion of purposive prehistory is not always popular. Many archeologists and historians still prefer the earlier version of timewalker colonization—either drifting around the world or simply relying on enough time to complete the process. However, I suspect them of double standards. While we would all object to a prehistoric past governed by incremental and progressive steps to the present, many still subscribe to the view that purpose came to *some* prehistoric societies with agriculture and the benefits of civilization. It is significant that this is as far back as our civilized roots are normally traced. These early civilizations are used to set an agenda for the shape of the future world in the form of bureaucracies, cities, armies, religions, law codes, and all the other accoutrements of state power. By actively changing the world and the relations between people, these ancient societies become the foundation stones for history. They were surrounded by other societies which showed little dramatic change; among these the fisher-gatherer-hunters apparently showed less than others. Our timewalkers are often portrayed as outside the active process of history which societies with more elaborate remains or written records are thought to possess.[3] Dynamic culture, whether in the form of irrigation, wall building, conquest, or heroic poetry, is contrasted with passive adaptations to nature at the edge of the civilized world, past and present. This not only applied to fisher-gatherer-hunters. The boats of the Polynesians could be stocked with domestic plants and animals but this did not guarantee them a place in history.

In fact the situation is usually much worse. Most discussions of the first civilizations ignore their hinterlands in the best traditions of benign neglect. The anonymity of the peoples outside the sphere of civilized action is total (fig. 2.1). They only become relevant as the ancient empires expand, for example when barbarian Europe sheds its prehistoric labels and takes on the name tags bestowed by Roman adventurers such as Julius Caesar. This invention of history is the invention of our kind of history. We have known this for a long time, as Vico so clearly set out:

> 122 It is another property of the human mind that wherever men can form no idea of distant and unknown things, they judge them by what is familiar and at hand.

> 123 This axiom points to the inexhaustible source of all the errors about the beginnings of humanity that have been adopted by entire nations and by all the scholars. For when the former begin to take notice of them and the latter to investigate them, it was on the basis of their own enlightened, cultivated and magnificent times that they judged the origins of humanity, which must nevertheless by the nature of things have been small, crude and quite obscure.[4]

The double standards I accused archeologists and historians of practicing continue the error of inventing history by confusing human purpose with historical progress. We only have to look at the way the geographical expansion of early empires is still contrasted with fisher-gatherer-hunters to illustrate the purported difference between history with purpose and natural prehistory. The possibility of purpose and progress is only extended to a select few societies after new plant and animal resources have appeared. The double standard surfaces in the descriptions of many past societies, their peoples regarded as fully human but nonetheless set outside any reasonable definition of humanity because they are not expected to influence any changes they might experience. Nature is cited as the reason why change for the best, outside the historic centers, runs at a different pace and usually requires a push from outside. So, while rejecting the nineteenth-century view that prehistory is the science of progress, the advocates of this exclusionist scenario invariably come back to a form of argument that ensures its implicit continuation. The treatment of global colonization in prehistory is a case in point.

This treatment may also take the form of a look into the future. Indeed forecasting has always been a feature of such adaptive views, where the best triumphs through natural selection. A good example is Lubbock's opinion, based on what he saw as natural laws governing human evolution, that

> The most sanguine hopes for the future are justified by the whole experience of the past. It is surely unreasonable to suppose that a process which has been going on for so many thousand years, should have now suddenly ceased; and he must be blind indeed who imagines that our civilisation is unsusceptible of improvement, or that we ourselves are in the highest state attainable by man.[5]

His optimism still echoes in our ears. For example, J.H. Huxley concluded in his major synthesis on neo-Darwinian evolution that evolutionary progress could be objectively defined as "all round biological improvement."[6] Evolutionary developments could be traced through the notion of a dominant type that characterized periods of the paleontological past. Humans were just the latest dominant types to evolve in a long list that included dinosaurs. Human progress, he argued, involves the subjective criteria of human values but is demonstrated by the objective view of biology.

These optimists by nature are often the staunchest defenders of adaptive purpose in those societies which are considered to have made the major contribution to human history. Archeologists supply the ammunition. For example, Colin Renfrew has presented a model to explain patterns in the growth of early towns based on the potential rewards of living in larger centers, the ominous sounding "social benefits" arising from communal, civilized life.[7] In the scale of social benefits, the winners were no doubt a minority who gave their services in order to live in the palace on the hill. Such civilized centers, manned in Renfrew's model by an élite, are fit through their history for change and expansion. Those at the periphery are designed to remain in their place until summoned to participate. Adaptive roles must be understood and observed as General Pitt-Rivers noted (chap. 2) in his support of evolutionary theory to warn against revolutionary changes in society.

Purpose and prehistory

Most prehistorians now consider, quite rightly, that they have rejected progress as a guiding principle in the study of the deep past, before civilizations appeared. However, this is no reason to reject human purpose. Purpose and progress do not need to feed at the same trough. The issue becomes, who had purpose in the past? This leads to wider questions about the construction of early prehistory as an active rather than a passive process; for instance, the exaptive unfolding of culture rather than the adaptive response to the forces of nature.

I have shown how the general acceptance of the earlier human drift as a natural, random process sits uneasily alongside claims for a more dynamic human role in fashioning history once agriculture and civilization appear. This has been the legacy of the Neolithic Revolution so forcefully put by Gordon Childe in 1935 and subscribed to ever since.

I do not have to claim today that either process was progressive, driven by some natural law of improvement. I have argued instead that both depended on purposive action where humans constructed and changed their worlds. For those archeologists interested in Pleistocene hunters, the curtain of civilization might fall a little sooner, but it still falls. This is shown by the Moderns (table 1.2), some of whom have now been rehabilitated as complex rather than savage hunters. History, in the sense of self-determination, has been returned to them. They are now part of modern humanity, separated by a rapid transformation, seen by some as the Human Revolution, from the earlier Ancients and Pioneers. The division between "us" and "them," the ancestors, thus falls some fifty thousand rather than ten or even five thousand years ago.

It is this curtain which I have been trying to tweak aside throughout this book. Before it fell timewalkers were not everywhere. They had a restricted range within the Old World. They avoided many environments where the living was not easy (fig. 1.1). I have pointed to many differences in settlement and technology. I have inferred from this evidence very different scales of society and contact. But for all these differences should we attribute the values of cultural purpose to those on our side of the curtain, leaving our earlier ancestors at the whim of natural processes and biological laws? Does an active prehistory only begin with the Moderns? Put another way, why should we return history to one cohort of timewalkers just because, as Stuart Piggott once put it, they were "as handsome and as wise as us," only to deny it to those in earlier periods?[8] I can only conclude that finding and defining our origins requires a distinction to be drawn between ourselves and outsiders. This is a present purpose of our past. Nature can explain all, and has done frequently with race and culture. This is the burden of civilized history.

But what is the alternative? How can we return history to those periods and ancestors on the other side of the curtain where, for so long, it has been denied because they have been cast in many and varied dialogues with nature as her well-adapted, best-behaved children?

Initially, nothing more is needed than an acceptance of the thesis that hominids, timewalkers, humans—whatever we choose to call them—change themselves and their societies rather than being buffeted by climate and pushed along by basic passions, natural needs. It is at this point that my purposive prehistory, as revealed through the course of global colonization, becomes important. Throughout this book I have returned to the contrast between adaptive and exaptive effects and processes. Adaptation undoubtedly takes place as selection from the environment forces change along in the random, opportunistic pattern of evolution. It is comparatively easy to supply a narrative for such a rambling process. This is done by denying the element of human purpose. I have resisted doing this, for example with an account of how the Neanderthals were replaced or people voyaged to Australia. To do so suggests a purpose that can only lead to teleology of the following sort:

> Moderns were better adapted than Ancients because, for example, they were designed to:
> - fill up the habitable world;
> - unleash artistic potential because they had language;
> - provide the platform for developing agriculture;
> - know all the answers to the riddles of culture, thereby explaining why people do what they do.

Consequently, I have argued that the reason people were everywhere was the result of purpose, the process of deliberate action. This operated in ecological time as behavior was exapted to meet changing circumstances. There was no direction or goal to such behavior other than survival and the reproduction of existing social life, where purpose is provided by negotiation. The effect in evolutionary time was the extension of human range. This was accompanied by changes to the timewalkers. Many of these were adaptive, for example skin color, stature, and head size. They also acquired the many facets of their humanity in this process since they exaptively constructed social worlds to engage and transform new natural worlds. Human nature, as Vico pointed out almost three hundred years ago, is not fixed. Instead it is a social creation in continuous evolution determined in respect to human needs and utilities.[9] With the novelist's perspective, Carlos Fuentes has remarked on how the possibilities of our unfinished humanity still remain to be explored.[10] The process continues and remains infinite even in a world with no new lands to unveil, only whole continents of social geography, because in Kant's phrase, "Man is unsocially social." One rich avenue for exploration lies in the themes of the past. The substance of a world prehistory, to quote Grahame Clark, "is no less than the emergence and self-realisation of humanity itself."[11]

By using purpose I am not for a minute suggesting migrations in the historical sense. As I have said before, prehistoric colonization is not a remote equivalent of the history of recent maritime empires. The purpose I refer to returns to a point raised in the first chapter but studiously avoided ever since. This concerned issues of motivation and intention. My conclusion is that intention was not restricted to the Moderns. It preceded them. We can now see that intention in human action did not arrive with agriculture. Neither did it first appear with the earliest Moderns. What remains different is the scale of effects. It was only when these effects impinged on evolutionary time that we see radical changes. One of these was the colonization of the earth. The exaptive resources of behavior and physiology which made them possible then established their own independent trajectories of cultural change. Self-realization that I argued was so enhanced by the appearance of spoken language was just such an effect. Social and cultural diversity are not the end product of human evolution. They are rather the fellow travelers of an exaptive radiation undertaken by sentient, purposive, informed bipedal animals.

Prehistory, of whatever date and associated with whichever timewalkers, is in my view the result of such purposive actions. I can say this for the simple reason that we can now see they usually chose *not* to expand.

Viewed adaptively the long wait to get out of sub-Saharan Africa or across to Sahul is usually explained in terms of waiting for the right gifts to evolve to make it happen. This was Quatrefages' influential view in 1879. Humans were everywhere because of intelligence and industry; nothing more, nothing less. The resultant explanation for global colonization is one of adaptive inevitability, a natural rather than historical process. The timewalkers' story could be repeated on an endless wheel, unvarying in process or outcome. As Lyell wrote,

> Supposing the human genus were to disappear entirely, with the exception of a single family, placed either upon the Ocean of the New Continent, in Australia, or upon some coral island of the Pacific Ocean, we may be sure that its descendants would, in the course of ages, succeed in invading the whole earth, although they might not have attained a higher degree of civilisation than the Esquimaux or the South Sea Islanders.[12]

The implications raised by Lyell were profound and continue to be applied. They remove purpose by making all timewalkers wait passively for change from outside. Only then did they realize their purpose to expand. They became mere demiurges, skilled workers carrying out a divine or natural architect's ideal plan.

When looked at as exaptive rather than adaptive, these checks to expansion do not provide sufficient reason for the delay. The limits are instead set internally and so are integral to social life, however complicated. They obviously refer to the outside world where survival and expansion must take place. The inevitability of prehistoric global colonization then ceases to be outside history.

Returning prehistory to humanity

The eventual outcome of tracing changes in behavior through exaptive processes, at a global scale and through the theme of colonization, is to ask, what is world prehistory? Grahame Clark saw it clearly. Through prehistory, he wrote in 1968, we are creating a mythology that is adequate to the times in that *all* people can find fulfilment in it.[13] But what is the scale of our audience? Are we dealing with "images for the élites" as the political scientist W.J.M. Mackenzie once described the photographs of a small, round vulnerable earth taken from space?[14] Although we may not act on the common image and tend the planet accordingly we have, he was sure, absorbed it. But even so, "what proportion of mankind now has this image of mankind? I suspect that it is a small one, and that the requisite variety of mankind is still largely unexplored."[15]

A similar conclusion no doubt follows my world prehistory. I would hope it presents a theme to reflect upon our common, unfinished humanity, a comment on our present diversity.[16]

It is to the dispossessed that history must be returned. An example of this trend comes from the recent history of hunters and gatherers. Carmel Schrire has shown how bushmen peoples such as the !Kung San in the Kalahari Desert regularly herded sheep and cattle.[17] They switched between hunting and gathering to herding as the wider physical and economic climates of which they were part also shifted. This is likely to have been the pattern over the past 1,500 years. The "preservation" of hunters such as the !Kung San in such environments does not mean that they are representatives of a primeval human state which change passed by. Far from it. They were active participants in the flow of culture and human history in that region. But the heavy hand of nineteenth-century judgment keeps such stereotypes very much in play, still denies them their history, still expects peoples to conform to narrow social and economic categories.

My aim has been to attempt a similar act of restitution. I have tried to rehabilitate our ancestors by resurrecting a question relegated to minor status at the time prehistory was formulated as a subject: Why were people everywhere? The presentation of a world prehistory around this theme has followed Collingwood's idea that the study of the past is a source of human self-knowledge.

From our survey of world prehistory we now understand that there are few genetic rules for behavior and that, therefore, there is great variation in behavior (plasticity), which is the outcome of exaptation. The timing and direction of change and diversity can be understood but not predicted or forecast. The inevitability of the process is not sustainable since the motor of prehistory has been behavior, ungoverned by natural laws though constrained by the practicalities of existence. Its unforeseen consequences have been global colonization and social elaboration.

Previous world prehistories have been confused by the static nature of the past—many thousands of years when nothing seems to happen. The reality of prehistory lies in understanding human action in deep time. Part of the understanding stems from the knowledge that perspectives about time differ. They provide different explanations. They address different questions. Evolutionary time should not be judged by our experiences of ecological time. Difficult words such as intention and purpose should be qualified by the time perspective they address. Returning history to prehistory can only be accomplished if it is realized that the latter has a different reality to the former. This is not based solely on the different lines of evidence, texts versus objects. It is founded, as Tim Murray has perceptively argued, on the methods by which the prehistoric past was originally humanized in order to make it understandable.[18] One hundred and fifty years ago Lubbock did this by illustrating the antiquities of Europe with the customs and manners of indigenous societies around the world. Vico, as we have seen, warned us about this human trait in the eighteenth century.

We no longer humanize the past so directly through the peoples of the present. My aim in this prehistory of global colonization has been to establish a common process that allows us to understand diversity in the past as part of the legacy of today's commonly constructed world. If I am humanizing what are "distant and unknown things" through this concept it is to make the past intelligible through the diversity of present human activity.

Finally, in 1851 Latham remarked in his book on migrations that "A fly is a fly even when we wonder how it came into the amber; and men belong to humanity even when their origin is a mystery."[19]

A world prehistory must be interested in origins, but not the quest for origins, which by dating and classifying ancestors seeks to establish in deep time and from natural principles the antiquity of exclusion, the fragmentation of humanity for whatever purposes and motives. The goal should instead be an interest in diversity in order to encompass and enrich rather than reject. If we cannot acquire self-knowledge via the study of our ancestors then what hope for the study of other animals and habitats which we also use to construct our worlds? The global timewalkers are likewise trapped in amber so long as we regard them as natural rather than cultural beings. We do not need to humanize them in our form to understand them. We just have to ask different questions. The when, why, and how of global colonization pose many such questions for the future development of a truly world prehistory.

Let me finish where I began, in a wilderness. The few true wilderness and wildlife parks that still remain in the world are those where no timewalkers live today. They are strictly regulated for their own longterm future. We have the same longterm hopes for this intentional conservation as we do for the pictures of ourselves still being carried by *Voyager 2* into deep space. Entry to such wilderness areas is a privilege that has to be negotiated rather than a right to be exercised. You are told to leave nothing but your footprints. It is fortunate for the future of prehistory, and the understanding of ourselves it provides, that timewalkers in deep time left a little more as they navigated the world.

Notes

Preface

1. From the "Letter of Columbus to various persons describing the results of his first voyage and written on the return journey" translated by J.M. Cohen (1969: 121).

2. Keegan and Diamond 1987: 68.

Chapter 1

1. The account of Kerguelen's discovery (Dunmore 1965: 196–249) is a lesson to all explorers wishing to find what is not there.

2. Latham (1851: 89) spoke for all: "First as to its [human] universality. In this respect we must look minutely before we shall find places where Man is *not*. These, if we find them at all, will come under one of two conditions: the climate will be extreme, or the isolation excessive."

3. "Views of the past: drawing as a record of place" at British Museum, 1987. Resolution *and* Discovery *in Christmas Harbour, Kerguelen Island 1776* by John Webber (1750–1793), catalog number Additional MS 15513, f3 British Museum.

4. Cook's voyages are best approached through Beaglehole's magisterially edited volumes (1955, 1963).

5. Darwin 1839.

6. Wallace 1880: viii.

7. Sir Humphrey Davy (1830), the Duke of Argyll (1869), and J.W. Dawson (1883).

8. For example, in the Book of Mormon, revealed to and translated by the prophet Joseph Smith in 1829 in upper New York State.

9. Spencer 1876–96.

10. Landau 1991 for the fullest account of her approach based on the seminal work of Vladimir Propp (1928) *Morphology of the folktale*.

11. Landau 1984: 264.

12. Gould and Vrba 1982.

13. Ibid., 13.

14. Chatwin 1987: 162.

15. First defined by Osborn (1902).

16. Stringer and Gamble 1993.

17. Kelly 1983.

18. Foley 1987a: 13.

Chapter 2

1. Friedman 1981.

2. Beaglehole 1955: xxiv.

3. This did not stop Columbus reporting that in parts of Cuba which he had not visited "the people are born with tails" (Cohen 1969: 119).

4. Marshall and Williams 1982: 261.

5. See Friedman (1981) for illustrations of the fabulous races.

6. Beaglehole 1955: xxvii; Moseley (1983: 137) for Mandeville's treatment of Pliny's monstrous races each of which lived on one of the Andaman Islands.

7. Marshall and Williams 1982: 302.

8. Beaglehole 1955: 388.

9. Lubbock 1865: 476.

10. London 1919: 9.

11. Garrett 1988: 434–5.

12. Monday 6 April 1722: "it was seen by us that smoke rose from various places, from which it may with reason be concluded that the said island (Easter Island), although it appears sandy and barren, nevertheless is inhabited by people" (Sharp 1970).

13. Defoe's title page in 1719 recalled "The life and strange surprizing adventures of Robinson Crusoe, of York. Mariner: who lived eight and twenty years, all alone in an un-inhabited island on the coast of America, near the mouth of the Great River of Oroonoque."

14. Connolly and Anderson 1988.

15. See Grumley (1975) for an example of the wide-eyed, but no doubt lucrative genre approach to hairy things that go bump in the night and that sometimes have impressed impressionable archeologists (Shackley 1983).

16. Hartley 1953.
17. Meek 1973: 89.
18. Moore 1969: 63. For a discussion of these philosophical time travelers and their impact on archeology see Jones (1992).
19. Lubbock 1965: 336, probably the most oft-quoted and notorious judgment from the person who got parliament to give England its first Bank Holiday.
20. Darwin 1871: 404. Bowler (1992: 722–5), has pointed out that the intervening years between Darwin seeing and writing about the Fuegians had led him to alter his judgment. No doubt the influence of Lubbock (his friend and neighbor in Kent) and his contribution to the development of prehistory (1865) led Darwin to the later judgment.
21. Lowenthal 1985; Shennan 1989; Gathercole and Lowenthal 1993.
22. Bragg 1987: 75.
23. Worsaae 1849: 149–50. The link between nationalism, the rise of the middle classes, and an interest in prehistory is discussed by Kristiansen (1981).
24. Symons 1979; Bloch 1977.
25. Murray 1989.
26. Pitt-Rivers (1891: 116) quoted in Bowden (1984); Bowden (1991) puts this eccentric, but essential, Victorian into context.
27. Meek 1976. The same argument, supplemented by Dégerando's famous memorandum (note 18 above), was still being repeated two centuries later by Dawson (1883: 3): "Existing humanity, as it appears in the native American, is little else than a survival of primeval man in Europe. In short, the early voyagers who first met the American tribes really held conference with their own ancestors, or with men among whom still lived manners and customs extinct in Europe before the dawn of history." Dawson was an ardent believer in degeneration.
28. Marshall and Williams 1982: 191.
29. Ibid., 191.
30. Turgot 1751; Ferguson 1767; Millar 1771.
31. Bergin and Fisch 1948.
32. Meek 1976: 117. Smith's views were recorded as student notes taken during his lectures in Glasgow.
33. Marshall and Williams 1982: 221.
34. Morgan 1877: 18.
35. Tylor 1881: 20.
36. Boas 1928: 60–1.
37. For example, Clark, J.G.D. 1946; Childe 1951; Wymer 1982.
38. Latham 1851: 62.
39. Published in 1800.
40. Huxley (1863: 244), in characteristically entertaining fashion, described the "Caucasian mystery," innocently perpetrated by Blumenbach who selected a skull from Georgia (the best in his collection) as the exemplar of the Caucasian race. Then "by some strange intellectual hocus-pocus, grew up the notion that the Caucasian man is the prototypic 'Adamic' man, and his country the primitive centre of our kind" (ibid., 245).
41. Darwin 1871: 226.
42. Stocking 1968, 1987. Huxley (1863: 243) was very clear about the varieties of monogenism on offer. There were the Adamitic monogenists who believed in Adam and Eve and the Rational monogenists such as Linnaeus, Buffon, Blumenbach, Cuvier, and Prichard. While Huxley found fault with many of their propositions, he championed their cause, reserving his scorn for the misguided polygenists.
43. Grayson 1983: 147; Bowler 1984.
44. Grayson 1983: 161–2; Gould 1983: chapter 8.
45. Fitz-Roy 1839: 650.
46. Ellegard 1958: 308–9; Wallace 1864.
47. Leach 1982.
48. Prichard 1855: 657.
49. Stocking 1968: 46.
50. Prichard 1855: 662.
51. Wallace 1864: 169.
52. Coon 1962: 656.
53. Ingold 1986: 5.
54. Montagu 1972: 238.

Chapter 3

1. Latham 1851: 49.
2. Darwin 1859: 323.
3. Darwin 1871: 166.
4. Ibid., 160.
5. Latham 1851: 61; *protos* = first; *plastos* = formed.
6. Nelson (1983: 471) provides a survey of past cradles and claims.

7. Lyell 1863: 387.
8. Haeckel (1909), following Dubois' discoveries in Java which were intended to prove Haeckel's theory of an Asian cradle correct.
9. Grayson 1983: 142.
10. Latham 1851: 248.
11. Scheidt 1924–5: 365.
12. Haeckel (1876) *The history of creation*, in Count (1950: 129).
13. Haddon 1911: 15.
14. Matthew 1915: 180.
15. Ibid., 209.
16. I. Kant (1775) *On the different races of man*, in Count (1950: 23).
17. The Duke of Argyll, a firm believer in degeneration, paints a particularly dismal picture of the inhabitants at the uttermost ends of the earth (1869: 173). See also the selected papers in Count 1950.
18. Darwin 1871: pt i: 199.
19. Darwin 1859: 373.
20. Matthew 1915: fig. 7, and many of the figures in Black's detailed study of the fossil primates.
21. Osborn 1910: 68.
22. Quoted in Macalister (1921: 572).
23. Laing (1895) reviewed the cases for both California (ibid., 416) and Spitsbergen (ibid., 412). Amenghino (1880) put forward the Pampas of Argentina as the human cradle. The evidence was refuted by Hrdlickca (1912).
24. Spencer (1990) for a full account and an intriguing suggestion for this perennial whodunnit.
25. Landau 1984: 63.
26. Latham 1851: 61.
27. In Count 1950: 23.
28. Thiselton-Dyer 1909: 308.
29. Darlington 1957: 566.
30. Ibid., 553.
31. Darwin 1859: 371.
32. Matthew 1915: 214.
33. Gamble (1992a) for a discussion of the impact of the uttermost ends ethnography on the development of archeology.
34. Osborn 1919: 19.
35. Simpson 1940: 162.
36. José de Acosta 1589; (see Fagan 1987: 28).
37. Huxley 1863: 251.
38. The Comte de Buffon put forward this view in 1749, *Histoire naturelle, générale et particulière*, in Count (1950: 15).
39. Scheidt 1924–5: 367.
40. Malthus 1798: 39.
41. Lyell 1863: 386.
42. Quatrefages 1879: 177.
43. Duke of Argyll 1869: 161.
44. Lyell 1863: 387.
45. Duke of Argyll 1869: 162.
46. Lubbock 1865: 474.
47. Ibid., 475.
48. Ratzel 1896: 10.
49. Ibid., 9.
50. Marett 1927: 13. He was professor of anthropology at Oxford.
51. Bergson 1911: 361.
52. To be more precise, "upon the entrance of the night preceding the twenty third day of October," 4004 B.C. Ussher made his calculation in 1658 (Grayson 1983: 27).
53. 1983: 211 and xi. The problem of demonstrating antiquity and measuring the age of the earth was highlighted by Lord Kelvin's influential criticisms of Darwin and Lyell (Burchfield 1975).
54. Simpson 1952: 174.
55. Bailey 1981: 10.
56. Eldredge and Gould (1972), and their reply to criticisms, Gould and Eldredge (1977).
57. Bailey 1983: 103.
58. MacArthur and Wilson 1967; see also Wilson 1975; Pianka 1978.
59. For an excellent account of scientific dating methods, see Aitken (1990).
60. Darwin also favored the ice ages as a mechanism to drive evolution.
61. Hampson 1968: 237; Marshall and Williams 1982.
62. Tylor 1881: 113.
63. Keith 1931: 61.
64. In Count 1950: 23.
65. Gould (1988) for an extended discussion of these competing and oddly complementary metaphors for time.
66. A subject well covered by Grayson (1983).
67. Imbrie and Imbrie 1979.
68. Wallace 1880: 228–9.
69. For example, Zeuner 1959.
70. Milankovich 1941; Imbrie and Imbrie 1979: 108.
71. Shackleton and Opdyke 1973; Roberts 1984.

72. Tarling and Tarling 1971.
73. Ruddiman and Raymo 1988: 2.
74. Hays, Imbrie, and Shackleton 1976; Berger et al. 1984.
75. Ruddiman and Raymo 1988.
76. Mix 1987.
77. Prentice and Denton 1988: 390.

Chapter 4

1. Clark 1975: 176.
2. Full accounts of work in Africa can be found in Reader 1981; Clark 1982a; Gowlett 1984; Foley 1987; Lewin 1989; Klein 1989. Two important edited volumes (Delson 1985; Grine 1988) cover recent work on the Australopithecines and early *Homo*. Aiello and Dean (1990) provide an invaluable guide to human anatomy and its evolution.
3. The ages of the geological periods relevant to human evolution are shown in table 3.2. The dating of the boundaries between the geological epochs is not always agreed. For example, the start of the Pleistocene is set by some at 1.6–1.7 Myr while others have argued that it should correspond to the appearance of the first ice sheets 2.5 Myr ago.
4. Hall et al. 1985.
5. Grove 1983.
6. Koobi Fora is located to the east of Lake Turkana (formerly Lake Rudolf) in northern Kenya. Hence a fossil classified as KNMER 1470 can be deciphered as Kenya National Museum East Rudolf specimen number 1470. West Turkana, as the name implies, lies on the other side of the lake. As the fossil discoveries were made here after the change of name they are classified as WT 15000 = West Turkana specimen number 15000.
7. A measure of the pace and quantity of fossil discoveries in these areas can be gained by comparing the successive editions of Michael Day's *Guide to fossil man* (1965, 1967, 1977, 1986).
8. Dart 1925.
9. Day 1967.
10. These are known as cratons.
11. Butzer and Cooke 1982.
12. Bonnefille 1984; Malone 1987.
13. van Couvering and van Couvering 1976: 163; Grove 1983: 126.
14. Ibid., fig. 1.
15. Brain (1975, 1981c) describes how caverns form and fauna and hominids are incorporated in them.
16. Aitken (1990) provides a comprehensive account of scientific dating techniques. See also Renfrew and Bahn 1992.
17. Klein 1983: 26.
18. Grove 1983: 124.
19. Pilbeam 1984: 65.
20. van Couvering 1980: 293–4.
21. Roberts 1984: fig. 2.4; Brooks and Robertshaw 1990.
22. Flenley 1979: fig. 1.8.
23. Grove 1983: 118.
24. Goudie 1977.
25. Goodman 1963.
26. Sarich and Wilson 1967.
27. The genetic maps of the great apes are discussed by Lowenstein and Zihlman (1988). For a critique of the molecular clock see Lewin (1990).
28. McGrew 1992.
29. Linden 1976.
30. Byrne and Whiten 1988.
31. Gribbin and Cherfas 1982.
32. Robin Dunbar; personal communication with tongue firmly in cheek! Diamond (1991) takes it more seriously.
33. Ridley (1986) provides a thorough critique while Conroy (1990) presents cladistics and other schemes of classification in a very understandable manner. Groves (1989a; see Kimbel 1991), Wood (1992a), and Skelton and McHenry (1992) show what can be achieved with a cladistic approach to human evolution.
34. Sarich 1982: 513.
35. Pilbeam 1984.
36. Sarich 1982.
37. Andrews and Cronin 1982; Andrews 1982; Pilbeam 1985.
38. Foley 1987a.
39. Eldredge and Tattersall 1982: 126.
40. A tempo to evolution known as Darwinian gradualism.
41. Quoted as the dedication in Gribbin and Cherfas (1982).
42. Bowler (1986) provides a comprehensive description of the protagonists and theories in human evolution. See Bowler (1984) and Stocking (1987) for the wider context of evolutionary studies.

43. The controversy over who committed the Piltdown hoax continues to fascinate. For a full update see Spencer (1990; where Sir Arthur Keith, the doyen of British anatomists at the time, is implicated) and the resulting debate in Tobias (1992).

44. A point well illustrated by the many ancestral trees drawn up. See e.g. Sir Arthur Keith's positioning of the Piltdown and Neanderthal branches, soon to wither, in his frontispiece to *The antiquity of man* (1915).

45. Reader 1981.

46. Foley 1987a: 238, fig. 9.8.

47. The story of Dubois' sensational discoveries is well told in von Koenigswald (1956) and by Reader (1981).

48. Shapiro 1976.

49. Leakey, Tobias, and Napier 1964.

50. A recent authoritative review of *Homo habilis* can be found in Wood (1992a).

51. Walker and Leakey 1978.

52. A case has been put by Hill et al. (1992) that a skull fragment from Lake Baringo should be classified as *Homo*. If supported this would make it the oldest *Homo* at 2.4 Myr old. As Wood (1992b: 678) notes, there is no unambiguous evidence for *Homo* from Koobi Fora or elsewhere in East Africa that is older than 2 Myr.

53. Chamberlain and Wood 1987.

54. Groves 1989a and b, Clarke 1990. See Kimbel (1991) for an extensive review.

55. Wood (1992a: fig. 4) points to the closer links in premolar and molar reduction between *H. ergaster*, *H. erectus* and *H. sapiens*. According to this scheme *H. habilis* is "intermediate" in form between *H. ergaster* and *H. rudolfensis* (Wood 1991).

56. Johanson et al. 1987.

57. Wood 1992a: 785.

58. Brain 1981.

59. Cladistic analysis was developed by Hennig (1966). See Ridley 1986 and Conroy 1990.

60. Chamberlain and Wood 1987.

61. The program that does the sorting is known as PAUP (Phylogenetic Analysis Using Parsimony) and was written by David Swofford.

62. Chamberlain and Wood 1987: 119.

63. Groves' (1989a) careful cladistic study of the material leads him to the conclusion that *Homo habilis* cannot be an ancestor for later humans. In strict cladistic terms the *Homo habilis* fossils from Olduvai (OH 7) possess some unique derived characters not found in later hominids. The same would be the case with 1470 (*Homo rudolfensis*). His preferred scheme puts the 1813 skull, which he calls *Homo ergaster*, in the important ancestral position at 1.9 Myr. Not everyone agrees. Clark Howell and Bernard Wood, for example, both regard 1813 as *Homo habilis*.

64. The geographical patterns of hominid distribution are drawn up by Foley (1987a: fig. 9.15).

65. Johanson and White 1979. The type specimen, however, comes from Laetoli and not Hadar.

66. Kimbel 1988.

67. Leakey and Harris 1987.

68. Oxnard 1987; see Aiello and Dean (1990: 244–74) for a discussion of locomotion in primates and hominids.

69. Stern and Susmann 1983.

70. McHenry 1988.

71. Jungers 1988.

72. White 1980, Aiello and Dean 1990: 273–4.

73. Johanson and White 1979. See Johanson and Edey (1981) for a popular account of the discoveries and their interpretation.

74. e.g. Pilbeam 1984.

75. Delson 1986: 496; Howell 1982.

76. Brain 1981.

77. Walker et al. 1986. WT stands for West Turkana. For an alternative view of investigator bias, the personalities of human paleontology, and the crises of classification which this has produced for the Black Skull see Clark (1988).

78. Delson 1987.

79. Wood and Chamberlain 1988: 637. For a general review of the robust Australopithecines see the papers in Grine (1988) and comment by Delson (1987). A recent cladistic review is provided by Skelton and McHenry (1992).

80. Ibid., 340.

81. Foley 1987a: 29.

82. Brain 1981; Delson 1986.

83. This would change if the cladistic analysis favored by Groves (1989a, b) is followed (see note 63 above).
84. Brain 1981.
85. Bilsborough 1986: 212.
86. Brown et al. 1985.
87. Clarke 1990.
88. Groves 1989a.
89. Turner and Chamberlain 1989; Kimbel 1991.
90. Groves 1989a: 296.
91. McHenry 1988.
92. Jungers 1988.
93. Feldesman and Lundy 1988.
94. Brain 1981c: 232. Mann (1975) produced a minimum estimate of 113 individuals.
95. Ibid. Brain (1981c: 272) considers that the proportions in each age class are very similar despite the different estimates of the number of hominid individuals (see fn. 88).
96. Feldesman and Lundy 1988.
97. Johanson et al. 1987.
98. Lieberman et al. 1988.
99. McHenry 1988. Encephalization quotients were first proposed by Eisenberg (1981) to allow comparison between species with very different body and brain sizes (see Foley and Dunbar 1989). EQ provides an index of relative brain size. As we shall see in chapter 5, hominids have large brains relative to body size when compared with other primates.
100. McHenry 1988.
101. Foley 1991.
102. Harris 1983; Hall et al. 1985.
103. These core artifacts are often known as pebble tools, choppers, or chopping tools. Considerable debate has existed concerning their status as cores for the production of flakes and implements in their own right. It seems likely that they served both functions.
104. Hence the name chosen by Leakey et al. (1964) *Homo habilis* or handy man. At this time the production and use of technology were regarded as key elements in defining the transition from apelike ancestor (*Australopithecus*) to part of the human lineage (*Homo*).
105. The term Oldowan is derived from Olduvai. It is common archeological practice to name a collection of stone tools after a locality. When similar assemblages (as judged by technique of manufacture and implement shapes, or types) are found in other places these are often called industries or cultures. The term culture, as applied by archeologists to recurrent patterns among assemblages of stone tools, is often synonymous with the means of reconstructing a prehistoric people—their movements in space and their development, or not, through time (Childe 1929: v–vi).
106. Isaac (1982), Klein (1989) for overview.
107. Susman 1988: 783.
108. Referred to as expedient behavior by Binford (1973, 1977) where a distinction is drawn with curated behavior where repair, recycling, and planning ahead are part of the organization of technology.
109. The concept of living floors was used by Mary Leakey (1971) in her discussion of the Olduvai excavations. See Potts (1988) for a full discussion and interpretation of the evidence.
110. The critique of living floors has been most forcefully put by Binford (1981, 1983, 1988). Glynn Isaac was the first person to coin the term hydraulic jumbles for these collections of bones and stones (personal communication).
111. Binford 1981. Several of the interpretations of hominid behavior were foreshadowed by Brain (1975a and b, 1976), who pioneered many taphonomic approaches which aim to understand how cave fillings accumulate (1969, 1981).
112. Binford 1983.
113. O'Connell and Hawkes 1988; Binford 1978; Speth 1983.
114. Binford 1981.
115. Bunn et al. 1980; Isaac 1981a.
116. Potts and Shipman 1981; Bunn and Kroll 1986.
117. Davidson and Solomon 1990.
118. Ardrey 1961, 1976, Morgan 1982. For discussion and critique see Binford (1981), Knight (1992), and Richards (1987).
119. Blumenschine 1986, 1987. The capture of food might also have been improved, as Blumenschine suggests, by confrontational scavenging between early hominids and carnivores. Stone throwing, stick waving, and coordinated group behavior may have been important during

lean seasons when scavenged carcasses were rare. The literature on scavenging and hunting is now extensive (O'Connell et al. 1988, Speth 1987, Shipman 1986, Potts 1984, Isaac and Crader 1981, Read-Martin and Read 1975, Isaac 1971). Edited volumes by Harding and Telecki (1981), Clutton-Brock and Grigson (1983), Foley (1984), and Nitecki and Nitecki (1987) provide further studies.

120. Isaac 1982: table 3.10.
121. Ibid., 224; Blumenschine 1987.
122. Isaac 1981b; Foley (1981) for an off-site approach to the same question of reconstructing the behavior of mobile foragers.
123. Potts 1984: 158; 1988.
124. Isaac 1976, 1978. The food sharing hypothesis suggested that the most important human invention was the bag to carry food back to base camps.
125. Leakey, M. 1971.
126. Binford 1981, 1983.
127. Ibid., 1984.
128. Grine 1986.
129. This was Robinson's (1954) dietary hypothesis.
130. Brain (1981c) for full description. The other principal workers in the Krugersdorp caves of the Sterkfontein Valley are Broom and Robinson, Tobias and Hughes at Sterkfontein, Brain and Vrba at Swartkrans and Broom (briefly), Vrba and Brain at Kromdraai. Dart was primarily concerned with work at the Makapansgat limeworks and Taung quarries, northern Cape Province, outside the Krugersdorp.
131. Termed by Dart (1957) the Osteodontokeratic (bone–tooth–horn) culture.
132. Ardrey 1961, 1976.
133. Brain 1981c: fig. 221.
134. Foley 1987a: 387.
135. Brain 1981c: 270–1.
136. Ibid., 271.
137. Brain and Sillen 1988.
138. Gowlett et al. 1981.
139. Susman 1988.
140. Bilsborough 1986.
141. Leakey and Walker 1976.
142. Bilsborough 1986: 205.
143. Groves 1989a, Clarke 1990.
144. Brown et al. 1985.
145. Benyon and Dean 1988: 513.

Chapter 5

1. Boriskovsky 1978: 89.
2. Campbell and Bernor 1976.
3. Washburn 1967: 23; 1982.
4. Endler 1977: 7.
5. Darlington 1957.
6. Brown 1957; Groves 1990.
7. Ibid., 1989b.
8. Founder effect describes the derivation of a new population from either a single individual or a limited number of individuals. This can commonly occur on islands (see below, vicariance). The founder population represents a small sample of the gene pool of the parent population to which it once belonged. Therefore natural selection working on this more restricted genetic variety produces gene combinations different from those in the parent population (Allaby 1991: 184).
9. Mayr 1963: 539.
10. White 1978: 107.
11. Haffer (1969, 1982), Vrba (1988: fig. 25.2) for discussion of refugium on an evolutionary timescale and Endler (1982) for an alternative clinal theory to refuge theory.
12. Meggers 1978.
13. Mayr 1942, 1963.
14. White 1978: 18.
15. Allaby 1991: 507.
16. Endler 1977: 183.
17. Eldredge and Gould 1972; Gould and Eldredge 1977.
18. Dawkins 1986: 241–2.
19. Cronin et al. 1981: 122.
20. Public lecture, University of Southampton 1991.
21. Letter to the author, 23 September 1992.
22. Darwin 1859: 352ff. and 372.
23. Simpson 1952: 173.
24. Nelson 1983, Nelson and Rosen eds 1981, Nelson and Platinick 1981. Patterson (1983), Kirsch (1984), Horton (1984), Humphries and Parenti (1986) provide a succinct overview of the aims and claims of vicariance biogeography.
25. Croizat 1962; his full panbiogeographical treatment was published in 1958.
26. Ibid., 1962: i.
27. Ibid., 605.
28. Hutton 1795.

29. Lovelock 1982, 1989.
30. Croizat 1962: 5.
31. Ibid., 188, fig. 44; 1981.
32. Eldredge 1981: 34.
33. Erwin 1981: 195.
34. Humphries and Parenti 1986.
35. Odling-Smee 1988.
36. Ibid., 121.
37. Ruddiman and Raymo 1988: 419.
38. Hsü et al. 1977: 402.
39. Clark, J.D. 1975: 179.
40. Bigalke 1978: fig. 1.1.
41. Kingdon 1971: 81–2.
42. Ibid., 76.
43. Ibid., 81.
44. Bigalke 1978: 8.
45. Butzer 1982a.
46. Ibid., 296.
47. Vrba 1985a; Malone 1987; Butzer 1982b: 23–4.
48. Ghiglieri 1987: fig. 1.
49. Pilbeam 1985: 58.
50. Susman et al. 1984; Malone 1987.
51. Wheeler 1984, 1985, 1988. A more efficient use of water, possibly a limited and certainly a competed for resource on the East African savannahs, is also implied by this model for bipedalism.
52. Steele 1989.
53. Malone 1987: 479.
54. Wallace 1880.
55. Lull 1917: 693. Vrba (1985b) discusses the initiating causes which have been invoked for speciation and extinction.
56. Vrba 1985a: 70.
57. Janis 1982: 307.
58. For a full and illustrated review of the African herbivores see Dorst and Dandelot 1970; Kingdon 1971, 1974a and b, 1977; Bigalke 1978.
59. Vrba 1985a: 64.
60. Ibid., 1984: 131.
61. Paterson 1981, 1982, 1985, 1986; see also Geist (1971, 1974, 1978, 1985), Turner and Chamberlain (1989), Turner and Paterson (1991).
62. Vrba 1985a: fig. 3; 1988: fig. 25.1; see also Prentice and Denton (1988: fig. 24.7), Brain (1981a, b: fig. 11; 1981b).
63. Vrba 1985a; Brain 1981b: 16.
64. Vrba 1985a.
65. Ibid., 1984: 135.
66. Ibid. (1980, 1983) outlines the *effect hypothesis* whereby species arise not through adaptative selection for speciation itself but as an incidental effect of selection for organismal fitness in new habitats (see also Allaby 1991: 157); Stanely (1979) discusses the principles of macroevolution. A further example of explosive speciation which could better be accounted for by the effect hypothesis rather than species selection is provided by Miocene age horses (MacFadden and Hulbert 1988). The authors point to high rates of taxonomic evolution (new species but only moderate rates of morphological evolution (e.g. dentitions). The adaptive radiation took place over 3 Myr. The forcing agency, if one exists, to explain these events is as yet unclear but exaptive effect rather than adaptive selection of species seems in order.
67. Vrba 1980: 61.
68. Janis 1982.
69. Geist 1971, 1985; see also Schaller (1977). For a full discussion of contest behavior involving armaments such as horns see Maynard-Smith (1979).
70. Geist 1974, 1978.
71. Ibid., x.
72. Grubb 1982: 550.
73. Geist 1978, 1985: 21. The general principle that allows an understanding of the predictive relationship between ecology and social behavior is referred to as the Jarman-Bell principle (Geist 1974: 206). The principle as formulated by Geist states that "the body size and population biomass of ungulate species is a function of the fiber content (digestibility) and density of the forage they exploit" (1974: 206). Body size affects many aspects of ungulate behavior (Jarman 1974, Bell 1971, Clutton-Brock and Harvey 1978).
74. Wynne-Edwards 1962, 1986.
75. Ibid., 10.
76. Geist 1978: 15.
77. Vrba 1984: 75.
78. Parsons 1983.
79. Geist 1978: 13–16.
80. Parsons 1983: 20; see Wallace 1981.
81. Geist (1978: 259 and 352) makes the argument for neoteny in relation to *H. erectus* and modern humans.
82. Ibid., 15.
83. Vrba 1983.
84. Mayr 1963: 539.

85. Vrba 1985a: 64–5. See Shipman and Harris (1988) for critique and Vrba's (1988: 414–16) reply.
86. Stern and Susman 1983.
87. Langdon 1985: 625.
88. Vrba 1985a: 66.
89. As Allaby (1991: 268) notes, there is considerable disagreement over whether macroevolution results from the additive effect of small microevolutionary changes or whether it is uncoupled, as Vrba would argue, from them (1983, 1984: 116; Stanley 1979).
90. This important distinction between generalist and specialist can be clarified further for all species: "Generalists persist because of their tolerance of alternative environmental conditions that come and go with climatic cycles. Yet at the species level the specialist lineages score. And this cuts to the essence of the effect hypothesis: namely that the differential success of specialist species that proliferate much more quickly carries with it *no connotation whatsoever* that the specialist organisms within those species should in any way be superior to the generalist organisms in the sister clade" (Elisabeth Vrba, letter to the author 23 September 1992; my emphasis).
91. Delson 1986; Chamberlain and Wood 1987; Wood 1992; see chapter 4 for the radiation which produced *H. ergaster*, *H. rudolfensis*, *H. habilis*, *H. erectus* and Foley (1991) for a discussion of how many early hominid species there should be.
92. Vrba 1985a: 70.
93. Brain 1981c. The stimulus he recieved from Vrba's work (1980) is acknowledged.
94. Brain 1981b: 13.
95. Roberts 1984: 45.
96. Vrba 1985a: 70.
97. Ruddiman and Raymo 1988: fig. 6.
98. Prentice and Denton (1988) provide a major discussion of the issues of cyclicity, periodicity, and tectonic change. The climatic impact of such tectonic shifts can now be traced. Bloemendal and deMenocal (1989) have identitied in deep sea sediment cores the major shift in tropical climate cycles at 2.4 Myr possibly linked to changes in monsoon intensity which in turn may be linked to tectonic activity.
99. Vrba 1985a: 71.
100. Flenley 1979: 3.
101. Bellwood 1985: 14.
102. Croizat 1962: 579.
103. Bowler 1986: 146.
104. Boriskovsky 1978.
105. Dennell et al. 1988a and b.
106. Dzaparidze et al. 1989.
107. Clarke 1990.
108. Roberts 1984.
109. Geist 1978: 2.
110. Parsons 1983: 209.
111. Turner 1984: 205; 1990.

Chapter 6

1. Baker 1978: 76.
2. Foley (1985) provides a thorough review of optimality concepts in human evolution.
3. Baker 1978: 39.
4. Ibid., 21.
5. Wynne-Edwards 1964: 4.
6. Andrewartha and Birch 1984: 13.
7. Baker 1978: 44.
8. Chapter 3, note 35.
9. Passingham 1982: 89.
10. Ibid., 105.
11. Ibid., 120–1.
12. Ibid., 112–13.
13. A very full discussion of the issues can be found in Gould (1977).
14. Groves 1989: 306.
15. Vonnegut 1987: 32.
16. Franklin 1778; Oakley 1957; Leakey et al. (1964); Foley 1987: 391; Clarke (1990) and Groves (1989) link the Oldowan and Acheulean to different fossil species.
17. Goodall 1986.
18. Kortlandt 1986.
19. Boesch and Boesch 1981, 1984. Films of tool using behavior by wild living chimpanzees are now available.
20. Humphrey 1976.
21. Lovejoy 1981: 347.
22. Richards 1986.
23. Humphrey 1976.
24. The geneticist J.B.S. Haldane is reputed to have described the concept in the following graphic terms: that he would lay his life down for more than two brothers, or sisters, eight cousins, thirty-two second cousins and so on according to the proportion of his own genes shared by these relatives. The literature on inclusive fitness is vast (Wilson 1975).

25. Jolly 1966.
26. Byrne and Whiten 1988.
27. Whiten and Byrne 1988: 58.
28. de Waal 1982; see also de Waal 1991.
29. de Waal in Byrne and Whiten (1988: 130).
30. Ibid., 131.
31. Harcourt 1988: 135.
32. Whiten and Byrne 1988: 132.
33. Symons 1979: 37.
34. Knight (1992) provides a spirited view of the sexual revolution in prehistory and in particular the implications for the origins of culture.
35. Morris 1967; a brief glimpse at the Kinsey and Hite reports shows that monogamy, while set up as an ideal in Western society, is just that: an ideal. Monogamous marriage systems, worldwide, are heavily outnumbered by polygamous practices. Morris had made the fatal error of accepting that the *aspirations* of his own society were the *realities* of the rest of the world, past and present.
36. Lovejoy 1981.
37. Dunbar 1988; Wrangham 1979, 1980; Ghiglieri 1987.
38. Symons 1979: 133; Lovejoy 1981.
39. Symons 1979: 141.
40. Milton 1988: 300.
41. Steele 1989.
42. Landau 1991.
43. Groves 1989: 310.
44. Montesquieu 1748.
45. Kortlandt and van Zorn 1969.
46. Wynne-Edwards 1964: 5.
47. Diamond 1977: 252.
48. Baker 1978: 44.
49. Andrewartha and Birch 1984: 262.
50. Richards 1989: 245.
51. Lovejoy 1981.
52. See chapter 4; Jungers 1988; McHenry 1988.
53. Dunbar 1988: 319.
54. Wrangham 1980: 264; Foley and Lee 1989.
55. Indeed, variation must be expected. In a review of the conference "Understanding chimpanzees: diversity and survival," Gibbons (1992) has drawn attention to the diversity in chimpanzee life—including hunting, fighting, tool use, social interactions, and even language—as shown by detailed studies in different parts of Africa. The socioecology behind this variety may be understandable, but once recognized out go any ideas of standard, inflexible patterns of chimp behavior. In turn this suggests that models of hominid socioecology (Wrangham 1979; Ghiglieri 1987; Foley 1989; Foley and Lee 1989) need to be more flexible.
56. Baker 1982: 51. Among humans Baker (1978: 294) identifies different migration costs to individuals of different ages. Exploratory migration is highest between twelve and twenty-two years of age for females and twelve and twenty-six years of age for males. It has to be said that his data from industrial nations is very generalized and rather anecdotal. A more focused study by Hewlett et al. (1986) on the Aka Pygmies found that in "most exploratory activity, or at least the travel that is remembered by adults, takes place roughly between 10 and 25 years of age" (ibid., 69). This is significant since "a spouse is chosen from among people met in the range explored" (ibid., 78).
57. The importance of play behavior in human evolution, its effects in the archeological record, and its persistence as a neotenous trait in adult life have not received adequate attention. Human evolution and Pleistocene archeology have perhaps been seen as too serious to allow such ideas credence? Perhaps today the universal celebration of the qualities of children (less apparent to Victorian anthropologists who put "savages" at the mental level of children) will finally allow such notions to be explored.
58. de Waal in Byrne and Whiten (1988: 125).
59. Speth 1987: 21.
60. Blumenschine 1987.
61. Washburn and DeVore 1961: 94–5.
62. Dunbar 1988.
63. Strum and Mitchell 1986.
64. Ibid., 99.
65. Dunbar (1992) supplies evidence that group size is a function of the relative neocortical volume among primates. The more gray matter the larger the group sizes. He argues that the significance of differences in neocortical size is directly related to the capacity of different primates

to process information. A restraint is therefore imposed on group living at the level of intensity usually associated with primates.

66. Willner and Martin 1985.

67. Ibid., 2.

68. Harvey and Nee (1991) discuss Charnov's optimality theory of female mammalian life history evolution.

69. Fisher 1930; Trivers and Willard 1973; Williams 1979; recent review by Sieff 1990.

70. Johnson 1985, 1988.

71. Ghiglieri 1987.

72. Chapter 5, Vrba 1985.

73. Dunbar (1992) draws the interesting conclusion that primates will only be able to invade habitats that require larger groups for foraging or defense against predators (Steele 1989), if they evolve larger neocortices to sustain social life. The switch from grooming to language as a means of information processing was obviously critical. While a necessary development, the move to neocortical expansion came from social selection.

Chapter 7

1. See e.g., Pfeiffer 1978.

2. Binford 1985: 254.

3. Accounts of how Arctic hunters survive can be found in Binford (1978), Nelson (1969, 1973), and Grønnow (1986), Grønnow et al. (1983).

4. Oswalt 1976.

5. Torrence 1983, 1989a and b.

6. Bleed 1986.

7. Minc 1986.

8. Pfeiffer 1982.

9. Speth 1987: 19.

10. Ibid., 15.

11. See also Speth and Spielmann 1983; Speth 1983.

12. Lindstedt and Boyce 1985.

13. Geist 1978.

14. The recent, and continuing, excavation of the remains of more than twenty individuals at the Middle Pleistocene site of Atapuerca (near Burgos, Spain) will soon provide an adequate sample to examine both inter- and intra-population differences. The initial results, reported at a workshop on "Human evolution in Europe and the Atapuerca evidence" in 1992, point to considerable sexual dimorphism among these archaic *H. sapiens* which must be regarded as members of a once contemporary group. A study of the teeth reveals a greater sexual dimorphism than among living people (Bermudez de Castro et al. 1993: 54–5). How this relates to body size remains to be investigated with the rich collection of post-cranial remains from Atapuerca. Also, the size of the Atapuerca females relative to earlier females has yet to be explored. The Atapuerca finds are most likely 300,000 years old and possibly older (Aguirre, personal communication).

15. A flake describes any piece struck from a nodule. The flake may be the required object or it may be incidental to shaping a core tool such as a handaxe. As a result flakes vary in size and shape. They can vary from thin and regular to thick and irregular in outline. Many other combinations are possible. Most flakes will possess a razor sharp cutting edge. "Retouch" refers to the subsequent trimming of these edges, usually by small controlled removals. Retouch can change the angle of the edge from shallow to steep which may be an advantage for working different materials such as skins or wood. Retouch can also change the shape of the outline of the flake to make convergent edges (points) or concave and convex edged scrapers.

16. Keeley 1980.

17. Known as bifaces because they have been flaked on both faces (bifacially). However, not all bifacially flaked stone tools are handaxes. Many projectile points from the North American plains dating to *c.* 10,000 years ago show highly sophisticated bifacial work.

18. Frere 1800.

19. Clarke 1988.

20. Leakey 1971.

21. Isaac 1982; Butzer 1982.

22. Gowlett 1988.

23. Isaac 1982.

24. Bunn et al. 1980.

25. Potts 1984.

26. Arambourg 1963.

27. Day 1986.

28. Biberson 1961; Isaac 1982. The site of Ain Hanech in Algeria has also

produced chopper/pebble tools that may be 730,000 years old as determined by magnetic reversals in the sediments (Sahouni; personal communication).

29. Misra 1987; Klein 1989.

30. Howell 1982: 128.

31. Bar-Yosef 1980, 1987; Clark 1989.

32. Bar-Yosef 1980: 107.

33. Dzaparidze et al. 1989.

34. *Megantereon* cf. *megantereon*.

35. Wenban-Smith (personal communication).

36. Paddayya 1977, 1982.

37. Paddayya 1977: 354.

38. Ranov 1984, 1991.

39. Paddayya 1977, 1982.

40. Wendorf et al. 1987.

41. Bowler 1986: 35. For an assessment of Dubois' achievement see von Koenigswald (1956) and Reader (1981) for his subsequent disillusionment with paleontology.

42. Day 1986: 337. The skull was first named as *Anthropopithecus* in 1892.

43. Named by Raymond Dart as *Australopithecus africanus* in his note in *Nature* in 1925.

44. A review of the material can be found in Day (1986).

45. Energetically described by von Koenigswald (1956).

46. Jacob 1978.

47. Pope and Cronin 1984; Jacob 1978. Rightmire (1987) argues that the Pucangan fossils are 1 Myr old. Bellwood (1985: fig. 1.11) provides a useful summary of a complicated geological sequence.

48. Day 1986: 369; Oakley et al. 1975; Pei 1939; Shapiro 1976; Jia and Huang 1990.

49. Wu and Wang 1985; Zhang 1985.

50. Wu Rukang and Dong Xingren 1985; Brooks and Wood 1990.

51. Wu and Olsen 1985: table 15.3.

52. Li and Etler 1992. The two skulls, though badly distorted because of compression in the ground, are said to have both *Homo erectus* and "archaic" *H. sapiens* features. A full publication is eagerly awaited. The Middle Pleistocene faunas of southern China are characterized by the presence of *Stegodon* (extinct elephant) and *Ailuropoda* (giant panda).

53. Numerically the four main species in the cave are hyena (at least 2,000 specimens), an extinct thick-jawed deer (at least 2,000 specimens), Merck's rhino and another extinct deer the size of a red deer (both *c.* 1,000 specimens). Horse was also present throughout the sequence (Aigner 1978). The association of large numbers of hyena, rhino, and horse is very typical of hyena dens throughout Pleistocene Eurasia. See e.g., Gamble 1986; Stiner 1991. Before cannibalism can be invoked the possible agency of hyenas and other bone splitting predators must be discounted.

54. Zhang Yinyun 1985.

55. Sutcliffe 1985.

56. Han Defen and Xu Chunhua 1985: tables 15.2 and 15.3 and map compiled by Qiu Zhongiang.

57. Bellwood 1985.

58. Bartstra and Basoeki 1989.

59. Bartstra et al. 1988.

60. Movius 1948; Pope and Cronin 1984.

61. Movius 1948: 411.

62. Astakhov 1986.

63. Yi and Clark 1983.

64. James 1989; Binford and Stone 1986.

65. Pope and Cronin 1984: 834.

66. Bellwood 1985.

67. Chen Tiemei and Yuan Sixun 1988; Wu and Olsen 1985.

68. Jia and Huang 1985; Luchterhand 1979, 1984; Medway 1971.

69. Gansser 1982.

70. Wang et al. 1982.

71. Sharma 1984; Liu and Ding 1984.

72. Ibid.

73. Defining Europe has never been easy. The finds from Dmanisi in the Caucasus mountains of Georgia could be regarded, on political and historic grounds, as part of Europe. However, I have treated them here as part of the Middle East and Central Asia. Such arbitrary divisions underscore the impossibility of talking in absolute terms about the earliest inhabitants of an ill-defined peninsula such as Europe.

74. For a detailed discussion of the European sites see Gamble (1986), Champion et al. (1984), Klein (1989), Stringer and Gamble (1993).

75. The quarry site of Ca' Belvedere at Monte Poggliolo in Emilia Romagna, Italy, has produced simple flake artifacts that possibly date to 1 Myr B.P. (Peretto; personal communication).

76. Klein 1989.

77. Referred to by some as *Homo sapiens heidelbergensis* (Groves 1989).

78. So named by Schoetensack in 1908. Groves 1989: 284–5; Stringer and Gamble 1993. The difficulties that face taxonomists when they come to classify fossil material stem primarily from the isolated samples (one skull here, a mandible there), often separated by several hundred thousand years, yet coming from a genus, *Homo*, that is known to vary enormously in size and morphology at the species level. The recent discoveries at Atapuerca (Spain) of a group of Middle Pleistocene hominids that apparently died at the same time bears out the plasticity of these early hominid populations (Bermudez de Castro et al. 1993).

79. Turner 1991.

80. For example, the big hyenas were absent in Spain. Turner suggests that earlier sites may yet be found in this part of Europe. According to his model the earliest traces of intermittent occupation may well, as at Monte Poggliolo in Italy, go back to 1 Myr.

81. Svoboda 1987, 1989. The Clactonian of southern England is now regarded by some as a knapping component of the Acheulean.

82. The minimum definition of a blade as opposed to a flake is that it must be twice as long as it is wide. The age when Levallois techniques of flake and blade preparation first appear is still unclear. The most likely date is sometime between 300,000 and 200,000 B.P.

83. Santonja and Villa 1990; Perretto 1989; Mussi 1992.

84. Boriskovsky 1984.

85. Ranov 1984, 1991.

86. Astakhov 1986.

87. Howell 1965.

88. Shipman and Rose 1983.

89. Callow and Cornford 1986.

90. Scott 1980, 1986.

91. Roberts 1986.

92. Wymer 1968, 1985.

93. Roebroeks 1988.

94. Tuffreau and Sommé 1988; Auguste 1991.

95. Mania 1991.

96. Bosinski et al. 1986.

97. Kretzoi and Vértes 1965.

98. Wymer 1985. Of course, no sack has ever been found.

99. de Lumley 1969.

100. Villa 1982.

101. Guthrie (1990) provides a brilliant account of finds of frozen animals; how they came to be incorporated into the permafrost and how past landscapes and animal behavior can be reconstructed from these remains.

102. Speth 1987.

103. Gamble 1987; Stringer and Gamble 1993.

104. Of course, following deposition many of these stone artifacts were swept into the rivers and transported some distance to their final resting place. Even more were redeposited as the rivers changed course and eroded earlier gravel deposits. The heavily rolled edges of many handaxes show how solifluction and river action has moved them from their original site. However, the majority must have stayed within the drainage basin of the rivers and it is to this generalized concentration of artifacts that I refer.

105. Estimates of population are notoriously difficult to supply. However, a figure of 1 person per 30km^2 (not unreasonable for modern foragers in temperate mid latitude environments) would yield a population of 2,000 timewalkers for the unglaciated area of southern England. Longterm breeding success would have been guaranteed with a mating network of approximately five hundred since the minimum requirement has been established as 175 (Wobst 1974). Therefore, as a rough approximation, two such mating networks may have been centered on the Hampshire and Ouse basins respectively with perhaps a further two in the upper and lower reaches of the Thames. Together these would account for 2,000 people. Within these regions, people would have been highly mobile as they foraged for food and even, perhaps, separated from other groups by areas of much poorer resources and hence much lower population densities. Finally, at times of low sea level, the Channel probably formed an important, low lying area for settlement as well as providing

links to population in northern France and the Netherlands.
106. Binford 1989.

Chapter 8

1. Wells 1921: 686.
2. Ibid., 692.
3. Golding 1955; Kurtén 1980; Auel 1980.
4. For a full account see Stringer and Gamble (1993), Trinkaus and Shipman (1993), Trinkaus and Howells (1979). The scientific debate can be found in a number of volumes but principally Mellars and Stringer (1989), Mellars (1990), Trinkaus (1989, 1992). The reasons why physical anthropologists spend so much time studying Neanderthals are well spelt out by Trinkaus and Shipman (1993).
5. Wolpoff 1989.
6. Gamble 1991.
7. Howells 1976.
8. Weidenreich 1943, 1947; Coon 1962.
9. Weiner et al. 1953; Spencer 1990.
10. Straus and Cave 1957.
11. Coon's picture and its sentiments are often reproduced. They present an extreme form of the continuity model for modern human origins. However, the point is well taken that technology, haircuts, and clothing are only superficial (Trinkaus 1992).
12. Classified by some as *Homo sapiens heidelbergensis* (Groves 1989a).
13. Stringer and Gamble 1993.
14. Bartstra et al. 1988.
15. Stringer and Andrews 1988: 1,266.
16. All these dating techniques are authoritatively described by Aitken (1990).
17. Graves 1991.
18. Stringer and Andrews 1988: 1,263; Groves (1989: 283) for the anatomical characteristics.
19. Brown 1987.
20. Ibid., 43–4.
21. WLH = Willandra Lakes Hominid; date from Steve Webb (personal communication).
22. Thorne 1984.
23. Brown 1987.
24. Brace 1979.
25. Brace 1964.
26. Hammond 1982.
27. Boule 1911–13.
28. Wolpoff 1989; Wolpoff et al. 1988.
29. Stringer 1984; Stringer et al. 1984; Stringer and Gamble 1993.
30. Brose and Wolpoff 1971: 1178.
31. Stringer 1984: 74.
32. Trinkaus 1981; Stringer 1984: 68.
33. Beals et al. 1984.
34. Ibid., 307.
35. The earliest anatomically modern human skulls are 130,000 years old.
36. The tropical rain forests of the world do not appear to have been occupied until 10,000 years ago (chap. 10). The conclusion must be that modern humans have very plastic phenotypes that do not necessarily respond at the same rate to similar environmental selection.
37. Hublin 1992.
38. Stringer and Gamble 1993.
39. Grün and Stringer 1991 (using ESR dating).
40. Clark 1988: 289.
41. Leakey et al. 1969.
42. Garrod and Bate 1937; McCown and Keith 1939; Brothwell 1961.
43. Suzuki and Takai 1970.
44. Bar-Yosef et al. 1992.
45. Vandermeersch 1981.
46. Valladas et al. 1988; Stringer 1988; Grün and Stringer 1991; Stringer and Gamble 1993, Appx 1.
47. Thorne and Wolpoff 1981; Thorne 1981.
48. Thorne and Wolpoff 1981: 346.
49. Li and Etler (1992: 407) regard the crushed skulls from Yunxian as strong evidence for longterm continuity within this regional population.
50. Groves 1989.
51. Wu and Wu 1985.
52. Groves 1989b.
53. Rightmire 1987.
54. Day 1967: 247.
55. A feature first named by Weidenreich (1943).
56. Brown 1989.
57. Tattersall 1986.
58. Smith 1984; but not in western Europe (Stringer et al. 1984).
59. Cavalli-Sforza et al. 1988: 6005.
60. Turner and Chamberlain 1989.
61. Stringer and Andrews 1988: 1264.
62. Howells 1976.
63. Clark 1946: 18.

64. Clark and Lindly 1989; Clark 1992.

65. The title "The human revolution" has been used on several occasions. Montagu (1965) argued that the revolution began with the first tools in Africa. More recently, as shown by the papers in Mellars and Stringer (1989), the revolution has been seen as a package of elements—technology, language, anatomy, social organization, symbolism, art, and foraging economy. As with many "packages," not everything arrives at the same time.

66. See Bleed (1986) for discussion of maintainable technologies. Keeley (1982) discusses how we might recognize hafting with archeological materials.

67. Curation implies the recycling, repair, and reuse of materials and equipment. Consequently, where a broken tool was thrown away is not, necessarily, where it was made and used; a crucial point for the archeologist faced with the task of reconstructing patterns of movement from excavated remains. *Curated* technologies are contrasted with *expedient* technologies where tools are made, used, and thrown away all in one spot (Binford 1973; Torrence 1989). The degree to which curation (planning ahead) accounts for differences between the Middle and Upper Paleolithic in Europe has been the subject of much debate (Bordes 1972; Dibble 1987; Dibble and Rolland 1992; Rolland and Dibble 1990; Kuhn 1991, 1992; Gamble 1986, 1993; Roebroeks et al. 1988).

68. See Stringer and Gamble 1993.

69. Clark 1982b, 1988, 1989; see also Hublin (1992: 186) who points out the difficulties that have been encountered in securing absolute dates for this industry.

70. Allsworth-Jones 1986; Desbrosse and Kozlowski 1988.

71. von Koenigswald et al. 1974.

72. Boriskovsky 1984.

73. Campbell 1977.

74. Kozlowski 1982. This industry, once referred to as Bachokirian, is now considered as one of the earliest examples of a widespread early Upper Paleolithic, the Aurignacian, named after the type site of Aurignac in France.

75. The Spanish cave of l'Arbreda has also produced an average date of 38,500 B.P. for Aurignacian material (Bischoff et al. 1989).

76. Singer and Wymer 1982; McBurney 1967; Jelinek 1990; Ronen et al. (unpublished).

77. Carter excavated the Lesotho material and obtained dates of *c.* 40,000 B.P. for the microlithic component from sites in eastern Lesotho (Carter and Vogel 1974: 570).

78. Marks 1983, 1990.

79. Mercier et al. 1991. The St. Césaire find has been referred to as the last Neanderthal in France (ApSimon 1980). After 35,000 years the stage is exclusively held by anatomically modern looking humans. The St. Césaire Neanderthal does point to a change over of at least 5,000 years and probably longer—hardly a sudden revolution.

80. Harrold (1989) has made the convincing case that the classic southwest French Neanderthals made these Châtelperronian industries; possibly in imitation of the blade technologies, such as the Aurignacian, which appear to be the exclusive handiwork of anatomically modern humans.

81. Bordes 1961a and b, 1972.

82. These skeletons were dug up in 1908 by Hauser and 1914 by Peyrony. Hauser sent his finds to Germany where they were destroyed during World War II by Allied bombing.

83. Bordes 1968; 144–5.

84. An account of the excavations is provided by Bordes (1972).

85. Denis Peyrony (1869–1954); François Bordes (1919–81).

86. A detailed description of these rock shelters is provided by Laville et al. (1980).

87. Clark 1988.

88. Known as the functional explanation for variability among assemblages of Middle Paleolithic stone tools (Binford and Binford 1966, 1968).

89. Binford (1973, 1977) for an ethnographic example of curation.

90. Barton 1988. Dibble (1987) provides a model whereby the continual use and resharpening of Middle Paleolithic flint scrapers leads to a predictable "biography" for stone tools. They might start life as a particular one of Bordes' scraper types, but

change shape through constant retouch, thereby becoming another type in the list. This reduction, a sort of lithic *rite de passage*, is put down by Dibble (see also Dibble and Rolland 1992; Rolland and Dibble 1990) to the availability of abundant, good quality supplies of raw material; see Kuhn (1992), Gamble (1993a and b) for discussion.

91. Mellars 1989a and b.

92. White 1989.

93. Gamble 1986.

94. Deacon 1990.

95. Roebroeks et al. 1988; Geneste 1988a and b; Gamble 1993a.

96. GEPP 1983.

97. Movius 1966.

98. Simek 1987.

99. Praslov and Rogachev 1982.

100. Gargett 1989.

101. Bricker (1989) reminds us that the 1912 finds at the French site of La Ferrassie were excavated in front of a special commission of experts who then made their observations a matter of formal record by signing a report (*procès-verbal*). The text of the report was published twenty-seven years later. It recorded that "the existence of trenches artificially dug and later filled was shown absolutely . . . there is, then, in the clearest fashion, the proof of a funerary ritual" (Peyrony 1939). Smirnov (1989, 1991) has produced a detailed account of all the Neanderthal burials.

102. See e.g. the variety of burial practices in one island, Tasmania (Meehan 1971).

103. Mellars 1986; Stringer and Gamble 1993: Appx. 1.

104. Stiner 1991; White and Toth 1991.

105. Nowhere more so than at the cave site of Shanidar in Iraq. Here Solecki (1971) claimed a Neanderthal burial strewn with flowers. The association of the pollen, from which the garlands were reconstructed, and the body has since been called into question as we now realize that pollen can percolate through sediments very easily. It seems that, like Stonehenge, every age gets the Neanderthal it wants (Stringer and Gamble 1993). The 1970s appropriately got the first flower child. For a full account of the Shanidar skeletons see Trinkaus (1983).

106. Klima 1988. There is also a considerable quantity of intentional grave goods in this burial.

107. Webb 1989.

108. Pfeiffer 1982.

109. Dorn et al. (1988) using the cat-ion technique that dates the desert varnish that covers the engravings.

110. Davidson and Noble 1989: 128.

111. Mungo III (White and O'Connell 1982).

112. A recent breakthrough has been the direct dating of cave art by small particle AMS radiocarbon techniques. The charcoal used to draw the animals at Altamira, El Castillo (Cantabria, Spain) and Niaux (France) has given dates of 14,000 B.P., 12,900 B.P. and 12,890 B.P. respectively (Valladas et al. 1992). Pigment from the art site of Cougnac (France) gave a date of *c.*14,000 B.P.

113. Chaloupka 1985; Loy et al. 1990.

114. These include a piece from the Middle Paleolithic levels of Bacho Kiro with intentional, heavily scored zig-zag markings. Marshack (1990) has made a detailed study of this piece and made a case for its inclusion within a symbolically organized system of human action.

115. Chase and Dibble 1987. On the other hand continuity and an increase in symbolically organized behavior are argued by Lindly and Clark (1990).

116. Wiessner 1984; Gamble 1991.

117. Wobst (1977) develops the principle that since style contributes, through information exchange, to survival it will be under selection. He considers selection pressure in terms of the target groups which receive and decode messages. Only those groups who are socially distant will be targeted for visual information. Other groups "closer to home" can use language and the rituals of daily contact. For these groups the extra cost of maintaining styles when producing artifacts such as clothing, ornaments, and projectile points is unnecessary. The "creative explosion" therefore came when the category of strangers, outsiders, was socially defined and additional means of exchanging information (in this case using visual cues) were required to cope with such individuals.

118. Lieberman 1984, 1989; Lieberman and Crelin 1971; Laitman 1983; White 1985.
119. Arensburg et al. 1989. Lieberman et al. (1989) question how an unattached hyoid could be used to reconstruct the entire vocal tract. As they point out, even though Arensburg and his colleagues claim similarity in shape between modern and Neanderthal hyoids, this does not mean it had the same position in the neck.
120. Passingham 1982.
121. Falk 1989: 142.
122. Wynn 1988: 279.
123. Passingham 1982: 190.
124. Hewes 1973; Steklis 1985.
125. Jerison 1982.
126. Davidson and Noble 1989; Noble and Davidson 1991.
127. Davidson and Noble 1989: 137.
128. Whallon 1989. The ecological model is disputed by Roebroeks et al. (1992), although they provide no alternative to the timing and pattern of global colonization.
129. White 1985: 111.
130. Bickerton 1981.
131. Whallon 1989: 451.
132. Stringer and Andrews 1988.
133. Cavalli-Sforza et al. 1988; Cavalli-Sforza 1991.
134. Wainscoat et al. 1989.
135. Cann et al. 1987; Cann 1988; Stoneking and Cann 1989.
136. Vigilant et al. 1991.
137. The point at issue is the construction of the tree using a computer program written by David Swofford "Phylogenetic analysis using parsimony" (PAUP). PAUP draws up many possible ancestral trees from the same data. Rerunning the program many hundreds of times, as its author insists is necessary (Barinaga 1992: 687), produces several different outcomes for the sample of 147 people. Most important, not all these alternative trees start from Africa (Goldman and Barton 1992). However, for all the doubt now cast on the methods of the mitochondrial "Eve" hypothesis, it is still undisputed that Africans have greater diversity in their mtDNA than peoples in any other continent, suggesting that humans have lived there longer than anywhere else.
138. Wainscoat 1987.
139. Graham Richards (personal communication).
140. Smith et al. 1989: 215; Page 1989.
141. Hall and Muralidharan 1989; Page 1989.
142. Hall and Muralidharan 1989: 213.

Chapter 9

1. Lewin 1989.
2. Gamble 1991b.
3. Jones 1977a; Bowdler 1982.
4. Jones 1977a: 196–7.
5. Duke of Argyll 1869.
6. Rhys Jones, personal communication.
7. The history of the Tasmanians following European contact can be found in Ryan (1981).
8. Sollas (1911) was important in enshrining attitudes toward Tasmanians and other hunters and gatherers by holding them up as representatives of Paleolithic stages (Gamble 1992a and b; Bowler 1992; Murray 1992).
9. Haddon 1895: 9.
10. Gamble 1986: chapter 3; Sutcliffe 1985; Frenzel 1973; Butzer 1982.
11. Guthrie 1982.
12. Bosinski 1982.
13. Tode 1953.
14. Boriskovsky 1984.
15. Zavernyaev 1978.
16. Goretsky and Ivanova 1982; Goretsky and Tseitlin 1977; Ivanova and Tseitlin 1987; Praslov 1982.
17. These earlier populations thrived, as Olga Soffer has argued (personal communication), in areas where topography helped to pool the resources from several types of environment.
18. Boriskovsky 1984: 94–134; Stanko et al. 1989.
19. These would include the southern German uplands and the hills and limestone plateaux of Czechoslovakia.
20. See the relevant regional chapters in Soffer and Gamble (1990).
21. Praslov and Rogachev 1982.
22. Abramova 1967; Gvozdover 1989.
23. Delporte 1979.
24. Kozlowski and Kozlowski 1979; Praslov and Rogachev 1982.
25. Gamble 1982, 1991b.
26. Bader 1978.

27. Gamble and Soffer 1990: 16; Wobst 1990.
28. Soffer 1985 a and b.
29. Frison has pointed to the lack of suitable projectile points among the stone tools from the Russian and Ukraine plains. This is in marked contrast to the fluted points from North and South America which experiments have shown can pierce the hide of elephants very effectively (Frison 1989).
30. Kuzmina (personal communication).
31. Schmider 1990; Audouze 1987.
32. Bratlund (1991) has reconstructed from damaged reindeer bones (where, on occasion, the tips of flint arrows are still embedded) exactly how the animals were hunted in the Ahresburg tunnel valley.
33. Bosinski 1990.
34. Klein 1988.
35. Shackley 1985.
36. Szabo et al. 1989.
37. Ranov 1984.
38. Deacon 1990.
39. Klein (1988) bases his argument on his many reports from faunal remains in the region.
40. Parkington 1990.
41. Deacon 1990.
42. Close and Wendorf 1990.
43. Ibid. The same evidence for violent death can be found in the graveyards along the Murray River that flows through arid New South Wales, Australia (Pardoe 1990).
44. Marks and Freidel 1977.
45. Gilead and Grigson 1984.
46. Byrd and Garrard 1990; Garrard and Gebel 1988.
47. Edwards 1990.
48. Agrawal et al. 1990.
49. Direct radiocarbon dates now exist for a few of the classic sites (Valladas et al. 1992).
50. Henry 1985.
51. Deacon 1990.
52. Aptly described by Schaller as mountain monarchs (1977).
53. Gamble 1986.
54. Astakhov 1986.
55. Movius 1953.
56. Spahni 1964.
57. de Lumley 1972.
58. Kubiak and Nadachowski 1982.
59. Bailey and Gamble 1990.
60. Straus 1987.
61. Bagolini and Dalmeri 1987.
62. Davis 1987.
63. An Zhimin 1982.
64. Carter 1969.
65. Bailey et al. 1989; Fisher and Strickland 1991; Bellwood 1985; Politis (personal communication).
66. Brooks and Robertshaw 1990.
67. Bellwood 1985.
68. Cited in ibid.
69. McBurney 1950.
70. Gamble 1986. The extent to which the earlier occupants of northern Europe avoided forests is now being debated (Roebroeks et al. 1992).
71. E.g. at Pontnewydd cave in North Wales during the penultimate interglacial 200,000 years ago. But a comparison of the deepsea cores reveals that this saw lower sea levels and hence a less oceanic climate than the exaggerated conditions 130,000 years ago during substage 5e (chap. 3).
72. Helm 1981.
73. Goebel et al. 1993.
74. Astakhov 1986.
75. Boriskovsky 1984: 313; Larichev et al. 1992; Ackerman 1984.

Chapter 10

1. Larichev et al. 1987: 424; Vitebsky 1989.
2. Tseitlin; personal communication.
3. See Keith 1915: 225.
4. Smith 1926.
5. The Trenton Paleoliths were made by humans. The claims that they came from very ancient sediments has been disproved. The Calico Hills material is the result of natural fracture. These geofacts, comparable to the European eoliths, do come from ancient deposits.
6. Larichev et al. 1987a: 425. For reviews of Siberian archeology see Powers (1973), Larichev et al. (1987b, 1990, 1992), Dolitsky (1985).
7. Fagan 1987.
8. For a detailed description of Beringia see the papers in Hopkins et al. (1982).
9. Mochanov 1977.
10. Tseitlin 1979: 226.
11. Larichev et al. 1992: 464. However, Hoffecker et al. (1993: 50) notes that the

artifacts and mammoth remains come from different stratigraphic levels at the site. Occupation most probably took place sometime after 12,000 B.P.

12. Boriskovsky 1984: 314; Larichev et al. 1992.

13. Powers and Hoffecker 1989.

14. David Yesner and Kris Crossen; personal communication.

15. Hoffecker et al. 1993.

16. Hopkins 1982: 14.

17. Fladmark 1979.

18. Hopkins 1982.

19. Vereschagin and Baryshnikov 1982.

20. Warner et al. 1982.

21. Reeves 1983: 399.

22. Frison and Walker 1990.

23. Lynch 1990; Meltzer (1989) examines the reasons why we cannot establish the date of the earliest occupation of North America.

24. Adovasio n.d.

25. Meltzer 1988: 2; Fagan 1987.

26. Frison (1991: 150–5) excavated a mammoth skeleton associated with a Clovis point at the Colby site in the Bighorn basin, Wyoming. Very few examples of such undisputed big game hunting have ever been found.

27. Frison 1991; Stanford and Day 1992.

28. Meltzer 1988.

29. Martin 1967.

30. Martin 1973: 970; Terrell 1986: 188.

31. Allowing an interval of sixteen years per generation.

32. Dillehay et al. 1992; Borrero 1989.

33. Martin 1973: 972.

34. Adovasio et al. 1983; Fagan 1987.

35. Hoffecker et al. 1993: 51–2. The issue they raise is that finding the origins of the Paleoindian tradition, with its spectacular stone projectile points (Clovis, Folsom), has been rendered almost impossible by archeologists' expectations of how such artifacts should develop and how long the process should take. The Alaskan material has usually been regarded as complicating rather than resolving the problem. However, in Hoffecker's, Powers' and Goebels' opinion the Nenana complex is the right age, in the right place and contains enough similar artifacts and techniques of stone working to fit the bill. A bifacial point has also been found in the lowest levels at the broken mammoth site, Tanana Valley (Alaska) at an age > 11,000 B.P. (David Yesner; personal communication). A southern link can be made with Charlie Lake cave in British Columbia (Fladmark et al. 1988) where a fluted projectile point has an age of *c.* 10,500 B.P. The origin of the Paleoindian traditions of the lower forty-eight states can therefore be found in America's backyard—Beringia.

36. Equivalent to the Younger Dryas cold snap in the European sequence.

37. Kelly and Todd 1988.

38. Dillehay 1989; Dillehay and Collins 1988.

39. Kozlowski and Bandi 1981. Turner (1992) argues from linguistic, dental, and blood analyses that there have been three migrations into North America during the late Pleistocene.

40. Dikov 1983.

41. As part of the site was still in permafrost it was possible to recover wooden and skin artifacts (Grønnow 1988). These finds revealed a diverse hunting technology, as well as wooden bowls, spoons and even parts of a kayak.

42. Art, as we have seen in other parts of the Moderns' world, is not always present. The site of Qeqertasussuk, rich in bones, wood, and skin garments produced only half a dozen art pieces. This is typical of the Saqqara culture, widespread in the Arctic, to which the site belongs. However, the much later Thule culture, also widespread in the Arctic, has produced some of the masterpieces of Inuit art. This reminds us that there are always alternative solutions—cultural, social, technological—to survival even in a harsh environment such as the Arctic.

43. Hanihara 1986; Howells 1986; Turner (1992: 45) favors, based on dental and other physical anthropological work, a homeland for the indigenous North Americans east of Lake Baikal.

44. Dumond 1984: 74.

45. Maxwell 1985.

46. Ackerman 1984; Dikov 1988.

47. Ackerman 1984: 118; Fitzhugh and Crowell 1988.

48. Support for this view comes from the environmental history of the Arctic. It now

appears (Bjarne Grønnow; personal communication) that much of the Arctic was available for colonization some 2000 years before the process began. Marine and terrestrial mammals were there for hunting—and yet colonization did not take place, suggesting yet again that colonization is a social process rather than just a question of seizing an environmental opportunity to expand. When they first hit these new lands they were often rich in what at later times would be scarce resources. For example, the first settlements in western Greenland made use of the abundant supplies of driftwood, including trunks of spruce and larch which had washed in over several thousand years from Siberia.

49. Nygaard 1989.

50. Pitulko and Makeyev 1991.

51. Spitzbergen can only be reached by a voyage over the open sea. Paleo-Eskimo and Inuit use of the sea is restricted to coastal waters for fishing and whale hunting.

52. Clark 1989: 258. "The crude and rather colourless nature of this [stone] industry may serve to remind us that the original Australian aborigines issued from one of the most unenterprising parts of the Late Pleistocene world."

53. For general accounts of Australian prehistory see White and O'Connell (1982), Flood (1983, 1990), Mulvaney (1975). A very full bibliography and discussion of Pleistocene age sites is provided by Smith and Sharp (1993b).

54. Groube et al. 1986.

55. Roberts et al. 1990.

56. Allen (1989) suggests an archeologically "invisible phase" of Australian colonization before 40,000 B.P. The Malakunanja data provide some of the first evidence for low level colonization producing a near "invisible" archeological signature on the continent.

57. Birdsell (1953, 1957, 1958) has made several studies of the routes and demographic consequences of Australian colonization.

58. Irwin 1980: 327.

59. Rowland 1987.

60. White and O'Connell 1982: 216–17.

61. Birdsell 1953, 1957.

62. Gorecki et al. 1984.

63. Bowdler 1977.

64. Ibid., 227.

65. Allen 1990; Ross et al. 1992.

66. Wright 1971.

67. Smith 1987: 710.

68. Gould 1980.

69. Jones and Bowler 1980.

70. Jones 1985.

71. Chaloupka 1985; see chapter 8.

72. Jones 1977b: 345.

73. Ibid.

74. Rowland 1984.

75. The practice of burning and its implications for land management by hunters and gatherers can be found in Jones (1969), Lewis (1982), Mellars (1976).

76. White and O'Connell 1982: 56.

77. Mountain 1991.

78. Hope et al. 1983: 46.

79. Beaton 1982.

80. Rosenfeld et al. 1981.

81. Flood 1980.

82. Cosgrove 1989, 1991.

83. Cosgrove et al. 1990.

84. Jones 1990.

85. Reviewed in ibid.

86. Cosgrove et al. 1990.

87. Bowdler 1981; Lourandos 1985; Williams 1987.

88. Webb 1987: 394. See also the debate over population in Meehan and White (1990).

89. Pardoe 1988, 1990.

90. McBryde 1988; Mulvaney 1976; Gamble (1993a) for discussion.

91. Kirch 1984; Terrell 1986.

92. Allen and Gosden 1991.

93. Allen et al. 1988.

94. Wickler and Spriggs 1988.

95. Suzuki and Hanihara 1982.

96. Serizawa 1978, 1986.

97. Reynolds 1985.

98. Aikens and Higuchi 1982: 88.

99. Barnes and Reynolds 1984; Reynolds and Kaner 1990; Aikens and Higuchi 1982: 89.

100. Suzuki and Hanihara 1982.

101. Suzuki 1974.

102. Irwin et al. 1990.

103. White n.d.

104. Irwin et al. 1990: fig. 2.

105. Torrence 1986.

106. Terrell 1986: 248.
107. Groube 1971; Kirch 1984: fig. 19.
108. Crawford 1852.
109. Levison et al. 1973.
110. Irwin et al. 1990.
111. Heyerdhal 1950: 230.
112. Irwin et al. 1990: 41.
113. Kirch 1984: 88, 133.
114. Flannery et al. n.d.
115. Anderson (1989) investigates the ins and outs surrounding the extinction of these prodigious birds.
116. Kirch 1984: 90.
117. Graves 1990.
118. Irwin 1989.
119. MacArthur and Wilson 1967.
120. Their failure to consider human extinctions highlights the poverty of history in their model (White n.d.; Flannery et al. n.d.) and so questions its applicability to questions of species diversity and change.
121. Keegan and Diamond 1987.
122. Ibid., 68.
123. Kirch 1984.
124. Quoted in ibid., 278.
125. Terrell 1986: 105.
126. Johnson 1983.
127. *Mammuthus exilis* and *Mammuthus columbi* (Azzaroli 1981).
128. Johnson 1983.
129. Simpson 1940: 154.
130. Cary and Wormington 1963.
131. Stoddart 1984: 1. Opinions differ over the earliest settlement. MacPhee and Burney (1991) have recovered cut-marked pygmy hippo femurs dated to between A.D. 0 and A.D. 300 while an age of A.D. 500 for settlement is often quoted (Mack 1986: 20). One trend is clear—all the ages are very recent.
132. Bailey et al. 1989: 69.
133. Andrews 1900: 19.
134. Magnusson 1987.
135. Foote and Wilson 1970: 52.
136. Woodman 1986: 8.
137. Onrubia Pintado 1987.
138. Cawkell et al. 1960.
139. Johnson 1771.
140. Rouse 1986.
141. Rouse and Allaire 1978: 474.
142. Cherry (1981, 1990) for a full account of Mediterranean settlement.
143. Simmons 1988; Rees 1989; Davis 1985.

Chapter 11

1. Clark 1961: 3.
2. Collingwood 1945: 65.
3. For a review of attitudes to those at the "uttermost ends of the earth" see Gamble (1992 a & b).
4. Bergin and Fisch 1948.
5. Lubbock 1865: 490–1.
6. Huxley 1963: 567.
7. Renfrew and Parton 1979: 441.
8. Piggott 1965.
9. Bergin and Fisch 1948: Axiom 141; discussed by Hampson (1968: 235).
10. The *Guardian*, 24 February 1989.
11. Clark 1989: 255.
12. Lyell (1863) quoted in Quatrefages (1879: 212–13).
13. Clark 1989: 433.
14. Mackenzie 1978.
15. Ibid., 88.
16. In this sense the aspiration of constructing a world prehistory is comparable to the aim of historians such as Eric Wolf where, in *Europe and the people without history* (1982), he skilfully wove together the elements that construct a common world. It was this process of construction which had been hijacked by national histories. Wolf reversed this by redefining culture, taking it away from the narrow concerns of nationalism. Prehistory needs to do the same (chap. 2). In Wolf's world history the European voyages of discovery provided a unified stage for human action that linked all peoples in a web of social, religious, and economic ties. These linkages, as I have stressed in my world prehistory, are more important for understanding the past than are entities.
17. Schrire 1984; Wilmsen 1989; Mazel (1992) presents an archeological rather than anthropological view.
18. Murray 1987, 1992.
19. Latham 1851: 249. He was referring to the Hottentots, his example of living prehistory.

Bibliography

Abramova, Z.A. 1967. Palaeolithic art in the USSR. *Arctic Anthropology* 4: 1–179.

Ackerman, R. 1984. Prehistory of the Asian Eskimo zone. In D. Damas (ed.) *Handbook of North American Indians*. Volume 5: *Arctic*, 106–118. Washington, Smithsonian Institution.

Acosta, José de. 1589. *Historia Natural y Moral de las Indias.*

Adovasio, J.M. n.d. Pre-Clovis populations in the New World. Paper presented at the Soviet-American Archaeological Field Symposium, Leningrad 1989.

Adovasio, J.M., J. Donahue, J.E. Guilday, R. Stuckenrath, J.D. Gunn, W.C. Johnson 1983. Meadowcroft rockshelter and the peopling of the New World. In P. Masters and N. Fleming (eds.) *Quaternary coastlines*, 413–40. London, Academic Press.

Agassiz, L. 1840. *Études sur les glaciers*. Neuchâtel, Jent et Gassman.

Agrawal, D.P., R. Dodia and M. Seth 1990. South Asian climate and environment at *c.*18 000 BP. In Gamble and Soffer 1990a, Vol. 2, 231–60.

Aiello, L. and C. Dean 1990. *An introduction to human evolutionary anatomy*. London, Academic Press.

Aigner, J.S. 1978. Pleistocene faunal and cultural stations in South China. In F. Ikawa-Smith (ed.) 1978. *Early Palaeolithic in South and East Asia*, 129–62. The Hague, Mouton.

Aikens, M. and T. Higuchi 1982. *Prehistory of Japan*. New York, Academic Press.

Aitken, M. 1990. *Science based dating in archaeology*. London, Longman.

Allaby, M. (ed.) 1991. *The concise Oxford dictionary of zoology*. Oxford, Oxford University Press.

Allen, H. 1990. Environmental history in southwestern New South Wales during the late Pleistocene. In Gamble and Soffer 1990a, 296–321.

Allen, J. 1989. When did humans first colonize Australia? *Search* 20 (5), 149–54.

Allen, J. and C. Gosden (eds.) 1991. *Report of the Lapita homeland project*. Department of Prehistory, Research School of Pacific Studies, Australian National University, Canberra.

Allen, J., C. Gosden, R. Jones and J.P. White 1988. Pleistocene dates for the human occupation of New Ireland, northern Melanesia. *Nature* 331: 707–9.

Allsworth-Jones, P. 1986. *The Szeletian and the transition from Middle to Upper Palaeolithic in central Europe*. Oxford, Clarendon Press.

Amenghino, F. 1880. *La antigueded del hombre en La Plata*. Paris and Buenos Aires, Imprenta Coni.

Anderson, A. 1989. *Prodigious birds: moas and moa-hunting in prehistoric New Zealand.* Cambridge, Cambridge University Press.

Andrewartha, H.G. and L.C. Birch 1984. *The ecological web: more on the distribution and abundance of animals*. Chicago, University of Chicago Press.

Andrews, C.W. 1900. *A monograph of Christmas Island (Indian Ocean)*. London, Longmans.

Andrews, P. 1982. Hominoid evolution. *Nature* 295: 185–6.

Andrews, P. and J.E. Cronin 1982. The relationship of *Sivapithecus* and *Ramapithecus* and the evolution of the orang-utan. *Nature* 297: 541–6.

An Zhimin 1982. Palaeoliths and microliths from Shenja and Shuanghu, northern Tibet. *Current Anthropology* 23: 493–99.

ApSimon, A.M. 1980. The last Neanderthal in France? *Nature* 287: 271–2.

Arambourg, C. 1963. Le gisement de Ternifine. *Archives de l'Institut Paléontologie Humaine* 31: 1–190.

Ardrey, R. 1961. *African genesis*. New York, Dell.
Ardrey, R. 1976. *The hunting hypothesis*. London, Collins.
Arensburg, B. et al. 1989. A Middle Palaeolithic human hyoid bone. *Nature* 338: 758–60.
Argyll, Duke of 1869. *Primeval man: an examination of some recent speculations*. London, Strahan.
Astakhov, S.N. 1986. *The Palaeolithic of Tuva*. St. Petersburg, Nauka.
Audouze, F. 1987. The Paris basin in Magdalenian times. In O. Soffer (ed.) *The Pleistocene Old World: regional perspectives*, 183–200. New York, Plenum.
Auel, J. 1980. *The clan of the cave bear.* Toronto, Bantam Books.
Auguste, P. 1991. Les grands mammiferes du site pleistocene moyen de Biache-Saint-Vaast (Pas-de-Calais, France): Nouvelles données paléthnographiques. In A. Tuffreau (ed.) *Paléolithique et Mésolithique du Norde de la France: Nouvelles recherches*. 11., 35–40. Centre d'Etudes et de Recherches Préhistoriques Université des Sciences et Technologies de Lille 3.
Azzaroli, A. 1981. Pygmy elephants. *Quaternary Research* 16: 423–5.
Bader, O.N. 1978. *Sunghir*. Moscow, Nauka.
Bagolini, B. and G. Dalmeri 1987. I siti mesolitici di Colbricon (Trentino). Analisi spaziale e fruizione del territorio. *Preistoria Alpina* 23: 7–188.
Bailey, G. and C. Gamble 1990. The Balkans at 18 000 BP: the view from Epirus. In Soffer and Gamble 1990, 148–67.
Bailey, G.N. 1981. Concepts of resource exploitation: continuity and discontinuity in palaeoeconomy. *World Archaeology* 13: 1–15.
Bailey, G.N. 1983. Concepts of time in Quaternary prehistory. *Annual Review of Anthropology* 12: 165–92.
Bailey, R.C., G. Head, M. Jenike, B. Owen, R. Rechtman and E. Zechenter 1989. Hunting and gathering in tropical rain forest: is it possible? *American Anthropologist* 91: 59–82.
Baker, R.R. 1978. *The evolutionary ecology of animal migration*. London, Hodder & Stoughton.
Baker, R.R. 1982. *Migration: paths through time and space*. London, Hodder & Stoughton.
Barinaga, M. 1992. "African Eve" backers beat a retreat. *Science* 255: 686–7.
Barnes, G.L. and T.E.G. Reynolds 1984. The Japanese Palaeolithic: a review. *Proceedings of the Prehistoric Society* 50: 49–62.
Barton, C.M. 1988. *Lithic variability and middle Palaeolithic behaviour*. BAR International Series 408, Oxford.
Bartstra, G-J. and Basoeki 1989. Recent work on the Pleistocene and the Palaeolithic of Java. *Current Anthropology* 30: 241–4.
Bartstra, G.-J., S. Soegondho, and A. van der Wijk, 1988. Ngandong man: age and artifacts. *Journal of Human Evolution* 17: 325–37.
Bar-Yosef, O. 1980. Prehistory of the Levant. *Annual Review of Anthropology* 9: 101–33.
Bar-Yosef, O. 1987. Pleistocene connexions between Africa and Southwest Asia: an archaeological perspective. *African Archaeological Review* 5: 29–38.
Bar-Yosef, O., B. Vandermeersch, B. Arensburg, A. Belfer-Cohen, P. Goldberg, H. Laville, L. Meignen, Y. Rak, J.D. Speth, E. Tchernov, A.-M. Tillier and S. Weiner 1992. The excavations in Kebara Cave, Mt Carmel. *Current Anthropology* 33: 497–550.
Beaglehole, J.C. (ed.) 1955. *The journals of Captain James Cook on his voyages of discovery*. 1: *The voyage of the Endeavour 1768–1771*. Cambridge, Cambridge University Press.
Beaglehole, J.C. 1963. *The Endeavour journal of Joseph Banks 1768–1771*. 2 vols. (2nd edn.). Sydney, Angus & Robertson.
Beals, K.L., C.L. Smith and S.M. Dodd 1984. Brain size, cranial morphology, climate and time machines. *Current Anthropology* 25: 301–30.

Beaton, J.M. 1982. Fire and water: aspects of Australian aboriginal management of cycads. *Archaeology in Oceania* 17: 51–8.
Bell, R.H.V. 1971. A grazing system in the Serengeti. *Scientific American* 225: 86–93.
Bellwood, P. 1985. *Prehistory of the Indo-Malaysian Archipelago*. Sydney, Academic Press.
Benyon, A.D. and M.C. Dean 1988. Distinct dental development patterns in early fossil hominids. *Nature* 335: 509–14.
Berger, A., J. Imbrie, J. Hays, G. Kukla, B. Saltzman, (eds.) 1984. *Milankovitch and climate: understanding the response to astronomical forcing* 2 vols. Dordrecht, D. Reidel.
Bergin, T.G. and M.H. Fisch 1948. *The new science of Giambattista Vico* (translated from 3rd edn. 1744). Ithaca, Cornell University Press.
Bergson, H. 1911. *Creative evolution*. London, Macmillan.
Bermudez de Castro, J.M., A.I. Durand and S.L. Ipine 1993. Sexual dimorphism in the human dental sample seom the SH site (Sierra de Atapuerca, Spain): a statistical approach. *Journal of Human Evolution* 24: 43–56.
Biberson, P. 1961. *Le Paléolithique inférieur du Maroc Atlantique*. Publications du Service des Antiquités du Maroc 17.
Bickerton, D. 1981. *The roots of language*. Ann Arbor, Karoma.
Bigalke, R.C. 1978. Present day mammals of Africa. In V. Maglio and H.B.S. Cooke (eds.) *Evolution of African mammals*, 1–16. Cambridge, Mass., Harvard University Press.
Bilsborough, A. 1986. Diversity, evolution and adaptation in early hominids. In G. Bailey and P. Callow (eds.) *Stone age prehistory: studies in memory of Charles McBurney*, 197–220. Cambridge, Cambridge University Press.
Binford, L.R. 1973. Interassemblage variability – the Mousterian and the 'functional' argument. In C. Renfrew (ed.) *The explanation of culture change*, 227–54. London, Duckworth.
Binford, L.R. 1977. Forty-seven trips. In R.V.S. Wright (ed.) *Stone tools as cultural markers*, 24–36. Canberra, Australian Institute of Aboriginal Studies.
Binford, L.R. 1978. *Nunamiut ethnoarcheology*. New York, Academic Press.
Binford, L.R. 1981. *Bones: ancient men and modern myths*. New York, Academic Press.
Binford, L.R. 1983. *In pursuit of the past*. London, Thames and Hudson.
Binford, L.R. 1984. *Faunal remains from Klasies River Mouth*. New York, Academic Press.
Binford, L.R. 1985. Human ancestors: changing views of their behaviour. *Journal of Anthropological Archaeology* 4: 292–327.
Binford, L.R. 1988. Fact and fiction about the Zinjanthropus floor: an analysis of data, arguments, and interpretations. *Current Anthropology* 29: 123–35.
Binford, L.R. 1989. Isolating the transition to cultural adaptations: an organizational approach. In Trinkaus 1989, 18–41.
Binford, L.R. and S.R. Binford 1966. A preliminary analysis of functional variability in the mousterian of Levallois facies. *American Anthropologist* 68(2): 238–95.
Binford, L.R. and N. Stone 1986. Zhoukoudian: a closer look. *Current Anthropology* 27: 453–75.
Binford, S.R. and L.R. Binford 1969. Stone tools and human behavior. *Scientific American* 220: 70–84.
Birdsell, J.B. 1953. Some environmental and cultural factors influencing the structuring of Australian Aboriginal populations. *American Naturalist* 87: 171–207.
Birdsell, J.B. 1957. Some population problems involving Pleistocene man. *Cold Spring Harbor Symposia on Quantitative Biology* 22: 47–70.
Birdsell, J.B. 1958. On population structure in generalized hunting and gathering populations. *Evolution* 12: 189–205.
Bischoff, J.L., N. Soler, J. Maroto and R. Julia 1989. Abrupt Mousterian/Aurignacian

boundary at c.40 ka bp: accelerator 14C dates from L'Arbreda cave (Catalunya, Spain). *Journal of Archaeological Science* 16: 563–76.
Black, D. 1925. Asia and the dispersal of primates. *Bulletin of the Geological Society of China* 4: 133–83.
Bleed, P. 1986. The optimal design of hunting weapons. *American Antiquity* 51: 737–47.
Bloch, M. 1977. The past and the present in the past. *Man* 12: 278–92.
Bloemendal, J. and P. deMenocal 1989. Evidence for a change in the periodicity of tropical climate cycles at 2.4 Myr from whole core magnetic susceptibility measurements. *Nature* 342: 897–99.
Blumenbach, J.F. 1776 (1865) *On the natural varieties of mankind*. Trans. Thomas Bendyshe. London.
Blumenschine, R. 1986. *Early hominid scavenging opportunities: implications of carcass availability in the Serengeti and Ngorongoro ecosystems*. Oxford, British Archaeological Reports International Series 238.
Blumenschine, R. 1987. Characteristics of an early hominid scavenging niche. *Current Anthropology* 28: 383–407.
Boas, F. 1928. *Anthropology and modern life*. New York, Norton (1962 edn).
Boesch, C. and H. Boesch 1981. Sex differences in the use of natural hammers by wild chimpanzees. *Journal of Human Evolution* 10: 583–93.
Boesch, C. and H. Boesch 1984. Mental map in wild chimpanzees: an analysis of hammer transports for nut cracking. *Primates* 25: 160–70.
Bonnefille, R. 1984. Cenozoic vegetation and environments of early hominids in East Africa. In R.O. Whyte (ed.) *The evolution of the East Asia environment*. Volume 2, 579–605. Hong Kong, Centre of Asian Studies, University of Hong Kong.
Bordes, F. 1961a. Typologie du Paléolithique ancien et moyen. *Publications de l'Institut de Préhistorie de l'Université de Bordeaux*, Mémoire No. 1, 2 vols.
Bordes, F. 1961b. Mousterian cultures in France. *Science* 134: 803–10.
Bordes, F. 1968. *The old stone age*. London, Weidenfeld & Nicolson.
Bordes, F. 1972. *A tale of two caves*. New York, Harper & Row.
Boriskovsky, P.I. 1978. Some problems of the Palaeolithic of south and southeast Asia. In F. Ikawa-Smith (ed.) *Early Palaeolithic in South and East Asia*, 87–96. The Hague, Mouton.
Boriskovsky, P.I. (ed.) 1984. *Palaeolithic of the U.S.S.R.* (in Russian). Moscow, Nauka.
Borrero, L.A. 1989. Replanteo de la Arquologia Patagonica. *Interciencia* 14: 127–35.
Bosinski, G. 1982. The transition from Lower/Middle Palaeolithic in northwestern Germany. In A. Ronen (ed.) *The transition from Lower to Middle Palaeolithic and the origin of modern man*, 165–75. Oxford, BAR 151.
Bosinski, G. 1990. *Homo sapiens: l'historie des chasseurs du Paléolithique supérieur en Europe (40 000–10 000av. J.C.)*. Paris, Editions Errance.
Bosinski, G., K. Kröger, J. Schäfer and E. Turner 1986. Alsteinzeitliche Siedlungsplätze auf den Osteifel-Vulkanen. *Jahrbuch des Römisch-Germanischen Zentralmuseums* 33: 97–130.
Boule, M. L'homme fossile de la Chapelle-aux-Saints. *Annales de Paléontologie* (1911) 6: 1–64; (1912) 7: 65–208; (1913) 8: 209–79.
Bowden, M. 1984. *General Pitt Rivers: the father of scientific archaeology*. Salisbury, Salisbury and South Wiltshire Museum.
Bowden, M. 1991. *Pitt Rivers: the life and archaeological work of Lieutenant-General Augustus Henry Lane Fox Pitt Rivers, DCL, FRS, FSA*. Cambridge, Cambridge University Press.
Bowdler, S. 1977. The coastal colonization of Australia. In J. Alklen, J. Golson and R. Jones (eds.) *Sunda and Sahul: prehistoric studies in South East Asia, Melanesia and Australia*, 205–46. London, Academic Press.
Bowdler, S. 1981. Hunters in the highlands: aboriginal adaptations in the eastern Australian uplands. *Archaeology in Oceania* 16: 99–111.

Bowdler, S. 1982. Prehistoric archaeology in Tasmania. In F. Wendorf and A. Close (eds.) *Advances in World Archaeology* 1: 1–49.

Bowler, P.J. 1984. *Evolution: the history of an idea*. Berkeley, University of California Press.

Bowler, P.J. 1986. *Theories of human evolution: a century of debate 1844–1944*. Oxford, Basil Blackwell.

Bowler, P.J. 1992. From 'savage' to 'primitive': Victorian evolutionism and the interpretation of marginalized peoples. *Antiquity* 66: 721–9.

Brace, C.L. 1964. The fate of the 'classic' Neanderthals: a consideration of hominid catastrophism. *Current Anthropology* 5: 3–43.

Brace, C.L. 1979. Krapina, "Classic" Neanderthals, and the evolution of the European face. *Journal of Human Evolution* 8: 527–550.

Bragg, M. 1987. *The maid of Buttermere*. London, Hodder & Stoughton.

Brain, C.K. 1969. The contribution of Namib Desert Hottentots to an understanding of australopithecine bone accumulations. *Scientific Papers of the Namib Desert Research Station* 39: 12–22.

Brain, C.K. 1975a. An interpretation of the bone assemblage from the Kromdraai australopithecine site, South Africa. In R.H. Tuttle (ed.) *Palaeoanthropology, morphology and palaeoecology*, 225–43. The Hague, Mouton.

Brain, C.K. 1975b. An introduction to the South African australopithecine bone accumulations. In A.T. Clason (ed.) *Archaeozoological studies*, 109–19. Amsterdam, Elsevier.

Brain, C.K. 1976. Some principles in the interpretation of bone accumulations associated with man. In G. Isaac and M. McCown (eds.) *Human origins: Louis Leakey and the East African evidence*, 96–116. Menlo Park, Benjamin.

Brain, C.K. 1981a. Hominid evolution and climatic change. *South African Journal of Science* 77: 104–5.

Brain, C.K. 1981b. The evolution of man in Africa: was it a consequence of cainozoic cooling? *Geological Society of South Africa Alex L. du Toit Memorial Lectures* No. 17.

Brain, C.K. 1981c. *The hunters or the hunted?* Chicago, Chicago University Press.

Brain, C.K. and A. Sillen 1988. Evidence from the Swartkrans cave for the earliest use of fire. *Nature* 336: 464–6.

Bratlund, B. 1991. The study of hunting lesions containing flint fragments on reindeer bones at Stellmoor, Schleswig-Holstein, Germany. In N. Barton, A.J. Roberts and D.A. Roe (eds.) *The late-glacial in north west Europe*, 193–207. London, CBA Research Report 77.

Bricker, H.M. 1989. Comment on Gargett. *Current Anthropology* 30: 177–8.

Broca, P. 1868. *Mémoire sur les crânes des Basques*. Paris, Masson.

Brooks, A. and P. Robertshaw 1990. The glacial maximum in tropical Africa 22 000–12 000 BP. In Gamble and Soffer 1990, 121–69.

Brooks, A. and B. Wood 1990. The Chinese side of the story. *Nature* 344: 288–9.

Brose, D. and M. Wolpoff 1971. Early upper Paleolithic man and late middle paleolithic tools. *American Anthropologist* 73: 1156–94.

Brothwell, D.R. 1961. The people of Mount Carmel: a reconsideration of their position in human evolution. *Proceedings of the Prehistoric Society* 27: 155–9.

Brown, F., J. Harris, R. Leakey and A. Walker 1985. Early *Homo erectus* skeleton from west Lake Turkana, Kenya. *Nature* 316: 788–92.

Brown, P. 1987. Pleistocene homogeneity and Holocene size reduction: the Australian human skeletal evidence. *Archaeology in Oceania* 22: 41–67.

Brown, P. 1989. *Coobool Creek: a morphological and metrical analysis of the crania, mandibles and dentitions of a prehistoric Australian population*. Canberra, Department of Prehistory, Research School of Pacific Studies, Australian National University. *Terra Australis* 13.

Brown, W.L. 1957. Centrifugal speciation. *Quarterly Review of Biology* 32: 247–77.

Buffon, G.L.C. Comte de 1749–67. *Histoire naturelle, générale et particulière*. 15 vols., Paris.

Bunn, H.T. and E.M. Kroll 1986. Systematic butchery by Plio/Pleistocene hominids at Olduvai Gorge, Tanzania. *Current Anthropology* 27: 431–52.

Bunn, H.T., J.W.K. Harris, G. Isaac, Z. Kaufulu, E. Kroll, K. Schick, N. Toth and A.K. Behrensmeyer 1980. FxJj50: an early Pleistocene site in northern Kenya. *World Archaeology* 12: 109–36.

Burchfield, J.D. 1975. *Lord Kelvin and the age of the earth*. New York, Science History Publications.

Butzer, K.W. 1982. *Archaeology as human ecology*. Cambridge, Cambridge University Press.

Butzer, K. and H.B.S. Cooke 1982. The palaeo-ecology of the African continent: the physical environment of Africa from the earliest geological to the Later Stone Age Times. In Clark 1982, Vol. 1, 1–69.

Byrd, B. and A.N. Garrard 1990. The last glacial maximum in the Jordanian desert. In Gamble and Soffer 1990a, 78–96.

Byrne, R.W. and A. Whiten (eds.) 1988. *Machiavellian intelligence: social expertise and the evolution of intellect in monkeys, apes and humans*. Oxford, Clarendon Press.

Callow, P. and J.M. Cornford (eds.) 1986. *La Cotte de St. Brelade 1961–1978. Excavations by C.B.M. McBurney*. Norwich, Geo Books.

Campbell, B.G. and R.L. Bernor 1976. The origin of the Hominidae: Africa or Asia? *Journal of Human Evolution* 5: 441–54.

Campbell, J.B. 1977. *The Upper Palaeolithic of Britain*. 2 vols. Oxford, Oxford University Press.

Cann, R. 1988. DNA and human origins. *Annual Review of Anthropology* 17: 127–43.

Cann, R., M. Stoneking and A. Wilson 1987. Mitochondrial DNA and human evolution. *Nature* 325: 31–6.

Carter, P.L. 1969. Moshebi's shelter: excavation and exploitation in eastern Lesotho. *Lesotho* 8: 1–11.

Carter, P.L. and J.C. Vogel 1974. The dating of industrial assemblages from stratified sites in eastern Lesotho. *Man* 9: 557–78.

Cary, M. and E.H. Wormington 1963. *The ancient explorers*. Harmondsworth, Pelican (first published 1929).

Cavalli-Sforza, L.-L. 1991. Genes, peoples and languages. *Scientific American* 265: 71–8.

Cavalli-Sforza, L.-L., A. Piazza, P. Menozzi and J. Mountain 1988. Reconstruction of human evolution: bringing together genetic, archeological, and linguistic data. *Proceedings of the National Academy of Sciences of the USA* 85: 6002–6.

Cawkell, M.B.R., D.H. Maling and E.M. Cawkell 1960. *The Falkland Islands*. London, Macmillan.

Chaloupka, G. 1985. Chronological sequence of Arnhem Land plateau rock art. In Jones 1985, 269–80.

Chamberlain, A.T. and B.A. Wood 1987. Early hominid phylogeny. *Journal of Human Evolution* 16: 119–33.

Champion, T.C., C.S. Gamble, S.J. Shennan and A.W. Whittle 1984. *Prehistoric Europe*. London, Academic Press.

Chase, P. and H.L. Dibble 1987. Middle Palaeolithic symbolism: a review of current evidence and interpretations. *Journal of Anthropological Archaeology* 6: 263–96.

Chatwin, B. 1987. *The songlines*. London, Picador.

Chen Tiemei and Yuan Sixun 1988. Uranium-series dating of bones and teeth from Chinese Palaeolithic sites. *Archaeometry* 30: 59–76.

Cherry, J.F. 1981. Pattern and process in the earliest colonization of the Mediterranean islands. *Proceedings of the Prehistoric Society* 47: 41–68.

Cherry, J.F. 1990. The first colonization of the Mediterranean islands: a review of recent research. *Journal of Mediterranean Archaeology* 3: 145–221.

Childe, V.G. 1929. *The Danube in prehistory*. Oxford, Oxford University Press.
Childe, V.G. 1935. Changing methods and aims in prehistory. *Proceedings of the Prehistoric Society* 1: 1–15.
Childe, V.G. 1951. *Social evolution*. London, Watts.
Clark, G.A. 1988. Some thoughts on the black skull: an archaeologist's assessment of WT–17000 (*A. boisei*) and systematics in human paleontology. *American Anthropologist* 90: 357–71.
Clark, G.A. 1992. Continuity or replacement? Putting modern human origins in an evolutionary context. In H.L. Dibble and P. Mellars (eds.) *The Middle Paleolithic: adaptation, behavior, and variability*, 183–206. Philadelphia, University Museum.
Clark, G.A. and J. Lindly 1989. The case for continuity: observations on the biocultural transition in Europe and western Asia. In P. Mellars and C. Stringer (eds.) *The human revolution: behavioural and biological perspectives in the origins of modern humans*, 626–76. Edinburgh, Edinburgh University Press.
Clark, J.D. 1975. Africa in prehistory: peripheral or paramount? *Man* 10: 175–98.
Clark, J.D. (ed.) 1982a. *The Cambridge History of Africa*. Volume 1: *From the earliest times to c.500 B.C.* Cambridge, Cambridge University Press.
Clark, J.D. 1982b. The cultures of the Middle Palaeolithic/Middle Stone Age. In Clark 1982b, 248–341.
Clark, J.D. 1988. The Middle Stone Age of East Africa and the beginnings of regional identity. *Journal of World Prehistory* 2: 235–305.
Clark, J.D. 1989. The origin and spread of modern humans: a broad perspective on the African evidence. In Mellars and Stringer 1989, 565–88.
Clark, J.G.D. 1946. *From savagery to civilization*. London, Cobbetts Press.
Clark, J.G.D. 1961. *World prehistory in new perspective*. Cambridge, Cambridge University Press.
Clark, J.G.D. 1989. *Economic prehistory*. Cambridge, Cambridge University Press.
Clarke, R.J. 1988. Habiline handaxes and Paranthropine pedigree at Sterkfontein. *World Archaeology* 20: 1–12.
Clarke, R.J. 1990. The Ndutu cranium and the origin of *Homo sapiens*. *Journal of Human Evolution* 19: 699–736.
Close, A.E. 1990. Living on the edge: Neolithic herders in the eastern Sahara. *Antiquity* 64: 79–96.
Close, A.E. and F. Wendorf 1990. North Africa at 18 000 BP. In Gamble and Soffer 1990a: 41–57.
Clutton-Brock, J. and C. Grigson (eds.) 1983. *Animals and archaeology*. Volume 1: *Hunters and their prey*. Oxford, British Archaeological Reports International Series 163.
Clutton-Brock, T.H. and P.H. Harvey 1978. Mammals, resources and reproductive strategies. *Nature* 273: 191–5.
Cohen, J.M. 1969. *The four voyages of Christopher Columbus*. London, Cresset Library.
Collingwood, R.G. 1945. *The idea of history*. Oxford, Oxford University Press.
Connolly, B. and R. Anderson 1988. *First contact*. London, Penguin Books.
Conroy, G.C. 1990. *Primate evolution*. New York, W.W. Norton.
Coon, C.S. 1954. *The story of man*. New York, Knopf.
Coon, C.S. 1962. *The origin of races*. New York, Knopf.
Cosgrove, R. 1989. Thirty thousand years of human colonization in Tasmania: new Pleistocene dates. *Science* 243: 1706–8.
Cosgrove, R. 1991. *The illusion of riches: issues of scale, resolution and explanation of Pleistocene human behaviour*. Unpublished Ph.D., La Trobe University.
Cosgrove, R., J. Allen and B. Marshall 1990. Palaeo-ecology and Pleistocene human occupation in south central Tasmania. *Antiquity* 64: 59–78.
Count, E.W. (ed.) 1950. *This is race*. New York, HenSchuman.
Crawford, J. 1852. *A dissertation on the affinities of the Malayan languages*. London, Smith Elder.

Croizat, L. 1958. *Panbiogeography or an introductory synthesis of zoogeography, phytogeography, and geology; with notes on evolution, systematics, ecology, anthropology, etc.* 2 vols. Caracas, published by the author.

Croizat, L. 1962. *Space, time, form: the biological synthesis*. Caracus, published by the author.

Croizat, L. 1981. *Biogeography: past, present, and future*. In Nelson and Rosen 1981, 501–23.

Croll, J. 1875. *Climate and time in their geological relations: a theory of secular changes of the earth's climate*. Edinburgh, Black.

Cronin, J.E., N.Y. Boaz, C.B. Stringer and Y. Rak 1981. Tempo and mode in hominid evolution. *Nature* 292: 113–22.

Cziesla, E. 1990. Zur Erhaltung von Oberflächen in ariden Gebieten: eine Betrachtung anhand ausgewählter Archäologischer Fundstellen aus der Ostsahara. *Berliner Geographische Studien* 30: 143–68,.

Cziesla, E. and R. Kuper 1989. Sitra – das Orakel der Steine. *Archäologie in Deutschland* 2: 10–14.

Darlington, P.J. 1957. *Zoogeography: the geographical distribution of animals*. New York, Wiley.

Dart, R.A. 1925. *Australopithecus africanus*: the man-ape of South Africa. *Nature* 115: 195–99.

Dart, R.A. 1957. The osteodontokeratic culture of *Australopithecus prometheus. Transvaal Museum Memoirs* 10: 1–105.

Darwin, C. 1839. *Narrative of the surveying voyages of His Majesty's Ships Adventure and Beagle between the years 1826 and 1836*. Volume 3: *Journals and remarks 1832–1836*. London, Coulburn.

Darwin, C. 1859. *On the origin of species by means of natural selection, or the preservation of favoured races in the struggle for life*. Harmondsworth, Penguin (1968 edn.).

Darwin, C. 1871. *The descent of man, and selection in relation to sex*. London, John Murray.

Davidson, I. and W. Noble 1989. The archaeology of perception: traces of depiction and language. *Current Anthropology* 30: 125–155.

Davidson, I. and S. Solomon 1990. Was OH7 the victim of a crocodile attack? In S. Solomon, I. Davidson and D. Watson (eds.) *Problem solving in taphonomy. Tempus* 2: 229–39.

Davis, R. 1987. Regional perspectives on the Soviet Central Asian Paleolithic. In O. Soffer (ed.) *The Pleistocene Old World: regional perspectives*, 121–33. New York, Plenum.

Davis, S. 1985. Tiny elephants and giant mice. *New Scientist* 1437: 25–7.

Davy, Sir Humphrey 1830. *Consolations in travel or the last days of a philosopher*. London, John Murray.

Dawkins, R. 1986. *The blind watchmaker*. Harmondsworth, Penguin.

Dawson, J.W. 1883. *Fossil men and their modern representatives. An attempt to illustrate the characters and condition of pre-historic men in Europe, by those of the American races*. 2nd edn. London, Hodder & Stoughton.

Day, M.H. 1965. *Guide to fossil man,* 1st edn; 2nd edn. 1967; 3rd edn. 1977, 4th edn. 1986. London, Cassell.

Deacon, J. 1990. Changes in the archaeological record in South Africa at 18 000 BP. In Gamble and Soffer 1990a, 170–88.

Delporte, H. 1979. *L'image de la femme dans l'art préhistorique*. Paris, Picard.

Delson, E. 1986. Human phylogeny revised again. *Nature* 322: 496–7.

Delson, E. 1987. Evolution and palaeobiology of robust *Australopithecus*. *Nature* 327: 654–5.

Dennell, R.W., H. Rendell and E. Hailwood 1988a. Late Pliocene artifacts from Northern Pakistan. *Current Anthropology* 29: 495–8.

Dennell, R.W., H. Rendell and E. Hailwood 1988b. Early tool making in Asia: two million year old artefacts in Pakistan. *Antiquity* 62: 98–106.
Desbrosse, R. and J. Kozlowski 1988. *Hommes et climats à l'age du mammoth: le Paléolithique supérieur d'Eurasie centrale*. Paris, Masson.
Diamond, J. 1977. Colonization cycles in man and beast. *World Archaeology* 8: 249–61.
Diamond, J. 1991. *The rise and fall of the third chimpanzee: how our animal heritage affects the way we live*. London, Vintage.
Dibble, H.L. 1987. The interpretation of middle Paleolithic scraper morphology. *American Antiquity* 52: 109–17.
Dibble, H.L. and N. Rolland 1992. On assemblage variability in the Middle Paleolithic of western Europe: history, perspectives, and a new synthesis. In H.L. Dibble and P. Mellars (eds.) *The Middle Paleolithic: adaptation, behavior, and variability*, 1–28. Philadelphia, University Museum.
Dikov, N.N. 1983. The stages and routes of human occupation of the Beringian land bridge based on archaeological data. In P. Masters and N. Fleming (eds.) *Quaternary coastlines*, 347–64. London, Academic Press.
Dikov, N.N. 1988. The earliest sea mammal hunters of Wrangel Island. *Arctic Anthropology* 25: 80–93.
Dillehay, T.D. (ed.) 1989. *Monte Verde: a late Pleistocene settlement in Chile*. Volume 1: *Paleoenvironment and site context*. Washington, Smithsonian Institution Press.
Dillehay, T.D. and M.B. Collins 1988. Early cultural evidence from Monte Verde in Chile. *Nature* 332: 150–2.
Dillehay, T.D., G.A. Calderón, G. Politis and M. Beltrao 1992. Earliest hunters and gatherers of South America. *Journal of World Prehistory* 6: 145–204.
Dolitsky, A.B. 1985. Siberian Palaeolithic archaeology: approaches and analytic methods. *Current Anthropology* 26: 361–78.
Dorn, R.I., M. Nobbs and T. Cahill 1988. Cation-ratio dating of rock-engravings from the Olary Province of arid South Australia. *Antiquity* 62: 681–9.
Dorst, J. and P. Dandelot 1970. *A field guide to the large mammals of Africa*. London, Collins.
Dumond, D. 1984. In D. Damas (ed.) *Handbook of North American Indians*. Volume 5: *Arctic*, 106–18. Washington, Smithsonian Institution.
Dunbar, R.I.M. 1988. *Primate social systems*. London, Croom Helm.
Dunbar, R.I.M. 1992. Neocortex size as a constraint on group size in primates. *Journal of Human Evolution* 20: 469–93.
Dunmore, J. 1965. *French explorers in the Pacific*. Volume 1: *The eighteenth century*. Oxford, Oxford University Press.
Dzaparidze, V. et al. 1989. Der altpaläolithische Fundplatz Dmanisi in Georgien (Kaukasus). *Jahrbuch des Römisch-Germanischen Zentralmuseums Mainz* 36: 67–116.
Edwards, P.C. 1990. Kebaran occupation at the last glacial maximum in Wadi al-Hammeh, Jordan Valley. In Gamble and Soffer 1990a, 97–118.
Eisenberg, J. 1981. *The mammalian radiations: an analysis of trends in evolution, adaptation and behaviour*. London, Athlone Press.
Eldredge, N. 1981. Discussion. In G. Nelson and D.E. Rosen 1981, 34–8.
Eldredge, N. and S.J. Gould 1972. Punctuated equilibria: an alternative to phyletic gradualism. In T.J.M. Schopf (ed.) *Models in paleobiology*, 82–115. San Francisco, Freeman Cooper.
Eldredge, N. and R. Tattersall 1982. *The myths of human evolution*. New York, Columbia University Press.
Ellegard, A. 1958. *Darwin and the general reader: the reception of Darwin's theory of evolution in the British periodical press, 1859–1872*. Göteborg.
Endler, J.A. 1977. *Geographic variation, speciation, and clines*. New Jersey, Princeton University Press.
Endler, J.A. 1982. Pleistocene forest refuges: fact or fancy? In G.T. Prance (ed.)

Biological diversification in the tropics, 641–57. New York, Columbia University Press.

Erwin, T.L. 1981. Taxon pulses, vicariance, and dispersal: an evolutionary synthesis illustrated by Carabid beetles. In Nelson and Rosen 1981, 159–96.

Fagan, B. 1987. *The great journey: the peopling of ancient America*. London, Thames and Hudson.

Falk, D. 1989. Comment on Davidson and Noble. *Current Anthropology* 30: 141–2.

Feldesman, M.R. and J.K. Lundy 1988. Stature estimation for some African Plio-Pleistocene fossil hominids. *Journal of Human Evolution* 17: 583–96.

Ferguson, A. 1767. *An essay on the history of civil society*. Edinburgh.

Fisher, J.W. and H.C. Strickland 1991. Dwellings and fireplaces: keys to Efe Pygmy campsite structure. In C. Gamble and W.A. Boismier, *Ethnoarchaeological approaches to mobile campsites: hunter-gatherer and pastoralist case studies*, 215–36. International Monographs in Prehistory, Ethnoarchaeological Series 1.

Fisher, R.A. 1930. *The genetical theory of natural selection*. Oxford, Clarendon Press.

Fitzhugh, W.W. and A. Crowell 1988. *Crossroads of continents: cultures of Siberia and Alaska*. Washington, Smithsonian Institution Press.

Fitz-Roy, R. 1839. *Narrative of the surveying voyages of His Majesty's Ships Adventure and Beagle between the years 1826 and 1836*. Volume 2: *Proceedings of the second expedition*. London, Coulburn.

Fladmark, K.R. 1979. Routes: alternate migration corridors for early man in north America. *American Antiquity* 44: 55–69.

Fladmark, K.R., J.C. Driver and D. Alexander 1988. The paleoindian component at Charlie Lake Cave (HbRf 39), British Columbia. *American Antiquity* 53: 371–84.

Flannery, T.F., J.P. White, J. Allen and C. Gosden n.d. Prehistoric faunal changes, apparently anthropogenic, in the Bismarck archipelago. MSS. University of Sydney.

Flenley, J. 1979. *The equatorial rain forest: a geological history*. London, Butterworths.

Flood, J.M. 1980. *The moth hunters*. Canberra, Australian Institute of Aboriginal Studies.

Flood, J.M. 1983. *The archaeology of the Dreamtime*. London, Collins.

Flood, J.M. 1990. *The riches of ancient Australia: a journey into prehistory*. Brisbane, University of Queensland Press and Australian Heritage Commission.

Foley, R.A. 1981. *Off-site archaeology and human adaptation in eastern Africa*. Oxford, British Archaeological Reports 97.

Foley, R.A. (ed.) 1984. *Hominid evolution and community ecology*. New York, Academic Press.

Foley, R.A. 1985. Optimality theory in anthropology. *Man* 20: 222–42.

Foley, R.A. 1987a. *Another unique species*. Longman, London.

Foley, R.A. 1987b. Hominid species and stone-tool assemblages: how are they related? *Antiquity* 61: 380–92.

Foley, R.A. 1989. The evolution of hominid social behaviour. In V. Standen and R.A. Foley (eds.) *Comparative socioecology*, 473–94. Oxford, Blackwell Scientific Publications.

Foley, R.A. 1991. How many species of hominid should there be? *Journal of Human Evolution* 20: 413–27.

Foley, R.A. and R. Dunbar 1989. Beyond the bones of contention. *New Scientist* 14 (October): 37–41.

Foley, R.A. and P.C. Lee 1989. Finite social space, evolutionary pathways, and reconstructing hominid behaviour. *Science* 243: 901–6.

Foote, P.G. and D.M. Wilson 1970. *The Viking achievement*. London, Book Club Associates.

Fox, B. 1984. Ecological principles in zoogeography. In M. Archer and G. Clayton (eds.) *Vertebrate zoogeography and evolution in Australia*, 31–43. Perth, Hesperian Press.

Franklin, B. 1778. Quoted in Boswell's *Life of Johnson*.

Frenzel, B. 1973. *Climatic fluctuations of the Ice Age*. Cleveland, Case Western Reserve University.

Frere, J. 1800. Account of flint weapons discovered at Hoxne in Suffolk. *Archaeologia* 13: 204–5.

Friedman, J.B. 1981. *The monstrous races in medieval art and thought*. Cambridge, Mass. Harvard University Press.

Frison, G.C. 1989. Experimental use of Clovis weaponry and tools on African elephants. *American Antiquity* 54: 766–84.

Frison, G.C. 1991. *Prehistoric hunters of the High Plains*. 2nd edn. San Diego, Academic Press.

Frison, G.C. and D.N. Walker 1990. New World palaeoecology at the last glacial maximum and the implications for New World prehistory. In Soffer and Gamble 1990, 312–30.

Galton, F. 1883. *Inquiries into human faculty and its development*. London.

Gamble, C.S. 1982. Interaction and alliance in palaeolithic society. *Man* 17: 92–107.

Gamble, C.S. 1986. *The Palaeolithic settlement of Europe*. Cambridge, Cambridge University Press.

Gamble, C.S. 1987. Man the shoveller: alternative models for middle Pleistocene colonization and occupation in northern latitudes. In O. Soffer (ed.) *The Pleistocene Old World: regional perspectives*, 81–98. New York, Plenum.

Gamble, C.S. 1991a. The social context for European palaeolithic art. *Proceedings of the Prehistoric Society* 57: 3–15.

Gamble, C.S. 1991b. Raising the curtain on modern human origins. *Antiquity* 65: 412–7.

Gamble, C.S. (ed.) 1992a. The uttermost ends of the earth. Special section. *Antiquity* 66: 710–83.

Gamble, C.S. 1992b. Archaeology, history and the uttermost ends of the earth – Tasmania, Tierra del Fuego and the Cape. *Antiquity* 66: 712–20.

Gamble, C.S. 1993a (in press). Exchange, foraging and local hominid networks. *Proceedings of the Prehistoric Society*.

Gamble, C.S. 1993b (in press). People on the move: interpretations of regional variation in Palaeolithic Europe. In J. Chapman and P. Dolhukhov (eds.) *CITEE Conference Proceedings*. Glasgow, Worldwide Archaeology.

Gamble, C.S. and O. Soffer 1990a. *The world at 18 000 BP*. Volume 2: *Low latitudes*. London, Unwin Hyman.

Gamble, C.S. and O. Soffer 1990b. Pleistocene polyphony: the diversity of human adaptations at the last glacial maximum. In Gamble and Soffer 1990a, 1–23.

Gansser, A. 1982. The morphogenic phase of mountain building. In K.J. Hsü (ed.) *Mountain building processes*, 221–8. London, Academic Press.

Gargett, R. 1989. Grave shortcomings: the evidence for Neanderthal burial. *Current Anthropology* 30: 157–90.

Garrard, A.N. and H.G. Gebel (eds.) 1988. *Prehistory of Jordan*. Oxford, BAR International Series 396.

Garrett, W.E. 1988. The peopling of the earth. *National Geographic* 174(4): 434–7.

Garrod, D.A.E. and D.M.A. Bate 1937. *The stone age of Mount Carmel*: vol 1. Oxford, Clarendon Press.

Gathercole, P. and D. Lowenthal (eds.) 1993. *The politics of the past*. London, Routledge.

Geist, V. 1971. *Mountain sheep: a study in behavior and evolution*. Chicago, Chicago University Press.

Geist, V. 1974. On the relationship of social evolution and ecology in ungulates. *American Zoologist* 14: 205–20.

Geist, V. 1978. *Life strategies, human evolution, environmental design*. New York, Springer.

Geist, V. 1985. On evolutionary patterns in the Caprinae with comments on the punctuated mode of evolution, gradualism and a general model of mammalian evolution. In S. Lovari (ed.) *The biology and management of mountain ungulates*, 15–30. London, Croom Helm.

Geneste, J.-M. 1988a. Systemes d'approvisionnement en matières premières au paléolithique moyen et au paléolithique supérieur en Aquitaine. *L'Homme de Néandertal* 8: 61–70.

Geneste, J.-M. 1988b. Les Industries de la Grotte Vaufrey: technologie du débitage, économie et circulation de la matière première lithique. In J.-P. Rigaud (ed.) La Grotte Vaufrey à Cenac et Saint-Julien (Dordogne), Paleoenvironments, chronologie et activités humaines. *Mémoires de la Société Préhistorique Française* 19: 441–518.

G.E.P.P. (Grupo para o Estudo do Paleolítico Portugues) 1983. A estaçao paleolitíca de Vilas Ruivas (Ródao) Campanha de 1979. *O Arquelólogo Portugues* IV: 15–38.

Ghiglieri, M.P. 1987. Sociobiology of the great apes and the hominid ancestor. *Journal of Human Evolution* 16: 319–57.

Gibbons, A. 1992. Chimps: more diverse than a barrel of monkeys. *Science* 255: 287–8.

Gilead, I. and C. Grigson 1984. Far'ah II: a Middle palaeolithic open-air site in the Northern Negev, Israel. *Proceedings of the Prehistoric Society* 50: 71–97.

Goebel, T., A.P. Derevianko and V.T. Petrin 1993 (in press). Dating the Middle to Upper Palaeolithic transition at Kara-Bom, Siberia. *Current Anthropology*.

Golding, W. 1955. *The inheritors*. London, Faber and Faber.

Goldman, N. and N.H. Barton 1992. Genetics and geography. *Nature* 357: 440–1.

Goodall, J. 1986. *The chimpanzees of Gombe*. Cambridge, Mass., Harvard University Press.

Goodman, M. 1963. Serological analysis of the systematics of recent hominids. *Human Biology* 35: 377–424.

Gorecki, P.P., D.R. Horton, N. Stern and R.V.S. Wright 1984. Coexistence of humans and megafauna in Australia: improved stratified evidence. *Archaeology in Oceania* 19: 117–19.

Goretsky, G.I. and I.K. Ivanova 1982. *Molodova I: unique Mousterian settlement in the middle Dniestr*. Moscow, Nauka.

Goretsky, G.I. and S.M. Tseitlin (eds.) 1977. *The multilayer Palaeolithic site Korman IV on the middle Dniestr*. Moscow, Nauka.

Goudie, A. 1977. *Environmental change*. Oxford, Clarendon Press.

Gould, R.A. 1980. *Living archaeology*. Cambridge, Cambridge University Press.

Gould, S.J. 1977. *Ontogeny and phylogeny*. Cambridge, Mass., Harvard University Press.

Gould, S.J. 1983. *The panda's thumb*. Harmondsworth, Pelican.

Gould, S.J. 1984. *The mismeasure of man*. Harmondsworth, Penguin.

Gould, S.J. 1988. *Time's arrow, time's cycle: myth and metaphor in the discovery of geological time*. London, Penguin Books.

Gould, S.J. 1991. Exaptation: a crucial tool for an evolutionary psychology. *Journal of Social Issues* 47: 43–65.

Gould, S.J. and N. Eldredge 1977. Punctuated equilibria: the tempo and mode of evolution reconsidered. *Paleobiology* 3: 115–51.

Gould, S.J. and E.S. Vrba 1982. Exaptation – a missing term in the science of form. *Palaeobiology* 8(1): 4–15.

Gowlett, J.A.J. 1984. *Ascent to civilisation*. London, Collins.

Gowlett, J.A.J. 1988. A case of developed Oldowan in the Acheulean? *World Archaeology* 20: 13–26.

Gowlett, J.A.J., J.W.K. Harris, D. Walton and B.A. Wood 1981. Early archaeological sites, hominid remains and traces of fire from Chesowanja, Kenya. *Nature* 294: 125–9.

Graves, P.M. 1990. The biological and the social in human evolution. Unpublished Ph.D. dissertation, University of Southampton.

Graves, P.M. 1991. New models and metaphors for the Neanderthal debate. *Current Anthropology* 32: 513–41.

Grayson, D.K. 1983. *The establishment of human antiquity*. New York, Academic Press.

Gribbin, J. and J. Cherfas 1982. *The monkey puzzle: are apes descended from man?* London, Bodley Head.

Grine, F.E. 1986. Dental evidence for dietary differences in Australopithecus and Paranthropus; a quantitative analysis of permanent molar microwear. *Journal of Human Evolution* 15: 783–822.

Grine, F.E. (ed.) 1988. *The evolutionary history of the robust Australopithecines.* New York, Aldine.

Groube, L.M. 1971. Tonga, Lapita pottery, and Polynesian origins. *Journal of the Polynesian Society* 80: 278–316.

Groube, L.M., J. Chappell, J. Muke and D. Price 1986. A 40 000 year old human occupation site at Huon Peninsula, Papua New Guinea. *Nature* 324: 453–5.

Grove, A.T. 1983. Evolution of the physical geography of the east African Rift Valley region. In R.W. Sims, J.H. Price and P.E.S. Whalley (eds.) *Evolution, time and space: the emergence of the biosphere*, 115–55. London, Academic Press.

Groves, C.P. 1989a. *A theory of human and primate evolution.* Oxford, Oxford University Press.

Groves, C.P. 1989b. Natural selection and intelligent ancestors. *Mankind* 19: 76–82.

Groves, C.P. 1990. The centrifugal pattern of speciation in Melanesian rainforest mammals. *Memoirs of the Queensland Museum* 28: 325–8.

Grubb, P. 1982. Refuges and dispersal in the speciation of African forest mammals. In G.T. Prance (ed.) *Biological diversification in the tropics*, 537–53. New York, Columbia University Press.

Grumley, M. 1975. *There are giants in the earth.* St. Albans, Panther.

Grün, R. and C.B. Stringer 1991. Electron spin resonance dating and the evolution of modern humans. *Archeometry* 33: 153–99.

Grønnow, B. 1986. Recent archaeological investigations of West Greenland caribou hunting. *Arctic Anthropology* 23, 57–80.

Grønnow, B. 1988. Prehistory in permafrost: investigations at the Saqqaq site, Qeqertasussuk, Disco Bay, West Greenland. *Journal of Danish Archaeology* 7: 24–39.

Grønnow, B., M. Meldgaard and J.B. Nielsen 1983. Aasivissuit – the great summer camp: archaeological, ethnographical and zoo-archaeological studies of a caribou-hunting site in West Greenland. *Meddelelser om Grønland, Man & Society* 5.

Guthrie, R.D. 1982. Mammals of the mammoth steppe as palaeoenvironmental indicators. In Hopkins et al. 1982, 307–26.

Guthrie, D. 1990. *Frozen fauna of the mammoth steppe.* Chicago, Chicago University Press.

Gvozdover, M.D. 1989. The typology of female figurines of the Kostenki palaeolithic culture. *Soviet Anthropology and Archaeology* 27 (4): 32–94.

Haddon, A.C. 1895. *Evolution in art: as illustrated by the life-histories of designs.* London, Walter Scott.

Haddon, A.C. 1911. *The wanderings of peoples.* Cambridge, Cambridge University Press.

Haeckel, E. 1876. *The history of creation: or the development of the earth and its inhabitants by the action of natural causes.* 2 vols. London, H.S. King.

Haeckel, E. 1909. *Natürliche Schöpfungs-Geschichte.* 11th edn. Berlin, G. Reimer.

Haffer, J. 1969. Speciation in Amazonian forest birds. *Science* 165: 131–7.

Haffer, J. 1982. General aspects of the refuge theory. In G.T. Prance (ed.) *Biological diversification in the tropics*, 6–24. New York, Columbia University Press.

Hall, C.M., R.C. Walter and D. York 1985. Tuff above "Lucy" is over 3ma old. *Eos* 66: 257.

Hall, H.G. and K. Muralidharan 1989. Evidence from mitochondrial DNA that African honey bees spread as continuous maternal lineages. *Nature* 339: 211–13.

Hammond, M. 1982. The explusion of the Neanderthals from human ancestry: Marcellin Boule and the social context of scientific research. *Social Studies of Science* 12: 1–36.

Hampson, N. 1968. *The Enlightenment, an evaluation of its assumptions, attitudes and values.* Harmondsworth, Penguin.

Han Defen and Xu Chunhua 1985. Pleistocene mammalian faunas of China. In Wu Rukang and Olson 1985, 267–86.

Hanihara, K. 1986. The origin of the Japanese—relation to other ethnic groups in East Asia. In R.J. Pearson, G.L. Barnes and K.L. Hutterer (eds.) *Windows on the Japanese past: studies in archeology and prehistory*, 75–83 Ann Arbor, Center for Japanese Studies.

Harcourt, A. 1988. Alliances in contests and social intelligence. In Byrne and Whiten 1988, 132–52.

Harding, R.S.O. and G. Teleki (eds.) (1981). *Omnivorous primates: gathering and hunting in human evolution*. New York, Columbia University Press.

Harris, J.W.K. 1983. Cultural beginnings: Plio-Pleistocene archaeological occurrences from the Afar, Ethiopia. *African Archaeological Review* 1: 3–31.

Harrold, F.J. 1989. Mousterian, Chatelperronian and early Aurignacian in western Europe: continuity or discontinuity? In Mellars and Stringer 1989, 677–713.

Hartley, L.P. 1953. *The go-between*. London, Hamish Hamilton.

Harvey, P.H. and S. Nee 1991. How to live like a mammal. *Nature* 350: 23–4.

Hays, J.D., J. Imbrie and N. Shackleton 1976. Variations in the earth's orbit: pacemaker of the ice ages. *Science* 194: 1121–32.

Helm, J. 1981. *Handbook of North American Indians*. Volume 6: *Subarctic*. Washington, Smithsonian Institute.

Hennig, W. 1966. *Phylogenetic Systematics*. Urbana, University of Illinois Press.

Henry, D.O. 1985. Preagricultural sedentism: the Natufian example. In T.D. Price and J.A. Brown (eds.) *Prehistoric hunter-gatherers: the emergence of cultural complexity*, 365–84. Orlando, Academic Press.

Hewes, G.W. 1973. Primate communication and the gestural origin of language. *Current Anthropology* 14: 5–24.

Heyerdhal, T. 1950. *Kon-Tiki: across the Pacific by raft*. Chicago, Rand McNally.

Hiw, A., S. Ward, A. Deino, G. Curtis and R. Drake 1992. Earliest *Homo*. *Nature* 355: 719–22.

Hoffecker, J.F., W.R. Powers and T. Goebel 1993. The colonization of Beringia and the peopling of the New World. *Science* 259: 46–53.

Hope, G., J. Golson and J. Allen 1983. Paleoecology and prehistory in New Guinea. *Journal of Human Evolution* 12: 37–60.

Hopkins, D.M. 1982. Aspects of the Palaeogeography of Beringia during the late Pleistocene. In D.M. Hopkins et al. (eds.) *Paleoecology of Beringia*, 3–28. New York, Academic Press.

Hopkins, D.M., J.V. Matthews, C.E. Scweger and S.B. Young (eds.) 1982. *Paleoecology of Beringia*. New York, Academic Press.

Horton, D. 1984. Dispersal and speciation: Pleistocene biogeography and the modern Australian biota. In M. Archer and G. Clayton (eds.) *Vertebrate zoogeography and evolution in Australia*, 113–18. Perth, Hesperian Press.

Howell, F.C. 1965. *Early man*. London, Time Life Books.

Howell, F.C. 1982. Origins and evolution of African Hominidae. In Clark 1982a, Vol. 1, 70–156.

Howells, W.W. 1976. Explaining modern man: evolutionists versus migrationists. *Journal of Human Evolution* 5: 477–95.

Howells, W.W. 1986. Physical anthropology of the prehistoric Japanese. In R.J. Pearson, G.L. Barnes and K.L. Hutterer (eds.) *Windows on the Japanese past: studies in archeology and prehistory*, 85–99. Ann Arbor, Center for Japanese Studies.

Hrdlicka, A. 1912. Early man in South America. *Bulletin of American Ethnology* 52, Washington D.C.

Hsü, K.J., L. Montadert, D. Bernoulli, M.B. Cita, A. Erikson, R.E. Garrison, R.B. Kidd, F. Mèlierés, C. Müller and R. Wright 1977. History of the Mediterranean salinity crisis. *Nature* 267: 399–403.

Hublin, J.-J. 1992. Recent human evolution in northwestern Africa. *Philosophical Transactions of the Royal Society of London B* 337: 185–91.

Humphrey, N.K. 1976. The social function of intellect. In P.P.G. Bateson and R.A. Hinde (eds.) *Growing points in ethology*, 303–17. Cambridge, Cambridge University Press (reprinted in Byrne and Whiten 1988, 13–26).

Humphries, C.J. and L.R. Parenti 1986. *Cladistic biogeography*. Oxford, Clarendon Press.

Hutton, J. 1795. *Theory of the earth with proofs and illustrations*. Edinburgh, William Creech.

Huxley, J.S. 1963. *Evolution the modern synthesis*. 2nd edn. London, Allen & Unwin.

Huxley, T.H. 1863 (1911). *Man's place in nature and other anthropological essays*. London, Macmillan.

Imbrie, J. and K.P. Imbrie 1979. *Ice ages: solving the mystery*. London, Macmillan.

Ingold, T. 1986. *Evolution and social life*. Cambridge, Cambridge University Press.

Irwin, G. 1980. The prehistory of Oceania: colonization and cultural change. In A. Sherratt (ed.) *The Cambridge encyclopedia of archaeology*, 324–32. Cambridge, Cambridge University Press.

Irwin, G. 1989. Against, across and down the wind: a case for the systematic exploration of the remote Pacific islands. *Journal of the Polynesian Society* 98: 167–206.

Irwin, G., S. Bickler and P. Quirke 1990. Voyaging by canoe and computer: experiments in the settlement of the Pacific Ocean. *Antiquity* 64: 34–50.

Isaac, G. 1971. The diet of early man. *World Archaeology* 2: 279–98.

Isaac, G. 1976. The activities of early African hominids: a review of archaeological evidence from the time span two and a half to one million years ago. In G. Isaac and E. McCown (eds.) *Human origins: Louis Leakey and the East African evidence*, 483–514. Menlo Park, Benjamin.

Isaac, G. 1978. The food sharing behavior of proto-human hominids. *Scientific American* 238: 90–108.

Isaac, G. 1981a. Archaeological tests of alternative models of early hominid behaviour: excavation and experiments. *Philosophical Transactions of the Royal Society of London B* 292: 177–88.

Isaac, G. 1981b. Stone age visiting cards: approaches to the study of early land use patterns. In I. Hodder et al. (eds.) *Pattern of the past: studies in honour of David Clarke*, 131–55. Cambridge, Cambridge University Press.

Isaac, G. 1982. The earliest archaeological traces. In Clark 1982a, Vol. 1, 157–247.

Isaac, G. and D.C. Crader 1981. To what extent were early hominids carnivorous? An archeological perspective. In R.S.O. Harding and G. Teleki (eds.) *Omnivorous primates: gathering and hunting in human evolution*, 37–103. New York, Columbia University Press.

Ivanova, I.K. and S.M. Tseitlin (eds.) 1987. *The multilayered paleolithic site of Molodova. Stone Age men and environment*. Moscow, Nauka.

Jacob, T. 1978. New finds of Lower and Middle Pleistocene Hominines from Indonesia and an examination of their antiquity. In F. Ikawa-Smith (ed.) 1978. *Early Palaeolithic in South and East Asia*, 13–22. The Hague, Mouton.

James, S.R. 1989. Hominid use of fire in the Lower and Middle Pleistocene. *Current Anthropology* 30: 1–26.

Janis, C. 1982. Evolution of horns in ungulates: ecology and paleoecology. *Biological Reviews* 57: 261–318.

Jarman, P.J. 1974. The social organization of antelope in relation to their ecology. *Behaviour* 48: 215–67.

Jelinek, A. 1990. The Amudian in the context of the Mugharan tradition at the Tabun Cave (Mount Carmel), Israel. In Mellars 1990, 81–90.

Jerison, H. 1982. The evolution of biological intelligence. In R.J. Sternberg (ed.) *Handbook of human intelligence*, 723–92. Cambridge, Cambridge University Press.

Jia Lanpo and Huang Weiwan 1985. On the recognition of China's Palaeolithic cultural traditions. In Wu Rukang and Olsen 1985, 259–66.

Jia Lanpo and Huang Weiwan 1990. *The story of Peking man.* Oxford, Oxford University Press.

Johanson, D.C. and M. Edey 1981. *Lucy: the beginnings of humankind.* New York, Simon & Schuster.

Johanson, D.C., F.T. Masao, G.G. Eck, T.D. White, R.C. Walter, W.H. Kimbel, B. Asfaw, P. Manega, P. Ndessokia and G. Suwa 1987. New partial skeleton of *Homo habilis* from Olduvai Gorge, Tanzania. *Nature* 327: 205–9.

Johanson, D.C. and T.D. White 1979. A systematic assessment of early African hominids. *Science* 202: 321–30.

Johnson, C.N. 1985. Ecology, social behavior and reproductive success in a population of red-necked wallabies. Unpublished Ph.D. thesis, University of New England, Armidale.

Johnson, C.N. 1988. Dispersal and the sex ratio at birth in primates. *Nature* 332: 726–8.

Johnson, D.L. 1983. The California continental borderland: landbridges, watergaps and biotic dispersals. In P. Masters and N. Fleming (eds.) *Quaternary coastlines*, 481–527. London, Academic Press.

Johnson, S. 1771. Thoughts on the late transactions respecting Falkland's islands. In D.J. Greene (ed.) *Samuel Johnson: political writings*, 346–86. New Haven, Yale University Press.

Jolly, A. 1966. Lemur social behaviour and primate intelligence. *Science* 153: 501–6 (reprinted in Byrne and Whiten 1988: 27–33).

Jones, R. 1969. Firestick farming. *Australian Natural History* 16: 224–8.

Jones, R. 1977a. The Tasmanian paradox. In R.V.S. Wright (ed.) *Stone tools as cultural markers*, 189–204. Canberra, Australian Institute of Aboriginal Studies.

Jones, R. 1977b. Man as an element in the continental fauna: the case of the sundering of the Bassian bridge. In J. Allen, J. Golson and R. Jones (eds.) *Sunda and Sahul*, 317–86. London, Academic Press.

Jones, R. (ed.) 1985. *Archaeological research in Kakadu National Park.* Australian National Parks and Wildlife Service Special Publication 13. Canberra, Commonwealth of Australia.

Jones, R. 1990. From Kakadu to Kutikina: the southern continent at 18 000 years ago. In Gamble and Soffer 1990a, 264–95.

Jones, R. 1992. Philosophical time travellers. *Antiquity* 66: 744–57.

Jones, R. and J. Bowler 1980. Struggle for the savanna: northern Australia in ecological and prehistoric perspective. In R. Jones (ed.) *Northern Australia, options and implications*, 4–31. Canberra, Research School of Pacific Studies, Australian National University.

Jungers, W.L. 1988. New estimates of body size in Australopithecines. In Grine 1988, 115–25.

Kames, H.H. Lord 1774–5. *Sketches of the natural history of man.* Vols. 1–4. Dublin.

Keegan, W. and J. Diamond 1987. Colonization of islands by humans: a biogeographic perspective. *Advances in Archaeological Method and Theory* 10: 49–92.

Keeley, L.H. 1980. *Experimental determination of stone tool use: a microwear analysis.* Chicago, University of Chicago Press.

Keeley, L.H. 1982. Hafting and retooling: effects on the archeological record. *American Antiquity* 47: 798–809.

Keith, A. 1915. *The antiquity of man.* London, Williams and Norgate.

Keith, A. 1931. The evolution of human races, past and present. In G. Elliot Smith et al. (eds.) *Early man: his origin, development and culture*, 47–64. London, Benn.

Kelly, R. 1983. Hunter-gatherer mobility strategies. *Journal of Anthropological Research* 39: 277–306.

Kelly, R.L. and L.C. Todd 1988. Coming into the country: early paleoindian hunting and mobility. *American Antiquity* 53: 231–44.

Kimbel, W.H. 1988. Identification of a partial cranium of *Australopithecus afarensis* from the Koobi Fora formation, Kenya. *Journal of Human Evolution* 17: 647–56.

Kimbel, W.H. 1991. Species, species concepts and hominid evolution. *Journal of Human Evolution* 20: 355–71.
Kingdon, J. 1971. *East African mammals: an atlas of evolution in Africa.* Volume 1. London, Academic Press.
Kingdon, J. 1974a. *East African mammals: an atlas of evolution in Africa.* Volume 2a. London, Academic Press.
Kingdon, J. 1974b. *East African mammals: an atlas of evolution in Africa.* Volume 2b. London, Academic Press.
Kingdon, J. 1977. *East African mammals: an atlas of evolution in Africa.* Volume 3. London, Academic Press.
Kirch, P.V. 1984. *The evolution of the Polynesian chiefdoms.* Cambridge, Cambridge University Press.
Kirsch, J. 1984. Vicariance biogeography. In M. Archer and G. Clayton (eds.) *Vertebrate zoogeography and evolution in Australia*, 109–12. Perth, Hesperian Press.
Klein, R.G. 1983. The stone age prehistory of southern Africa. *Annual Review of Anthropology* 12: 25–48.
Klein, R.G. 1988. The archaeological significance of animal bones from Acheulean sites in southern Africa. *African Archaeological Review* 6: 3–25.
Klein, R.G. 1989. *The human career.* Chicago, University of Chicago Press.
Klima, B. 1988. A triple burial from the Upper Palaeolithic of Dolni Vestonice, Czechoslovakia. *Journal of Human Evolution* 16: 831–5.
Knight, C. 1992. *Blood relations: menstruation and the origins of culture.* New Haven, Yale University Press.
Kortland, A. 1986. Use of stone tools by wild-living chimpanzees and the earliest hominids. *Journal of Human Evolution* 15: 77–132.
Kortlandt, A. and J.C.J. van Zorn 1969. The present state of research on the dehumanization hypothesis of African ape evolution. *Proceedings 2nd International Congress on Primatology* 3: 10–13.
Kozlowski, J.K. (ed.) 1982. *Excavation in the Bacho Kiro cave, Bulgaria (Final Report).* Warsaw, Paristwowe Wydarunictwo, Naukowe.
Kozlowski, J.K. and H.-G. Bandi 1981. Le Problème des racines asiatiques du premier peuplement de l'Amérique. *Société Suisse des Américanistes Bulletin* 45: 7–42.
Kozlowski, J.K. and S.K. Kozlowski 1979. *Upper Palaeolithic and Mesolithic in Europe: taxonomy and palaeohistory.* Warsaw, Polska Akademia Nauk.
Kretzoi, N. and L. Vértes 1965. Upper Biharian (intermindel) pebble-industry site in western Hungary. *Current Anthropology* 6: 74–87.
Kristiansen, K. 1981. A social history of Danish archaeology. In G. Daniel (ed.) *Towards a history of archaeology*, 20–44. London, Thames and Hudson.
Kubiak, H. and A. Nadachowski 1982. The artiodactyla. *Excavation in the Bacho Kiro cave, Bulgaria (Final Report)*, 61–6. Warsaw, Paristwowe Wydarunictwo, Naukowe.
Kuhn, S.L. 1991. "Unpacking" reduction: lithic new material economy in the Mousterian of west-central Italy. *Journal of Anthropological Archaeology* 10: 76–106.
Kuhn, S.L. 1992. On planning and curated technologies in the Middle Palaeolithic. *Journal of Anthropological Research* 48: 185–214.
Kurtén, B. 1980. *The dance of the tiger.* London, Abacus/Sphere.
Laing, S. 1895. *Human origins.* London, Chapman & Hall.
Laitman, J.T. 1983. The evolution of the hominid upper respiratory system and implications for the origins of speech. In E. de Grolier (ed.) *Glossogenetics*, 63–90. New York, Harwood Academic Publishers.
Landau, M. 1984. Human evolution as narrative. *American Scientist* 72: 262–7.
Landau, M. 1991. *Narratives of human evolution.* New Haven, Yale University Press.
Langdon, J.H. 1985. Fossils and the origins of bipedalism. *Journal of Human Evolution* 14: 615–35.
Larichev, V., U. Khol'ushkin and I. Larichev 1987. The Lower and Middle Palaeolithic of

Northern Asia: achievements, problems, and perspectives. Northeastern Siberia and the Russian Far East. *Journal of World Prehistory* 1: 415–64.

Larichev, V., U. Khol'ushkin and I. Laricheva 1988. The Upper Palaeolithic of Northern Asia: achievements, problems, and perspectives I. Northeastern Siberia and the Russian Far East. *Journal of World Prehistory* 2: 359–96.

Larichev, V., U. Khol'ushkin and I. Larichev 1990. The Upper Palaeolithic of Northern Asia: achievements, problems, and perspectives II. Northeastern Siberia and the Russian Far East. *Journal of World Prehistory* 4: 347–85.

Larichev, V., U. Khol'ushkin and I. Laricheva 1992. The Upper Palaeolithic of Northern Asia: achievements, problems, and perspectives III. Northeastern Siberia and the Russian Far East. *Journal of World Prehistory* 6: 441–76.

Latham, R.G. 1851. *Man and his migrations*. London, John van Voorst.

Laville, H., J.-P. Rigaud and J.R. Sackett, 1980. *Rock shelters of the Périgord*. New York, Academic Press.

Lawrence, W. 1828. *Lectures on physiology, zoology, and the natural history of man, delivered at the Royal College of Surgeons*. Salem, Foote and Brown.

Leach, E.R. 1982. *Social Anthropology*. Glasgow, Collins.

Leakey, L.S.B., P.V. Tobias and J.R. Napier 1964. A new species of the genus *Homo* from Olduvai Gorge. *Nature* 202: 308–12.

Leakey, M.D. 1971. *Olduvai Gorge: excavations in Beds I and II 1960–1963*. Cambridge, Cambridge University Press.

Leakey, M.D. and J.M. Harris (eds.) 1987. *Laetoli: a Pliocene site in northern Tanzania*. Oxford, Clarendon Press.

Leakey, R., K.W. Butzer and M. Day 1969. Early *Homo sapiens* remains from the Omo River region of south-west Ethiopia. *Nature* 222: 1132–8.

Leakey, R.E.F. and A. Walker 1976. *Australopithecus, Homo erectus* and the single species hypothesis. *Nature* 261: 572–4.

Levison, M., R.G. Ward and J.W. Webb 1973. *The settlement of Polynesia: a computer simulation*. Minneapolis, University of Minnesota Press.

Lewin, R. 1989. *Bones of contention: controversies in the search for human origins*. Harmondsworth, Penguin.

Lewin, R. 1990. Molecular clocks run out of time. *New Scientist* (10 February): 38–41.

Lewis, H.T. 1982. Fire technology and resource management in Aboriginal North America and Australia. In N.M. Williams and E.S. Hunn (eds.) *Resource managers: North America and Australian hunter-gatherers*, 45–67. Canberra, Australian Institute of Aboriginal Studies.

Li Tianyuan and D.A. Etler 1992. New Middle Pleistocene hominid crania from Yunxian in China. *Nature* 357: 404–7.

Lieberman, D.E., D.R. Pilbeam and B.A. Wood 1988. A probabilistic approach to the problem of sexual dimorphism in *Homo habilis*: a comparison of KNM-ER 1470 and KNM-ER 1813. *Journal of Human Evolution* 17: 503–11.

Lieberman, P. 1984. *The biology and evolution of language*. Cambridge, Mass., Harvard University Press.

Lieberman, P. 1989. The origins of some aspects of human language and cognition. In Mellars and Stringer 1989, 391–414.

Lieberman, P. and E.S. Crelin 1971. On the speech of Neanderthal man. *Linguistic Inquiry* 11: 203–22.

Lieberman, P., J.T. Laitman, J.S. Reidenberg, K. Landhal and P.J. Gannon 1989. Folk psychology and talking hyoids. *Nature* 342: 486–7.

Linden, E. 1976. *Apes, men and language*. Harmondsworth, Penguin.

Lindly, J. and G.A. Clark 1990. Symbolism and modern human origins. *Current Anthropology* 31: 233–40.

Lindstedt, S.L. and M.S. Boyce 1985. Seasonality, fasting endurance, and body size in mammals. *American Naturalist* 125: 873–8.

Linnaeus, C. 1800. *A general system of nature* (trans. W. Turton from the last edition of the *Systema naturae* published by Gmelin). London, Lackington, Allen.

Liu Dongsheng and Ding Menglin 1984. The characteristics and evolution of the palaeoenvironment of China since the late Tertiary. In R. Whyte (ed.) *The evolution of the East Asian environment*, 11–40. University of Hong Kong, Centre of Asian Studies.

London, J. 1919. *The human drift*. London, Mills & Boon.

Lourandos, H. 1985. Intensification and Australian prehistory. In T.D. Price and J.A. Brown (eds.) *Prehistoric hunters and gatherers: the emergence of cultural complexity*, 385–423. New York, Academic Press.

Lovejoy, C.O. 1981. The origin of man. *Science* 211: 341–50.

Lovelock, J. 1982. *Gaia: a new look at life on earth*. Oxford, Oxford University Press.

Lovelock, J. 1989. *The ages of Gaia: a biography of our living earth*. Oxford, Oxford University Press.

Lowenstein, J. and A. Zihlman 1988. The invisible ape. *New Scientist* (3 December): 56–9.

Lowenthal, D. 1985. *The past is a foreign country*. Cambridge, Cambridge University Press.

Loy, T.H., R. Jones, D.E. Nelson, B. Meehan and J. Vogel 1990. Accelerator radiocarbon dating of human blood proteins in pigments from late Pleistocene art sites in Australia. *Antiquity* 64: 110–16.

Lubbock, J. 1865. *Pre-historic times, as illustrated by ancient remains and the manners and customs of modern savages*. London, Williams & Norgate.

Luchterhand, K. 1979. Late Cenozoic climate, mammalian evolutionary patterns and Middle Pleistocene human adaptation in eastern Asia. In L. Freeman (ed.) *Views of the past: essays in old world prehistory and palaeoanthropology*, 363–421. The Hague, Mouton.

Luchterhand, K. 1984. Mammalian endemism and diversity and Middle Pleistocene hominid distribution and adaptation in eastern Asia. In R. Whyte (ed.) *The evolution of the East Asian environment*, 848–63. University of Hong Kong, Centre of Asian Studies.

Lull, R.S. 1917. *Organic evolution*. London, Williams & Norgate.

Lumley, H. de 1969. A Paleolithic camp site at Nice. *Scientific American* 220: 42–50.

Lumley, H. de (ed.) 1972. *La Grotte moustérienne de l'Hortus*. Marseilles, Études Quaternaires, 1.

Lydekker, R. 1896. *A geographical history of mammals*. Cambridge, Cambridge University Press.

Lyell, C. 1863. *The geological evidences of the antiquity of man with remarks on theories of the origin of species by variation*. London, Murray.

Lynch, T.F. 1990. Glacial-age man in South America: a critical review. *American Antiquity* 55: 12–36.

Macalister, R.A.S. 1921. *A text-book of European archaeology*. Volume 1: *The Palaeolithic period*. Cambridge, Cambridge University Press.

MacArthur, R. and E.O. Wilson 1967. The theory of island biogeography. *Monographs in population biology* 1. Princeton, Princeton University Press.

McBryde, I. 1988. Goods from another country: exchange networks and the people of the Lake Eyre basin. In J. Mulvaney and P. White (eds.) *Archaeology to 1788*, 253–73. Waddon Associates, Sydney.

McBurney, C.B.M. 1950. The geographical study of the older Palaeolithic stages in Europe. *Proceedings of the Prehistoric Society* 16: 163–83.

McBurney, C.B.M. 1967. *The Haua Fteah (Cyrenaica) and the stone age of the south east Mediterranean*. Cambridge, Cambridge University Press.

McCown, T.D. and A. Keith 1939. *The stone age of Mount Carmel II: the fossil human remains from the Levalloiso-Mousterian*. Oxford, Clarendon Press.

MacFadden, B.J. and R.C. Hulbert, 1988. Explosive speciation at the base of the adaptive radiation of Miocene grazing horses. *Nature* 336: 466–8.

McHenry, H.M. 1988. New estimates of body weight in early hominids and their significance to encephalization and megadontia in "robust" Australopithecines. In Grine 1988, 133–48.

Mack, J. 1986. *Madagascar, island of the ancestors*. London, British Museum Publications.

Mackenzie, W.J.M. 1978. *Biological ideas in politics*. Harmondsworth, Pelican.

MacPhee, R.D.E. and D.A. Burney 1991. Dating of modified femora of extinct dwarf hippopotamus from southern Madagascar: implications for constraining human colonizing and vertebrate extinction events. *Journal of Archaeological Science* 18: 695–706.

Magnusson, M. 1987. *Iceland saga*. London, Bodley Head.

Malone, D. 1987. Mechanisms of hominoid dispersal in Miocene East Africa. *Journal of Human Evolution* 16: 469–81.

Malthus, T.R. 1798. *An essay on the principle of population as it affects the future improvement of society with remarks on the speculations of Mr. Godwin, M. Condorcet, and other writers*. London, J. Johnson (facsimile edn. 1966 London, Macmillan).

Mania, D. 1991. The zonal division of the Lower Palaeolithic open-air site Bilzingsleben. *Anthropologie* 29: 17–24.

Mann, A.E. 1975. *Paleodemographic aspects of the South African Australopithecines*. University of Pennsylvania Publications in Anthropology, 1. Philadelphia, University of Pennsylvania Press.

Marett, R.R. 1927. *Man in the making*. London, Benn.

Marks, A.E. 1983. The Middle to Upper Palaeolithic transition in the Levant. *Advances in World Archaeology* 2: 51–98.

Marks, A.E. 1990. The Middle and Upper Palaeolithic of the Near East and the Nile Valley: the problem of cultural transformations. In Mellars 1990, 56–80.

Marks, A.E. and D.A. Freidel 1977. Prehistoric settlement patterns in the Avdat/Aqev area. In A.E. Marks (ed.) *Prehistory and paleoenvironments in the central Negev, Israel*. Volume 2: *The Avdat/Aqev area, Part 2 and the Har Harif*, 131–59. Dallas, Southern Methodist University.

Marshack, A. 1990. Early hominid symbol and evolution of the human capacity. In Mellars 1990, 457–98.

Marshall, P.J. and G. Williams 1982. *The great map of mankind; British perceptions of the world in the age of enlightenment*. London, Dent.

Martin, P.S. 1967. Pleistocene overkill. In P.S. Martin and H.E. Wright (eds.) *Pleistocene extinctions: the search for a cause*, 75–120. New Haven, Yale University Press.

Martin, P.S. 1973. The discovery of America. *Science* 179: 969–74.

Matthew, W.D. 1915. Climate and evolution. *Annals of the New York Academy of Science* 24: 171–318.

Maxwell, M.S. 1985. *Prehistory of the eastern Arctic*. Orlando, Academic Press.

Maynard-Smith, J. 1979. Game theory and the evolution of behaviour. *Proceedings of the Royal Society of London B* 205: 475–88.

Mayr, E. 1942. *Systematics and the origin of species: from the viewpoint of a zoologist*. New York, Columbia University Press.

Mayr, E. 1963. *Animal species and evolution*. Cambridge, Mass., Harvard University Press.

Mazel, A.D. 1992. Changing fortunes: 150 years of San hunter-gatherer history in the Natal Drakensberg, South Africa. *Antiquity* 66: 758–67.

Medway, Lord 1971. The Quaternary mammals of Malesia: a review. In P. Ashton and M. Ashton (eds.) *Transactions of the second Aberdeen-Hull Symposium on Malesian ecology*, 63–98. University of Hull, Department of Geography, Miscellaneous Series 13.

Meehan, B. 1971. The form, distribution and antiquity of Australian Aboriginal mortuary practices. Unpublished M.A. thesis. University of Sydney.
Meehan, B. and N. White (eds.) 1990. *Hunter-gatherer demography: past and present.* Oceania Monographs, University of Sydney.
Meek, R.L. (ed.) 1973. *Turgot: on progress, sociology and economics.* Cambridge, Cambridge University Press.
Meek, R.L. 1976. *Social science and the ignoble savage.* Cambridge, Cambridge University Press.
Meggers, B.J. 1978. Vegetational fluctuations and prehistoric cultural adaptation in Amazonia: some tentative correlations. *World Archaeology* 8(3): 287–303.
Mellars, P.A. 1973. The character of the Middle–Upper Palaeolithic transition in south-west France. In C. Renfrew (ed.) *The explanation of culture change: models in prehistory*, 255–76. London, Duckworth.
Mellars, P.A. 1976. Fire ecology, animal populations and man: a study of some ecological relationships in prehistory. *Proceedings of the Prehistoric Society* 42: 15–45.
Mellars, P.A. 1986. A new chronology for the French Mousterian period. *Nature* 322: 410–11.
Mellars, P.A. 1989a. Technological changes at the Middle–Upper Palaeolithic transition: economic, social and cognitive perspectives. In Mellars and Stringer 1989, 338–65.
Mellars, P.A. 1989b. Major issues in the emergence of modern humans. *Current Anthropology* 30: 349–85.
Mellars, P.A. (ed.) 1990. *The emergence of modern humans.* Edinburgh, Edinburgh University Press.
Mellars, P.A. and C. Stringer (eds.) 1989. *The human revolution: behavioural and biological perspectives on the origins of modern humans.* Edinburgh, Edinburgh University Press.
Meltzer, D. 1988. Late Pleistocene human adaptations in eastern North America. *Journal of World Prehistory* 2: 1–52.
Meltzer, D. 1989. Why don't we know when the first people came to North America? *American Antiquity* 54: 471–90.
Mercier, N., H. Valladas, J.-L. Joron, J.-L. Reyss, F. Lévêque and B. Vandermeersch 1991. Thermoluminescence dating of the late Neanderthal remains from Saint Césaire. *Nature* 351: 737–9.
Milankovich, M.M. 1941. *Canon of insolation and the ice age problem.* Koniglich Servische Akademie, Beograd.
Millar, J. 1771. *Observations concerning the distinction of ranks in society.* Edinburgh.
Milton, K. 1988. Foraging behaviour and the evolution of primate intelligence. In Byrne and Whiten 1988, 285–305.
Minc, L. 1986. Scarcity and survival: the role of oral tradition in mediating subsistence crises. *Journal of Anthropological Archaeology* 5: 39–113.
Misra, V.N. 1987. Middle Pleistocene adaptations in India. In O. Soffer (ed.) *The Pleistocene Old World: regional perspectives*, 99–120. New York, Plenum.
Mix, A.C. 1987. Hundred-kiloyear cycle queried. *Nature* 327: 370.
Mochanov, U.A. 1977. *The earliest stages of the peopling of northeastern Asia* (in Russian). Novosibirsk, Nauka.
Montagu, A. 1965. *The human revolution.* Cleveland and New York, World Publishing.
Montagu, A. 1972. *Statement on race* (3rd edn.). New York, Oxford University Press.
Montesquieu, Baron de C.L.S. 1748. *De l'Esprit des lois.* Paris.
Moore, F.C.T. (trans.) 1969. *The observation of savage peoples by Joseph-Marie Degérando (1800).* London, Routledge & Kegan Paul.
Morgan, E. 1982. *The aquatic ape: a theory of human evolution.* London, Souvenir Press.
Morgan, L.H. 1877. *Ancient society.* New York, World Publishing.
Morris, D. 1967. *The naked ape.* London, Cape.

Morton, S.G. 1839. *Crania Americana or a comparative view of the skulls of various aboriginal nations of North and South America*. Philadelphia, John Pennington.

Moseley, C.W.R.D. (trans.) 1983. *The travels of Sir John Mandeville*. Harmondsworth, Penguin.

Mountain, M.-J. 1991. Highland New-Guinea hunter-gatherers from the Pleistocene Nombe rockshelter, Simbu. Unpublished Ph.D. dissertation, Australian National University.

Movius, H.L. 1948. The Lower Paleolithic cultures of southern and eastern Asia. *Transactions of the American Philosophical Society* 38.

Movius, H.L. 1953. The Mousterian cave of Teshik-Tash, southeastern Uzbekistan, Central Asia. *American School of Prehistoric Research* 17: 11–71.

Movius, H.L. 1966. The hearths of the Upper Perigordian and Aurignacian horizons at the Abri Pataud, Les Eyzies (Dordogne), and their possible significance. *American Anthropologist* 68: 296–325.

Mulvaney, D.J. 1975. *The Prehistory of Australia*. 2nd edn. Melbourne, Pelican.

Mulvaney, D.J. 1976. "The chain of connection": the material evidence. In N. Peterson (ed.) *Tribes and boundaries in Australia*, 72–94. Canberra, A.I.A.S.

Murray, T. 1987. Remembrances of things present: appeals to authority in the history and philosophy of archaeology. Unpublished Ph.D. dissertation, University of Sydney.

Murray, T. 1989. The history, philosophy and sociology of archaeology: the case of the Ancient Monuments Protection Act (1882). In V. Pinsky and A. Wylie (eds.) *Critical traditions in contemporary archaeology*, 55–67. Cambridge, Cambridge University Press.

Murray, T. 1992. Tasmania and the constitution of "the dawn of humanity." *Antiquity* 66: 730–43.

Mussi, M. 1992. *Il Paleolitico e il Mesolitico in Italia*. Popoli e civilta dell' l'Italia, vol. 10. Bologna, Biblioteca di Storia Patria.

Muzzolini, A. 1989. La 'Néolithisation' du nord de l'Afrique et ses causes. In O. Aurenche and J. Cauvin (eds.) *Néolithisations*, 145–86. Oxford, BAR International Series 516.

Nelson, G. 1983. Vicariance and cladistics: historical perspectives with implications for the future. In R.W. Sims, J.H. Price and P.E.S. Whalley (eds.) *Evolution, time and space: the emergence of the biosphere*, 469–92. London, Academic Press.

Nelson, G. and N. Platnick 1981. *Systematics and biogeography: cladistics and vicariance*. New York, Columbia University Press.

Nelson, G. and D.E. Rosen (eds.) 1981. *Vicariance biogeography: a critique*. New York, Columbia University Press.

Nelson, R.K. 1969. *Hunters of the northern ice*. Chicago, University of Chicago Press.

Nelson, R.K. 1973. *Hunters of the northern forest*. Chicago, University of Chicago Press.

Nitecki, M.H. and D.V. Nitecki (eds.) 1987. *The evolution of human hunting*. New York, Plenum.

Noble, W. and I. Davidson 1991. The evolutionary emergence of modern human behaviour: language and its archaeology. *Man* 26: 223–53.

Nygaard, S. 1989. The stone age of northern Scandinavia: a review. *Journal of World Prehistory* 3: 71–116.

O'Connell, J.F. and K. Hawkes 1988. Hadza hunting, butchering, and bone transport and their archaeological implications. *Journal of Anthropological Research* 44: 113–61.

O'Connell, J.F., K. Hawkes and N. Blurton-Jones 1988. Hadza scavenging: implications for Plio/Pleistocene hominid subsistence. *Current Anthropology* 29: 356–63.

Oakley, K. 1949. *Man the tool maker*. London, Natural History Museum.

Oakley, K. 1957. Tools makyth man. *Antiquity* 31: 199–209.

Oakley, K.P., B.G. Campbell and T.I. Molleson 1975. *Catalogue of fossil hominids*. Part III: *Americas, Asia, Australasia*. London, British Museum (Natural History).

Odling-Smee, J. 1988. Niche constructing phenotypes. In H.C. Plotkin (ed.) *The role of behavior in evolution*, 73–132. Cambridge, Mass., MIT Press.

Onrubia Pintado, J. 1987. Les cultures préhistoriques des Iles Canaries: état de la question. *Anthropologie* 91: 653–78.
Osborn, H.F. 1902. The law of adaptive radiation. *American Naturalist* 36: 353–63.
Osborn, H.F. 1910. *The age of mammals in Europe, Asia and North America.* New York, Macmillan.
Osborn, H.F. 1915/1919. *Men of the Old Stone Age, their environment, life and art.* 1st and 3rd edns. London, G. Bell.
Oswalt, W.H. 1976. *An anthropological analysis of food-getting technology.* New York, Wiley.
Oxnard, C.E. 1987. *Fossils, teeth and sex.* Seattle, University of Washington Press.
Paddayya, K. 1977. An Acheulian occupation site at Hunsgi, peninsular India: a summary of the results of two seasons of excavation (1975–6) *World Archaeology* 8: 344–55.
Paddayya, K. 1982. *The Acheulian culture of the Hunsgi valley (peninsular India): a settlement system perspective.* Poona, Deccan College Postgraduate and Research Institute.
Page, R.E. 1989. Neotropical African bees. *Nature* 339: 181–2.
Pardoe, C. 1988. The cemetery as symbol. *Archaeology in Oceania* 23: 1–16.
Pardoe, C. 1990. The demographic basis of human evolution in southeastern Australia. In Meehan and White (eds.) 1990, 59–70.
Parkington, J. 1990. A view from the south: southern Africa before, during and after the last glacial maximum. In Gamble and Soffer 1990a, 214–280.
Parsons, P. 1983. *The evolutionary biology of colonizing species.* Cambridge, Cambridge University Press.
Passingham, R.E. 1982. *The human primate.* San Francisco, W.H. Freeman.
Paterson, H.E.H. 1981. The continuing search for the unknown and the unknowable: a critique of contemporary ideas on speciation. *South African Journal of Science* 77: 113–19.
Paterson, H.E.H. 1982. Perspective on speciation by reinforcement. *South African Journal of Science* 78: 33–7.
Paterson, H.E.H. 1985. The recognition concept of species. In E.S. Vrba (ed.) *Species and speciation*, 21–9. Pretoria, Transvaal Museum Monograph 4.
Paterson, H.E.H. 1986. Environment and species. *South African Journal of Science* 82: 62–5.
Patterson, C. 1983. Aims and methods in biogeography. In R.W. Sims, J.H. Price and P.E.S. Whalley (eds.) *Evolution, time and space: the emergence of the biosphere*, 1–28. London, Academic Press.
Pei, W.C. 1939. A preliminary study of a new Palaeolithic station known as locality 15 within the Choukoutien region. *Bulletin of the Geological Society of China* 29: 147–87.
Penck, A. and E. Brückner 1909. *Die Alpen in Eiszeitalter.* Leipzig.
Perretto, C. 1989. Aspects et problems du premier peuplement d'Italie. *Proceedings of the 2nd International Congress of Human Paleontology, Turin 1987*, 373–7. Turin, Jaca Books.
Peyrony, D. 1939. Le Compte Begouën en Périgord. In *Mélanges de préhistoire et d'anthropologie offerts par ses collègues, amis et disciples au Professeur Compte H. Begouën*, 235–41. Toulouse, Editions du Museé.
Pfeiffer, J.E. 1978. *The emergence of man.* 3rd edn. New York, Harper & Row.
Pfeiffer, J.E. 1982. *The creative explosion.* New York, Harper & Row.
Pianka, E.R. 1978. *Evolutionary ecology.* 2nd edn. New York, Harper & Row.
Piggott, S. 1965. *Ancient Europe.* Edinburgh, Edinburgh University Press.
Pilbeam, D. 1984. The descent of hominoids and hominids. *Scientific American* 250: 60–9.
Pilbeam, D. 1985. Patterns of hominoid evolution. In E. Delson (ed.) *Ancestors: the hard evidence*, 51–9. New York, Alan R. Liss.
Pitt-Rivers, A.H.L. Fox 1891. Typological museums. *Journal of the Society of Arts* 40: 115–22.

Pitulko, V. and V. Makeyev 1991. Ancient arctic hunters. *Nature* 349: 374.
Pope, G.G. and J.E. Cronin 1984. The Asian *Hominidae*. *Journal of Human Evolution* 13: 377–96.
Potts, R. 1984. Hominid hunters? Problems of identifying the earliest hunter/gatherers. In R. Foley (ed.) *Hominid evolution and community ecology*, 129–66. New York, Academic Press.
Potts, R. 1988. *Early hominid activities at Olduvai*. New York, Aldine.
Potts, R. and P. Shipman 1981. Cutmarks made by stone tools on bones from Olduvai Gorge, Tanzania. *Nature* 291: 577–80.
Powers, W.R. 1973. Palaeolithic man in northeast Asia. *Arctic Anthropology* 10: 1–106.
Powers, W.R. and J.F. Hoffecker 1989. Late Pleistocene settlement in the Nenana Valley, central Alaska. *American Antiquity* 54: 263–87.
Praslov, N.D. (ed.) 1982. *Ketrosy*. Moscow, Nauka.
Praslov, N.D. and A.N. Rogachev (eds.) 1982. *Palaeolithic of the Kostenki-Borshevo area on the Don river, 1879–1979*. Leningrad, Nauka.
Prentice, M.L. and G.H. Denton 1988. The deep-sea oxygen record, the global ice sheet system and hominid evolution. In Grine 1988, 383–403.
Prichard, J.C. 1813. *Researches into the physical history of man*. Chicago, University of Chicago Press. Reprinted 1973.
Prichard, J.C. 1841. *The natural history of man*. 4th edn. London.
Propp, V. 1928. *Morphology of the folktale*. Austin, University of Texas Press. (Reprint 1968.)
Quatrefages, A. de 1879. *The human species*. London, C. Kegan Paul.
Ranov, V.A. 1984. Zentralasien. In neue Forschungen zur Altsteinzeit. *Forschungen zur Allgemeinen und Vergleichenden Archäologie* Band 4: 299–343.
Ranov, V.A. 1991. Les sites très anciens de l'âge de la pierre en U.R.S.S. In E. Bonifay and B. Vandermeersch (eds.) *Les premiers peuplements humains de l'Europe*, 209–16. Actes du 11 4e Congrès National de Sociétés Savantes, Paris.
Ratzel, F. 1896. *The history of mankind*. 3 vols. trans. A.J. Butler. London, Macmillan.
Read-Martin, C.E. and D.W. Read 1975. Australopithecine scavenging and human evolution: an approach from faunal analysis. *Current Anthropology* 16: 359–68.
Reader, J. 1981. *Missing links*. London, History Book Club.
Reese, D. 1989. Tracking the extinct pygmy hippopotamus of Cyprus. *Field Museum of Natural History Bulletin* 60: 22–9.
Reeves, B.O.K. 1983. Bergs, barriers and Beringia: reflections on the peopling of the New World. In P. Masters and N. Fleming (eds.) *Quaternary coastlines*, 389–411. London, Academic Press.
Renfrew, C. and P. Bahn 1991. *Archaeology*. London, Thames and Hudson.
Renfrew, C. and T. Poston 1979. Discontinuities in the endogenous change of settlement pattern. In C. Renfrew and K.L. Cooke (eds.) *Transformations: mathematical approaches to cultural change*, 437–61. New York, Academic Press.
Reynolds, T.E.G. 1985. The early Palaeolithic of Japan. *Antiquity* 59: 93–6.
Reynolds, T.E.G. and S.C. Kaner 1990. Japan and Korea at 18 000 BP. In Soffer and Gamble 1990, 296–311.
Richards, G. 1986. Freed hands or enslaved feet? *Journal of Human Evolution* 15: 143–50.
Richards, G. 1987. *Human evolution: an introduction for the behavioural sciences*. London, Routledge.
Richards, G. 1989. Human behavioural evolution: a physiomorphic model. *Current Anthropology* 30: 244–55.
Richter, J. 1989. Im Zeichen der Giraffe: Sammler-Jäger-Maler in Namibia. *Archäologie in Deutschland* 2: 38–41.
Ridley, M. 1986. *Evolution and classification: the reformation of cladism*. Oxford, Blackwells.

Rightmire, P. 1987. L'evolution des premier hominides en Asie du Sud-Est. *Anthropologie* 91: 455–66.

Roberts, M.B. 1986. Excavation of the Lower Palaeolithic site at Amey's Eartham Pit, Boxgrove, West Sussex: a preliminary report. *Proceedings of the Prehistoric Society* 52: 215–46.

Roberts, N. 1984. Pleistocene environments in time and space. In R. Foley (ed.) *Hominid evolution and community ecology*, 25–54. London, Academic Press.

Roberts, R.G., R. Jones and M.A. Smith 1990. Thermoluminescence dating of a 50 000 year old human occupation site in northern Australia. *Nature* 345: 153–6.

Robinson, J.T. 1954. Prehominid dentition and hominid evolution. *Evolution* 8: 324–34.

Roebroeks, W. 1988. *From find scatters to early hominid behaviour: a study of Middle Palaeolithic riverside settlements at Maastricht-Belvédère (The Netherlands)*. Analecta Praehistorica Leidensia 21, University of Leiden.

Roebroeks, W., N.J. Conard and T. van Kolfschoten 1992. Dense forests, cold steppes, and the Palaeolithic settlement of northern Europe. *Current Anthropology* 33: 551–86.

Roebroeks, W., J. Kolen and E. Rensink 1988. Planning depth, anticipation and the organization of Middle Palaeolithic technology: the "archaic natives" meet Eve's descendants. *Helinium* 28: 17–34.

Rolland, N. and H. Dibble 1990. A new synthesis of Middle Paleolithic variability. *American Antiquity* 55: 480–99.

Ronen, A., M. Lamdan and L. Shmookler n.d. Pre-upper Palaeolithic occurrences of isotope stage 5. Unpublished paper, Tel Aviv.

Rosenfeld, A., D.A. Horton and J. Winter 1981. Early man in North Queensland: art and archaeology in the Laura area. Canberra, Department of Prehistory, Research School of Pacific Studies, Australian National University. *Terra Australis* 6.

Ross, A., T. Donnelly and R. Wasson 1992. The peopling of the arid zone: human-environment interactions. In J. Dodson (ed.) *The naive lands: prehistory and environmental change in Australia and the southwest Pacific*, 76–114. Melbourne, Longman.

Rouse, I. 1986. *Migrations in prehistory: inferring population movement from cultural remains*. New Haven, Yale University Press.

Rouse, I. and L. Allaire 1978. The Caribbean. In R.E. Taylor and C.W. Meighan (eds.) *Chronologies in New World Archeology*, 432–81. New York, Academic Press.

Rowland, M.J. 1984. A long way in a bark canoe: Aboriginal occupation of the Percy Isles. *Australian Archaeology* 18: 17–31.

Rowland, M.J. 1987. The distribution of Aboriginal watercraft on the east coast of Queensland: implications for culture contact. *Australian Aboriginal Studies* 2: 38–45.

Ruddiman, W.F. and M.E. Raymo 1988. Northern Hemisphere climate régimes during the past 3Ma: possible tectonic connections. In N.J. Shackleton, R.G. West and D.Q. Bowen (eds.) *The past three million years: evolution of climatic variability in the North Atlantic region*, 1–20. London, Royal Society.

Ryan, L. 1981. *The aboriginal Tasmanians*. St. Lucia, University of Queensland Press.

Santonja, M. and P. Villa 1990. The lower Palaeolithic of Spain and Portugal. *Journal of World Prehistory* 4: 45–94.

Sarich, V. 1982. Comment on M.H. Wolpoff, *Ramapithecus* and hominid origins. *Current Anthropology* 23: 513–4.

Sarich, V.M. and A.C. Wilson 1967. Immunological time scale for hominid evolution. *Science* 158: 1200–3.

Schaller, G.B. 1977. *Mountain monarchs*. Chicago, University of Chicago Press.

Scheidt, W. 1924–5. The concept of race in anthropology and the divisions into human races from Linnaeus to Deniker. In Count 1950, 354–91.

Schmider, B. 1990. The last Pleniglacial in the Paris Basin (22 500–17 000 BP). In Soffer and Gamble 1990, 41–53.

Schötensack, O. 1901. Die Bedeutung Australiens für die Heranbildung des Menschen aus einer niederen Form. *Zeitschrift für Ethnologie* 33: 127.

Schrire, C. (ed.) 1984. *Past and present in hunter-gatherer studies*. London, Academic Press.

Scott, K. 1980. Two hunting episodes of Middle Palaeolithic age at La Cotte de Saint-Brelade, Jersey (Channel Islands). *World Archaeology* 12:137–52.

Scott, K. 1986. The bone assemblage from layers 3 and 6. In Callow and Cornford 1986, 159–84.

Serizawa, C. 1978. The stone age of Japan. *Asian Perspectives* 19: 1–14.

Serizawa, C. 1986. The Palaeolithic age of Japan in the context of East Asia: a brief introduction. In R.J. Pearson, G.L. Barnes and K.L. Hutterer (eds.) *Windows on the Japanese past: studies in archeology and prehistory*, 191–7. Ann Arbor, Center for Japanese Studies.

Shackleton, N.J. and N.D. Opdyke 1973. Oxygen isotope and palaeomagnetic stratigraphy of equatorial Pacific core V28–238. *Quaternary Research* 3: 39–55.

Shackley, M. 1983. *Wildmen: Yeti, Sasquatch and the Neanderthal enigma*. London, Thames and Hudson.

Shackley, M. 1985. Palaeolithic archaeology of the central Namib Desert. *Cimbebasia* Memoir 6.

Shapiro, H.L. 1976. *Peking Man*. London, Allen & Unwin.

Sharma, K.K. 1984. The sequence of uplift of the Himalaya. In R. Whyte (ed.) *The evolution of the East Asian environment*, 56–70. University of Hong Kong, Centre of Asian Studies.

Sharp, A. (ed.) 1970. *The journal of Jacob Roggeveen*. Oxford, Clarendon Press.

Shennan, S.J. (ed.) 1989. *Archaeological approaches to cultural identity*. London, Routledge.

Shipman, P. 1986. Scavenging or hunting in early hominids: theoretical framework and tests. *American Anthropologist* 88: 27–43.

Shipman, P. and J.M. Harris 1988. Habitat preference and paleoecology of *Australopithecus boisei* in eastern Africa. In Grine 1988, 343–82.

Shipman, P. and J. Rose 1983. Evidence of butchery and hominid activities at Torralba and Ambrona: an evaluation using microscopic techniques. *Journal of Archaeological Science* 10: 465–74.

Sieff, D. 1990. Explaining biased sex ratios in human populations. *Current Anthropology* 31: 25–48.

Simek, J. 1987. Spatial order and behavioural change in the French Palaeolithic. *Antiquity* 61: 25–40.

Simmons, A.H. 1988. Extinct pygmy hippopotamus and early man in Cyprus. *Nature* 333: 554–7.

Simpson, G.G. 1940. Mammals and land bridges. *Journal of the Washington Academy of Sciences* 30: 137–63.

Simpson, G.G. 1952. Probabilities of dispersal in geologic time. *Bulletin of the American Museum of Natural History* 99: 163–76.

Singer, R and J. Wymer, 1982. *The Middle Stone Age at Klasies River Mouth in South Africa*. Chicago, University of Chicago Press.

Skelton, R.R. and H.M. McHenry 1992. Evolutionary relationships among early hominids. *Journal of Human Evolution* 23: 309–49.

Smirnov, Y.A. 1989. Intentional human burial: Middle Palaeolithic (Last Glaciation) beginnings. *Journal of World Prehistory* 3: 199–233.

Smirnov, Y.A. 1991. *Mousterian burials* (in Russian). Moscow, Nauka.

Smith, D.R., O.R. Taylor and W.M. Brown 1989. Neotropical Africanized honey bees have African mitochondrial DNA. *Nature* 339: 213–15.

Smith, F. 1926. *Prehistoric man and the Cambridge gravels*. Cambridge, Heffers.

Smith, F.H. 1984. Fossil hominids from the Upper Pleistocene of central Europe and the origin of modern Europeans. In F.H. Smith and F. Spencer (eds.) *The origins of modern humans: a world survey of the fossil evidence*, 137–209. New York, Alan Liss.

Smith, G.E. 1929. *The migrations of early culture*. Manchester, Manchester University Press.

Smith, G.E. 1933. *The diffusion of culture*. London, Watts.

Smith, M.A. 1987. Pleistocene occupation in arid central Australia. *Nature* 328: 710–11.

Smith, M.A. 1989. The case for a resident human population in the central Australian ranges during full glacial aridity. *Archaeology in Oceania* 24: 93–105.

Smith, M.A. and N.D. Sharp 1993a (in press). A revised bibliography of Pleistocene archaeological sites in Australia, New Guinea and island Melanesia. In M.A. Smith (ed.) *Sahul in review*. Canberra, ANU Press.

Smith, M.A. and N.D. Sharp 1993b (in press). Pleistocene sites in Australia, New Guinea and island Melanesia: geographic and temporal structure of the archaeological record. In M.A. Smith (ed.) *Sahul in review*. Canberra, ANU Press.

Soffer, O. 1985a. Patterns of intensification as seen from the Upper Paleolithic of the central Russian plain. In T.D. Price and J.A. Brown (eds.) *Prehistoric hunter-gatherers: the emergence of cultural complexity*, 235–70. New York, Academic Press.

Soffer, O. 1985b. *The Upper Paleolithic of the central Russian plain*. New York, Academic Press.

Soffer, O., and C.S. Gamble (eds.) 1990. *The world at 18 000 BP*. Volume 1: *High latitudes*. London, Unwin Hyman.

Solecki, R.S. 1971. *Shanidar—the first flower people*. New York, Knopf.

Sollas, W.J. 1911. *Ancient hunters and their modern representatives*. London, Macmillan.

Spahni, J.-C. 1964. Les gisements à Ursus spelaeus de l'Autriche et leurs problèmes. *Bulletin de la Société Préhistorique Française* 61: 346–67.

Spencer, F. 1990. *Piltdown, a scientific forgery*. London, Natural History Museum.

Spencer, H. 1876–96. *The principles of sociology*. 3 vols. London.

Speth, J.D. 1983. *Bison kills and bone counts*. Chicago, University of Chicago Press.

Speth, J.D. 1987. Early hominid subsistence strategies in seasonal habitats. *Journal of Archaeological Science* 14: 13–29.

Speth, J.D. and K. Spielmann 1983. Energy source, protein metabolism and hunter-gatherer subsistence strategies. *Journal of Anthropological Archaeology* 2: 1–31.

Stanford, D.J. and J.S. Day (eds.) 1992. *Ice age hunters of the Rockies*. Denver, University Press of Colorado.

Stanko, V.N., G. Grigorieva and T. Shvlako 1989. *Anetovka II*. Kiev, Nauka Dumka.

Stanley, S.M. 1979. *Macroevolution: pattern and process*. San Francisco, W.H. Freeman.

Steele, J. 1989 Hominid evolution and primate social cognition. *Journal of Human Evolution* 18: 421–32.

Steklis, H. 1985. Primate communication, comparative neurology, and the origin of language reconsidered. *Journal of Human Evolution* 14: 157–73.

Stern, J.T. and R.L. Susman 1983. The locomotor anatomy of *Australopithecus afarensis*. *American Journal of Physical Anthropology* 60: 279–317.

Stiner, M.C. 1991. A taphonomic perspective on the origins of the faunal remains of Grotta Guattari (Latium, Italy). *Current Anthropology* 32: 103–17.

Stocking, G.W. 1968. *Race, culture, and evolution; essays in the history of anthropology*. New York, Free Press.

Stocking, G.W. 1987. *Victorian anthropology*. New York, Free Press.

Stoddart, D.R. 1984. Scientific studies in the Seychelles. In D.R. Stodder (ed.) *Biogeography and ecology of the Seychelles Islands*, 1–15. The Hague, Junk.

Stoneking, M. and R. Cann 1989. African origin of human mitochondrial DNA. In P. Mellars and C. Stringer (eds.) *The human revolution: behavioural and biological perspectives on the origins of modern humans*, 17–30. Edinburgh, Edinburgh University Press.

Stratz den Haag, C.H. 1904. The problem of classifying mankind into races. In Count 1950, 230–8.

Straus, L.G. 1987. Upper Palaeolithic ibex hunting in S.W. Europe. *Journal of Archaeological Science* 14: 163–78,

Straus, W.E. and A.J.E. Cave 1957. Pathology and posture of Neanderthal man. *Quarterly Review of Biology* 32: 348–63.

Stringer, C. 1984. Hominid evolution and biological adaptation in the Pleistocene. In R. Foley (ed.) *Human evolution and community ecology*, 55–83. London, Academic Press.

Stringer, C. 1988. The dates of Eden. *Nature* 331: 565–6.

Stringer, C. and P. Andrews 1988. Genetic and fossil evidence for the origin of modern humans. *Science* 239: 1263–68.

Stringer, C. and C. Gamble 1993. *In search of the Neanderthals: solving the puzzle of human origins*. London, Thames and Hudson.

Stringer, C.B., J.-J. Hublin and B.Vandermeersch 1984. The origin of anatomically modern humans in western Europe. In F.H. Smith and F. Spencer (eds.) *The origins of modern humans: a world survey of the fossil evidence*, 51–135. New York, Alan Liss.

Strum, S.C. and W. Mitchell 1986. Baboon models and muddles. In W.G. Kinzey (ed.) *The evolution of human behavior: primate models*, 87–104. Albany, State University of New York Press.

Susman, R.L. 1988. Hand of Paranthropus robustus from member 1, Swartkrans: fossil evidence for tool behaviour. *Science* 240: 781–4.

Susman, R.L., J.T. Stern and W.L. Jungers 1984. Arboreality and bipedality in the Hadar hominids. *Folia Primatologia* 43: 113–56.

Sutcliffe, A.J. 1985. *On the track of ice age mammals*. London, Natural History Museum.

Suzuki, H. and K. Hanihara 1982. *The Minatogawa man: the Upper Pleistocene man from the island of Okinawa*. University Museum, University of Tokyo Bulletin 19.

Suzuki, H. and F. Takai 1970. *The Amud man and his cave*. Tokyo, University of Tokyo Press.

Suzuki, M. 1974. Chronology of prehistoric human activity in Kanto, Japan. *Journal of the Faculty of Science University of Tokyo* 4: 395–469.

Svoboda, J. 1987. Lithic industries of the Arago, Vértesszöllös, and Bilzingsleben hominids: comparison and evolutionary interpretation. *Current Anthropology* 28: 219–27.

Svoboda, J. 1989. Middle Pleistocene adaptation in central Europe. *Journal of World Prehistory* 3: 33–70.

Symons, D. 1979. *The evolution of human sexuality*. New York, Oxford University Press.

Szabo, B.J., W.P. McHugh, G.G. Schaber, C.V. Haynes and C.S. Breed 1989. Uranium-series dated authigenic carbonates and Acheulian sites in southern Egypt. *Science* 243: 1053–6.

Tarling, D.H. and M.P. Tarling, 1971. *Continental drift, a study of the earth's moving surface*. London, G. Bell.

Tattersall, I. 1986. Species recognition in human palaeontology. *Journal of Human Evolution* 15: 165–75.

Taylor, G. 1927. *Environment and race: a study of the evolution, migration, settlement, and status of the races of man*. London, Oxford University Press.

Terrell, J. 1986. *Prehistory in the Pacific Islands*. Cambridge, Cambridge University Press.

Thistleton-Dyer, W. 1909. Geographical distribution of plants. In A.C. Seward (ed.) *Darwin and modern science*, 298–318. Cambridge, Cambridge University Press.

Thorne, A.G. 1981. The centre and the edge: the significance of Australian hominids to African palaeoanthropology. In R. Leakey and B.A. Ogot (eds.) *Proceedings of the 8th Panafrican Congress of Prehistory and Quaternary Studies*, 180–1. Nairobi, International Louis Leakey Memorial Institute for African Prehistory.

Thorne, A.G. 1984. Australia's human origins: how many sources? *American Journal of Physical Anthropology* 63: 227.

Thorne, A.G. and M.H. Wolpoff 1981. Regional continuity in Australasian Pleistocene hominid evolution. *American Journal of Physical Anthropology* 55: 337–49.

Tobias, P. 1992. Piltdown: an appraisal of the case against Sir Arthur Keith. *Current Anthropology* 33(3): 243–93.

Tode, A. (ed.) 1953. Die Untersuchung der paläolithischen Freilandstation von Salzgitter-Lebenstedt. *Eiszeitalter und Gegenwart* 3: 144–220.

Torrence, R. 1983. Time budgeting and hunter-gatherer technology. In G.N. Bailey (ed.) *Hunter-gatherer economy in prehistory*, 11–22. Cambridge, Cambridge University Press.

Torrence, R. 1986. *Production and exchange of stone tools*. Cambridge, Cambridge University Press.

Torrence, R. 1989a. Tools as optimal solutions. In R. Torrence (ed.) *Time, energy and stone tools*, 1–6. Cambridge, Cambridge University Press.

Torrence, R. 1989b. Retooling: towards a behavioural theory of stone tools. In R. Torrence (ed.) *Time, energy and stone tools*, 57–66. Cambridge, Cambridge University Press.

Trinkaus, E. 1981. Neanderthal limb proportions and cold adaptation. In C. Stringer (ed.) *Aspects of human evolution*, 187–224. London, Taylor & Francis.

Trinkaus, E. 1983. *The Shanidar Neanderthals*. New York, Academic Press.

Trinkaus, E. (ed.) 1989. *The emergence of modern humans: biocultural adaptations in the later Pleistocene*. Cambridge, Cambridge University Press.

Trinkaus, E. 1992. Paleontological perspectives on Neanderthal behaviour. In M. Toussaint (ed.) 5 millions d'années, l'aventure humaine. *Études et Recherches Archéologiques de l'Université de Liège* 56: 151–76.

Trinkaus, E. and W.W. Howells 1979. The Neanderthals. *Scientific American* 241: 94–105.

Trinkaus, E. and P. Shipman 1993. *The Neandertals*. New York, Knopf.

Trivers, R.L. and D.E. Willard 1973. Natural selection of parental ability to vary the sex ratio of offspring. *Science* 179: 90–2.

Tseitlin, S.M. 1979. *Geology of the Palaeolithic of northern Asia* (in Russian). Moscow, Nauka.

Tuffreau, A. and J. Sommé (eds.) 1988. Le gisement Paléolithique moyen de Biache-St.-Vaast (Pas-de-Calais). *Mémoirs de la Société Préhistorique Française* 21.

Turgot, A.R.J. 1751. On universal history. In R. Meek 1973, 61–118.

Turner, A. 1984. Hominids and fellow travellers: human migration into high latitudes as part of a large mammal community. In R. Foley (ed.) *Hominid evolution and community ecology*, 193–217. London, Academic Press.

Turner, A. 1990. The evolution of the guild of larger terrestrial carnivores during the Plio-Pleistocene in Africa. *Geobios* 23: 349–68.

Turner, A. 1992. Large carnivores and earliest European hominids: changing determinants of resource availability during the Lower and Middle Pleistocene. *Journal of Human Evolution* 22: 109–26.

Turner, A. and A.T. Chamberlain 1989. Speciation, morphological change and the status of African *Homo erectus*. *Journal of Human Evolution* 18: 115–30.

Turner, A. and H.E.H. Paterson 1991. Species and speciation: evolutionary tempo and mode in the fossil record reconsidered. *Geobios* 24: 761–9.

Turner, C. 1992. New World origins: new research from the Americas and the Soviet Union. In D.J. Stanford and J.S. Day (eds.) 1992. *Ice age hunters of the Rockies*, 7–50. Denver, University Press of Colorado.

Tylor, E.B. 1881. *Anthropology: an introduction to the study of man and civilisation*. London, Macmillan.

Udvardy, M.D.F. 1969. *Dynamic zoogeography, with special reference to land animals*. New York, Van Nostrand.

Valladas, H., H. Cachier, P. Maurice, F. Bernaldo de Quiros, J. Clottes, V. Cabrera Valdés, P. Uzquiano and M. Arnold 1992. Direct radiocarbon dates for prehistoric paintings at the Altamira, El Castillo and Niaux caves. *Nature* 357: 68–70.

Valladas, H., J.L. Reyss, J.L. Joron, G. Valladas, O. Bar-Yosef and B. Vandermeersch 1988. Thermoluminescence dating of Mousterian 'Proto-Cro-Magnon' remains from Israel and the origin of modern man. *Nature* 331: 614–16.

van Couvering, J.A.H. and J.A. van Couvering 1976. Early Miocene mammal faunas from East Africa: aspects of geology, faunistics and palaeoecology. In G. Isaac and E. McCown (eds.) *Human origins: Louis Leakey and the East African evidence*, 155–207. Menlo Park, Benjamin.

van Couvering, J.A.H. 1980. Community evolution in East Africa during the Late Cenozoic. In A.K. Behrensmeyer and A.P. Hill (eds.) *Fossils in the making*, 272–97. Chicago, Chicago University Press.

Vandermeersch, B. 1981. *Les hommes fossiles de Qafzeh (Isräel)*. Paris, C.N.R.S.

Vandermeersch, B. 1989. The evolution of modern humans: recent evidence from southwest Asia. In Mellars and Stringer 1989, 155–64.

Vereshschagin, N.K. and G.F. Baryshnikov 1982. Palaeoecology of the mammoth fauna in the Eurasian Arctic. In D.M. Hopkins et al. 1982, 1267–79.

Vigilant, L., M. Stoneking, H. Harpending, K. Hawkes and A.C. Wilson 1991. African populations and the evolution of human mitochondrial DNA. *Science* 253: 1503–7.

Villa, P. 1982. Conjoinable pieces and site formation processes. *American Antiquity* 47: 276–90.

Vitebsky, P. 1989. Siberian conference on the cradle of mankind. *Polar Record* 25: 149.

von Koenigswald, G.H.R. 1956. *Meeting prehistoric man*. London, Scientific Bookclub.

von Koenigswald, G.H.R., W. H.-J. Müller-Beck and E. Pressmar 1974. *Die Archäologie und Paläontologie in der Weinberghölen bei Mauern (Bayern)*. Tübingen, Archaeologica Venatoria 3.

Vonnegut, K. 1987. *Galápagos*. London, Grafton Books.

Vrba, E.S. 1980. Evolution, species and fossils: how does life evolve? *South African Journal of Science* 76: 61–84.

Vrba, E.S. 1983. Macroevolutionary trends: new perspectives on the roles of adaptation and incidental effect. *Science* 221: 387–9.

Vrba, E.S. 1984. Patterns in the fossil record and evolutionary processes. In M.W. Ho and P.S. Saunders (eds.) *Beyond neo-Darwinism*, 115–42. London, Academic Press.

Vrba, E.S. 1985a. Ecological and adaptive changes associated with early hominid evolution. In E. Delson (ed.) *Ancestors: the hard evidence*, 63–71. New York, Alan R. Liss.

Vrba, E.S. 1985b. Environment and evolution: alternative causes of the temporal distribution of evolutionary events. *South African Journal of Science* 81: 229–36.

Vrba, E.S. 1988. Late Pliocene climatic events and hominid evolution. In Grine 1988, 405–26.

Waal, F. de 1982. *Chimpanzee politics*. London, Cape (selection reprinted in Byrne and Whiten 1988, 122–31).

Waal, F. de 1991. *Peacemaking among primates*. London, Penguin.

Wainscoat, J.S. 1987. Out of the garden of Eden. *Nature* 325: 13.

Wainscoat, J.S., A.V.S. Hill, S.L. Thein, J. Flint, J.C. Chapman, D.J. Weatherall, J.B. Clegg and D.R. Higgs 1989. Geographic distribution of Alpha- and Beta-Globin gene cluster polymorphisms. In Mellars and Stringer 1989, 31–8.

Walker, A.C., R.E. Leakey, J.M. Harris and F.H. Brown 1986. 2.5–Myr *Australopithecus boisei* from west of Lake Turkana, Kenya. *Nature* 322: 517–22.

Walker, A.C. and R.E.F. Leakey 1978. The hominids of East Turkana. *Scientific American* 239: 54–66.

Wallace, A.R. 1864. The origin of human races and the antiquity of man deduced from the theory of "Natural Selection." *Journal of the Anthropological Society of London* 2: 158–87.

Wallace, A.R. 1876. *The geographical distribution of animals, with a study of the relations of living and extinct faunas as elucidating the past changes of the earth's surface*. London, Macmillan.

Wallace, A.R. 1880. *Island life: or the phenomena and causes of insular faunas and floras, including a revision and attempted solution of the problem of geological climates*. London, Macmillan.

Wallace, B. 1981. *Basic population genetics*. New York, Columbia University Press.

Wang Chiyuen, Shi Yaolin and Zhou Wenhu 1982. Dynamic uplift of the Himalaya. *Nature* 298: 553–6.

Warner, B.G., R.W. Mathewes and J.J. Clague 1982. Ice free conditions on the Queen Charlotte Islands, British Columbia, at the height of Late Wisconsin Glaciation. *Science* 218: 675–7.

Washburn, S.L. 1967. Behaviour and the origin of man. *Proceedings of the Royal Anthropological Institute of Great Britain and Ireland for 1967*: 21–7.

Washburn, S.L. 1982. The evolution of man. In L.F. Laporte (ed.) *The fossil record and evolution: readings from* Scientific American, 182–90. San Francisco, W.H. Freeman.

Washburn, S.L. and I. DeVore 1961. Social behaviour of baboons and early man. In S.L. Washburn (ed.) *Social life of early man*, 91–105. New York, Wenner Gren.

Webb, S.G. 1987. A palaeodemographic model of late Holocene central Murray Aboriginal society, Australia. *Human Evolution* 2: 385–406.

Webb, S.G. 1989. *The Willandra lakes hominids*. Canberra, Department of Prehistory, Research School of Pacific Studies, Australian National University. *Occasional Papers in Prehistory* 16.

Weidenreich, F. 1943. The skull of *Sinanthropus pekinensis*: a comparative study of a primitive hominid skull. *Palaeontologia Sinica* (n.s.D) 10. Beijing, Geological Survey of China.

Weidenreich, F. 1947. The trend of evolution. *Evolution* 1: 221–36.

Weiner, J.S., K.P. Oakley, W.E. Le Gros Clark 1953. The solution to the Piltdown problem. *Bulletin of the British Museum Natural History* 2: 141–6.

Wells, H.G. 1921. *The grisly folk*. (Reprinted in H.G. Wells *Selected short stories* 1958.) Harmondsworth, Penguin.

Wendorf, F., A.E. Close and R. Schild 1987. Recent work on the Middle Palaeolithic of the Eastern Sahara. *African Archaeological Review* 5: 49–63.

Whallon, R. 1989. Elements of cultural change in the later Palaeolithic. In Mellars and Stringer 1989, 433–54.

Wheeler, P.E. 1984. The evolution of bipedality and loss of functional body hair in hominids. *Journal of Human Evolution* 13: 91–8.

Wheeler, P.E. 1985. The loss of functional body hair in man: the influence of thermal environment, body form and bipedality. *Journal of Human Evolution* 14: 23–8.

Wheeler, P.E. 1988. Stand tall and stay cool. *New Scientist* (12 May): 62–5.

White, C. 1799. *An account of the regular gradation in man, and in different animals and vegetables; and from the former to the latter*. London.

White, J.P. 1987. Islands: mirrors or mirages of continents? Paper delivered at the S.A.A. Meetings, Toronto.

White, J.P. and J. O'Connell 1982. *A prehistory of Australia, New Guinea and Sahul*. Sydney, Academic Press.

White, M.J.D. 1978. *Modes of speciation*. San Francisco, W.H. Freeman.

White, R. 1985. Thoughts on social relationships and language in hominid evolution. *Journal of Social and Personal Relationships* 2: 95–115.

White, R. 1989. Production complexity and standardization in early Aurignacian bead and pendant manufacture: evolutionary implications. In Mellars and Stringer 1989, 366–90.

White, T.D. 1980. Evolutionary implications of Pliocene hominid footprints. *Science* 208: 175–6.

White, T.D. and N. Toth 1991. The question of ritual cannibalism at Grotta Guattari. *Current Anthropology* 32: 118–38.

Whiten, A. and R. Byrne 1988. Taking (Machiavellian) intelligence apart. In Byrne and Whiten 1988, 50–65.

Wickler, S. and M. Spriggs 1988. Pleistocene human occupation of the Solomon Islands, Melanesia. *Antiquity* 62: 703–6.

Wiessner, P. 1984. Reconsidering the behavioural basis for style: a case study among the Kalahari San. *Journal of Anthropological Archaeology* 3: 190–234.

Williams, E. 1987. Complex hunter-gatherers: a view from Australia. *Antiquity* 61: 310–21.

Williams, G.C. 1979. The question of adaptive sex ratio in outcrossed vertebrates. *Proceedings of the Royal Society of London* B 205: 567–80.

Willner, L.A. and R.D. Martin 1985. Some basic principles of mammalian sexual dimorphism. In J. Ghesquiere, R.D. Martin and F. Newcombe (eds.) *Human sexual dimorphism*, 1–42. Symposia of the Society for the Study of Human Biology, Volume 24. London and Philadelphia, Taylor & Francis.

Wilmsen, E.N. 1989. *Land filled with flies: a political economy of the Kalahari*. Chicago, University of Chicago Press.

Wilson, E.O. 1975. *Sociobiology: the new synthesis*. Cambridge, Mass., Bellknap Press.

Wobst, H.M. 1974. Boundary conditions for Paleolithic social systems: a simulation approach. *American Antiquity* 39: 147–78.

Wobst, H.M. 1977. Stylistic behavior and information exchange. In C.E. Cleland (ed.) *Papers for the director: research essays in honor of James B. Griffin*, 317–42. Anthropological papers, Museum of Anthropology, University of Michigan No. 61.

Wobst, H.M. 1990. Minitime and megaspace in the Palaeolithic at 18K and otherwise. In Soffer and Gamble 1990, 331–43.

Wolf, E. 1982. *Europe and the people without history*. London, University of California Press.

Wolpoff, M.H. 1982. *Ramapithecus* and hominid origins. *Current Anthropology* 23: 501ff.

Wolpoff, M.H. 1989. Multiregional evolution: the fossil alternative to Eden. In Mellars and Stringer 1989, 62–108.

Wolpoff, M.H. et al. 1988. Modern human origins. *Science* 241: 772–3.

Wood, B.A. 1987. Who is the 'real' *Homo habilis*? *Nature* 327: 187–8.

Wood, B.A. 1991. *Koobi Fora research project IV: hominid cranial remains from Koobi Fora*. Oxford, Clarendon Press.

Wood, B.A. 1992a. Origin and evolution of the genus *Homo*. *Nature* 355: 783–90.

Wood, B.A. 1992b. Old bones match old stones. *Nature* 355: 678–9.

Wood, B.A. and A.T. Chamberlain 1988. The nature and affinities of the "robust" Australopithecines: a review. *Journal of Human Evolution* 16: 625–41.

Wood, J.G. 1870. *The natural history of man; an account of the manners and customs of the uncivilized races of man*. Volume 2: *Australia, New Zealand, Polynesia, America, Asia, and Ancient Europe*. London, Routledge.

Woodman, P.C. 1986. Problems in the colonisation of Ireland. *Ulster Journal of Archaeology* 49: 7–17.

Worsaae, J.J.A. 1849. *The primeval antiquities of Denmark*. London, Parker.

Wrangham, R.W. 1979. On the evolution of ape social systems. *Social Science Information* 28: 335–68.

Wrangham, R.W. 1980. An ecological model of female-bonded primate groups. *Behaviour* 75: 262–99.

Wright, R.V.S. 1971. The archaeology of Koonalda cave. In D.J. Mulvaney and J. Golson (eds.) *Aboriginal man and environment in Australia*, 106–13. Canberra, Australian National University Press.

Wu Rukang and Dong Xingren 1985. *Homo erectus* in China. In Wu Rukang and Olson 1985, 79–90.

Wu Rukang and J.W. Olson (eds.) 1985. *Paleoanthropology and Paleolithic archeology in the People's Republic of China*. Orlando, Academic Press.

Wu Rukang and Wang Linghong 1985. Chronology in Chinese paleoanthropology. In Wu Rukang and Olsen 1985, 29–52.

Wu Xinzhi and Wu Maolin 1985. Early *Homo sapiens* in China. In Wu Rukang and Olsen 1985, 91–106.

Wymer, J.J. 1968. *Lower Palaeolothic archaeology in Britain, as represented by the Thames Valley*. London, John Baker.

Wymer, J.J. 1982. *The Palaeolithic age*. London, Croom Helm.

Wymer, J.J. 1985. *Lower Palaeolithic sites in East Anglia*. Norwich, Geo Books.

Wynn, T. 1988. Tools and the evolution of human intelligence. In Byrne and Whiten 1988, 271–84.

Wynne-Edwards, V.C. 1962. *Animal dispersion in relation to social behavior*. Edinburgh, Oliver & Bond.

Wynne-Edwards, V.C. 1964. Population control in animals. *Scientific American* 192: 2–8.

Wynne-Edwards, V.C. 1986. *Evolution through group selection*. Oxford, Basil Blackwell.

Yi, Senbok and G.A. Clark 1983. Observations on the Lower Palaeolithic of northeast Asia. *Current Anthropology* 24: 181–202.

Zavernyaev, T.M. 1978. *Xhotylevo*. Leningrad, Nauka.

Zeuner, F.E. 1959. *The Pleistocene period*. London, Hutchinson.

Zhang Yinyun 1985. *Gigantopithecus* and "*Australopithecus*" in China. In Wu Rukang and Olsen 1985, 69–78.

Zimmermann, E.A.W. 1778. *Geographic history of man and the universally distributed animals*. Brunswick.

Index

Page numbers in italic type refer to illustrations.

Abri Pataud, *153*, 166
Acheulean, 71, 123, 163, 189
adaptation, 4
adaptive radiation, 6, *7*, 83–4
Afar depression, 47, 49
Agassiz, L., 36, 40
agriculture, 197–8, 242, 244, 246
Aleutian Islands, 207
Allen, H., 217
Allen, J., 225, 229
Amud, 154, *153*
ancients, *8*, 12, 117, 121, 122, *126*, 129, 131–4, 139–41, 143, 164, 169, *169*, 178, 179, 194, 199, 215, 217, 244, 245
Apollo cave, *153*, 167, 192
Arctic colonization, 211–14
Ardrey, R., 67, 69
art and ornament, 167–70, 173, 180–1, 187, 192, 201, 214, 218, 220, 222, 225, 226, 235
artifacts, 190, 218; bone and ivory 164, 186; wood 217
Astakhov, S., 193
Atlantic Ocean, 237–9
Australia colonization, 214–27
australopithecines (Apiths), *8*, 11, 47–8, 52, 58–9, 61, *62*, 62–3, 68, 69, 70, 71, *72*, 73, 81, 92–3, 105, 107, 109, 110, 115, 116, 125, 127, 131, 155, 176, 193; *A. aethiopicus 60*, 61; *A. afarensis*, 58–9, *60*, 70, 88, 125; *A. africanus*, 53–4, 59, *60*, 61, 62, 69, 88, 89; *A. boisei*, *60*, 61–3, 88; *A. crassidens*, 62; *A. robustus*, 53, 59, *60*, 61, 62, 63, 69, 70, 88
Avdeevo, 186, 187

Bacho Kiro, *153*, 161, 194
Bailey, G., 37, 194
Bailey, R., 197
Baker, R.R., 96, 97–8, 106
Bandi, H.-G., 211
Bar-Yosef, O., 127, 128
behavior, 5, 6, 12, 38, 39, 41, 82, 95, 99, 100, 106–7, 111, 112, 116, 117, 143–4, 148, 156, 157, 182; behavioral diversity, 26
Bellwood, P., 198
Benyon, A.D., 73
Bergmann's rule, 121
Bergson, H., 36
Beringia, 184, 204, 205–7, 211, 213
Biache-St.-Vaast, 138
Bilzingsleben, 135, 136, 138
Binford, L.R., 66, 118, 158–9, 164
biological adaptation, climate, 121, 150; food shortage, 151
bipedalism, 84, 86, 87, 101, 105
Birch, L.C., 107
Birdsell, J., 216, 226
Bir Tarfawi, 129, 189
Black, D., 31
Blumenbach, J.F., 23, 25, 31, 111
Boas, F., 23, 27
Boker Tachtit, *153*, 162, 190
Bone cave, 225
Border cave, 181
Bordes, F., 163, 164
Bosinski, G., 184
Boule, M., 145, 150
Bowdler, S., 214–15, 217, 225
Bowler, J., 214
Bowler, P.J., 93
Boxgrove, 137
Brace, C.L., 149, 150, 157
Brain, C.K., 69–70, 89, *90*
British Columbia, 207
Brooks, A., 197
Brown, P., 149, 150, 154
Brown, W., 74
Brückner, E., 40–1, 42
Brunhes epoch, 44
Buffon, G.L.C., 31, 39
burials, 166–7, 187, 202, 205, 215, 217, 228; cremation, 215, 217

Cabrera-Valdés, V., 161
campsites, 213, 215, 217
Caribbean, 239
carnivore activity, 66–7, 69–70, 88, 135–6, 140, 194, 195, 196
Carter, P., 196
catastrophe theory, 76
Cavalli-Sforza, L.-L., 156, 174–5
Cave, A.J.E., 145
Chaloupka, G., 168, 218–20
Chamberlain, A.T., 56, *57*, 62, 156
Childe, G., 244
Clacton, 140
Clark, Desmond, 47, 79, 159, 164
Clark, Geoff, 157
Clark, Grahame, 157, 158, 241, 246, 247
Clarke, R.J., 61, 72, 93, 125
climate, 30, 32, 39, 41–5, 49, 81–2, 88–9, 93, *122*, 133–4, 150, 156, 164, 183–4, 185, 188, 203, 218, 220–2; climatic change, 40–1, *42*, 45, 49, 94; climatic cycles, 41, 50, 80, 83, 90, 123, 184; climatic optima, 189; climatic stress, 152
Cloggs cave, 222
Close, A.E., 129, 190
Clovis points, 208
Collingwood, R.G., 242, 247
colonization, adaptive, *94*; exaptive *95*, 181–2
Combe Grenal, *153*, 163
Coon, C.S., 27, 32, 33, 145, *146*, 147
Cosgrove, R., 223, 225
cradles, 30–3, 34, 39; Africa 74, 79, 92, 93; Asia, 91; distribution, 2
Croizat, L., 77–8, 92, *92*, 93, 145
Cro-Magnon, 144, 148, 150, 151
Croll, J., 40, 41, 44

Dali, *153*, 154
Darlington, P.J., 33, 74
Dart, R.A., 53, 69
Darwin, C., 29–30, 30, 31, 33, 36, 52, 76, 77–8, 92, 175, 179, 241
Darwin Crater, 224
Darwin glass, 224–5
dating, radiocarbon, 191, 201, 205, 208, 209, 211, 215, 218; thermoluminescence, 148, 215; uranium series, 148
Davidson, I., 66, 173–4
Davis, R., 193
Dawkins, R., 76
Deacon, J., 165, 189
deepsea cores, 42–5, 89–90
Dennell, R.W., 93
Diamond, J., 106, 234, 236
Dibble, H.L., 164, 168
diet, 68, 104, 112; dietary deficiency, 121
Dikov, N., 205, 211
Dillehay, T., 210
Djebel Irhoud, 152, *153*
DNA, mitochondrial, 176, 178; nuclear, 50, 176
Dolní Věstonice, *153*, 167, 186, 187
domestic food resources, 234, 239, 241, 242
Dorn, R.I., 218
dreamtime myths, 226–7
Dubois, E., 129, 133
Dumond, D., 211, 213
dwellings, 202, 214

Ehringsdorf, 138
Elandsfontein, 189
Eldredge, N., 37, 52, 76
Endler, J.A., 74, 75
environment, 10, 11, 30, 39, 78–9, 87, 95, 117, 121–3, 125, 133, 139, 142, 143, 215, 244; coastal, 218; seasonal wetlands, 218, 220
evolution, human, 2, 4, 6, 7–8, 10–11, 12, 13, 17, 27, 29–30, 37, 39, 46, 50–73, 74–5, 76, 82, 83, 89, 92, 100, 106, 107, 133, 157, 175, 243, 246; parallel, 145; progressive, 3, *24*, 25, 27; evolutionary change, 36, 82; ecology, 96; process, 41; risk, 96–7; stages, 20
exaptation, 5
exaptive radiation, 6, *7*, 83–4, 85
extinctions 232, 233; megafauna, 208–9, 210, 216

fire, evidence for, 132, 136–7, 140, 222
Fladmark, K.R., 207
Flood, J., 222–3
Foley, R.A., 64
food, 66, 67–8, 69, 81, 88, 108, 110, 117, 118, 123, 136–7, 185, 188, 235; foraging 139–41; processing, 190; resources, 11, 125, 133, 135, 140, 142; r selected, 225;

risk buffering, 119, 234; scavenging 140; stress, 120, 121
forests 197–202; Britain, 199–200; coniferous, 200–2; tropical, 197– 200
Franchthi cave, 239
Freidel, D.A., 190

Gansser, A., 134
Gargett, R., 166
Garrod, D., 152, 191
Geist, V., 84, 85, 86, 87, 94, 121
Gigantopithecus, 131
Gilead, I., 190
Goebel, T., 206, 211
Golson, J., 222
Gosden, C., 229
Gould, S.J., 5, 27, 37, 76
Graves, P., 148, 233
Greenland, 211, 213, 237
Grigson, C., 190
Grønnow, B., 211
Groube, L., 231
Groves, C.P., 56, 61, 62, 72, 75, 100, 105, 155

Hadar, 47, 58–9, 70, 125
Haeckel, E., 30, 31, 129
Haynes, V., 210
hearths, 138, 140, 165–6, 201, 205, 210, 218
Heyerdhal, T., 231
Hoffecker, J., 206, 211
home bases and campsites, 67–8, 112
hominids, 11, 57, 58, 130, 136, 179; early, 11, 88, 93, 112, 120, 121, 183; evolution, 52; origin, 47, 134, 189
Homo, 60, 88, 89, 96, 109, 115, 125; early, 62, 104, 107, 116; *Homo erectus*, 12, 27, 53, 59, 61–2, *62*, *71*, 71–3, *72*, 93, 96, 117, 121, 125, 127, 130, 131, 135, 145, *147*, 151, 154, 155–6; *Homo ergaster*, 61, *62*, 71; *Homo habilis*, 53–4, 59, 61, *62*, 65, 67, 70, 71, 72, 73, 88, 93, 116, 125, 142, 147, 151, 172; *Homo leakeyei*, 61, 71, 72; *Homo rudolfensis*, 51, 61, *62*, 71, 116; *Homo sapiens*, archaic, 12, 53, *62*, 71, 134, 135, 147, 147, 148, 154, 156, 157, 189; *Homo sapiens sapiens*, 12, 145, 149, 162, 180
houses, 205
Howiesons Poort, *161*, 162
Hoxne, 138
Hrdlicka, A., 32
hunting, 117–23, 136, 140, 189, 194–5, 213, 223, 232, 247; big game, 209, 210, 216; marine, 214
Huxley, T.H., 25, 34, 179
Huxley, J.H., 243

ice age, 39–41, 42, 44, 89, 118, 133, 156, 184; Cordilleran ice sheet, 207; last glacial maximum, 50, 191; Laurentide ice sheet, 207
Indian Ocean, colonization, 236–7
intelligence, 101–4, 106
Irwin, G., 230, 231, 232, 233
Isaac, G., 66–8
island colonization, 233–6

James, S.R., 132
Java, 53, 92–3, 121, 131–3, 154, 155
Jerison, H., 172
Johanson, D.C., 59, *60*
Jolly, A., 102, 114
Jones, R., 180, 220, 222
Jourdain, J., 236
Jungers, W.L., 58, 61, 62–3

Kakadu 153; rock art, *221*
Kant, 32, 39, 245
Kebara, 154
Keegan, W., 234, 236
Keith, Sir A., 39
Kelly, R., 11, 210
Keniff cave, 214, 216, 222
Kents Cavern, 159, *153*
Khol'ushkin, U., 203
Klasies River Mouth, 162, *153*, 179, 181
Klein, R., 188, 189
Klithi rock shelter, 194, 195, *195*
Koobi-Fora, 47, 49, 54, 56, 63, 65, 66, 67, 70, 71, *72*, 88, 125, 172
Koonalda cave, 218
Kortlandt, A., 101, 105
Kostenki, *153*, 185, 186, 187
Kromdraai cave, 69

La Chapelle aux Saints 145, *153*
La Cotte, 137, 138

Laetoli, 47, 58–9, 64, 152
La Ferrassie, 163
Lake Mungo, *153*, 167, 215
La Madeleine, 163
Landau, M., 3–6, 105
language, spoken, 170–5, 246
Latham, R.G., 23, 29, 30–2, 37, 46, 248
Laugerie Haute, 163
Leach, E.R., 26
Leakey, L., 53, 65, 101, 125
Leakey, M.D., 52, 58, 125
Leakey, R., 54
Leroi-Gourhan, A. *24*
Levallois technique, 136, 158, 159, 185
Leverrier, 40
Levison, M., 231
Lewin, R., 179
Lieberman, P., 170–1
Linnaeus, C., 23, 30
Locke, J., 20, 241
Lourandos, H., 225, 226
Lovejoy, C.O., 101, 104, 108
Lubbock, Sir J., 14, 16, 17, 19, 25, 35, 96, 124, 144, 241, 243, 248
Lyell, C.S., 25, 30, 35, 36, 40, 246

MacArthur, R., 37, 233
McBryde, I., 226
McBurney, C.B.M. 137, 199
McHenry, H.M., 58, *60*, 61, 63
MacKenzie, W.J.M., 247
Matenkupkum cave, 228, *229*
Malthus, T.R., 20, 209
Mandeville, Sir J., 241
Marett, R.R., 35
Marks, A.E., 162, 190
Marshall, B., 225
Martin, P., 209, 210
Matthew, W.D., 31, 33
Matuyama epoch, 44
Meadowcroft rock shelter, 208, 210, 211, 235
Mediterranean, 239–40
Meltzer, D., 207
migration, 6, *7*, 33–6, 97–8, *98*, 99, 106–7, 110, 116, 123, 143, 149; population, 225
Milankovitch, M.M., 41, 89
Milton, K., 104–5
Mochanov, Y., 203, 205
models, Darwin/Wallace, 30; watermelon model, 93, 129; migration behavior, 108–12, *108*; organic tree-of-life, 76; multiregional evolution, 144–57, 177; Out of Africa 2, 144–57, *146*, 175–7, 181, 215
modern humans, *8*, 144, 145, *147*, 148–52, 154, 156–8, 159, 164, *169*, 173, 175, 178, 179, 185, 190, 217, 245, 246; behavioral, 162; genetic, 175; Old World, *183*; voice box, *171*
Morgan, E., 67
Morgan, L.H. 20–1, 26, 35
Movius, H.L., 132, 166
Mulvaney, J., 214, 226
Murray Springs, 208

natural selection, 6, 27, 29–30, 36, 37, 79, 87, 94, 99, 100, 104
Neanderthals 12, 53, 130, 144, 145, 147, 147, 148, 150, 151, 152–4, 155, 157, 158, 159, 163, 164, 166, 167, 181, 194; speech, 170–1; voice box, *171*
neoteny, 85, 100, 104
Ngandong, 131, 153, 154, 155
North America, 203–11
Nukak, *198*

O'Connell, J., 216
Okladnikov, A.P. 193–4
Olding–Smee, J., 78
Olduvai Gorge, 47, 49, 52, 53–4, 56, 61, 63, 65, 66–7, 68, 70, 125
Old World, deserts/arid lands, 188–92; mountains, 192–7; plains 183–8
Omo Kibish, 170
Onrubia Pintado, J., 238
origins, human, 3–4, 30, 36, 46, 47, 52, 74, 93, 125, 130, 145, 148
oxygen isotopes, 42–5, *43*

Pacific colonization, 227–33
Paddayya, K., 128, 129
Papua New Guinea, 16, 215, 218, 225
Pardoe, C., 226
Parkington, J., 189
Parsons, P., 85, 94, 175
Passingham, R.E., 99, 100, 172
past, uses of, 18–19

Penck, A., 40–1, 42
Petralona, 135, 150
Peyrony, D., 163
Pfeiffer, J.E., 120, 167, 168, 170
phenotype, 86, *86*, 87, 142; development, 85; dispersal, 84, 85
phyletic gradualism, 76
Pianka, E.R., *38*
Piggott, S., 244
pioneer phase, 166, 167, 184, 185, 190, 193, 244
pit houses, 186
Pitt-Rivers, General A.H.L, 19–20, 124, 244
Pliny, 13
pollen, 50, 206, 222
Pontnewydd, 135
Pope, G., 133
Powers, R., 206, 211
pre-Clovis tradition, 209
pre-sapiens theory, 223
Prichard, J.C., 26
punctuated equilibria, 76

Qafzeh, 162, *153*, 154, 170, 175, 181
Quatrefages, A. de, 31, 39, 203, 246
Quirke, P., 231

Ramapithecus, 52
Ranov, V.A., 128, 196
Raposo, L., 166
Ratzel, F. 33, 35
Raymo, M.E., 45, 90
red ocher, 128, 192
Reeves, B.O.K., 207
Renfrew, C., 244
Rensink, E., 165
Rift Valley, 48, 50, 80
Rightmire, P., 155
Riwat, 93
Roberts, N., 89, 90, 93
Robertshaw, P., 197
Roebroeks, W., 165
Rosny-Aîné, 137
Ruddiman, W.F., 45, 90, 133

Sahul, 215–16, 217, 222, 228; shelf, *153*
St. Césaire, *153*, 162, 164
Sangiran, 130, 154, 155
Schild, R., 129
Schötensack, O., 32
Schrire, C., 247
seasonality, 50, 88, 119, 120, 121, 122, 142, 143, 197
selection, *38*, 116, 178; K selection, 37–8, *38*, 85, 106, 107; pressure, 117; *r* selection, 37–8, *38*, 85, 106, 223
settlement, 188, 202, 210
sexual dimorphism, 113, *114*; apiths, 59–61, 122, 151
Sharma, K.K., 134
Simek, J., 189
Simpson, G.G., 37, 77, 98
Sivapithecus, 52
Skhul cave, 152, 154, 162, 170, 181
Skjolsvold, A., 235
Smith, Reverend F., 203
Smith, G.E., 32, 34
Smith, M., 218
social communication, 107, 110, 111
social Darwinism, 3, 33
Soffer, O., 188
Sokchang-ni, 132
Sollas, W.J., 21
Solo River, 129–30, 131, 154
speciation, 74–8, 80–5, 87, 89, 90–1, *90*, 93, 94, 98
Spencer, H., 3, 27, 29
Speth, J.D., 111, 120, 139
Spriggs, M., 229
Stadel cave, *153*, 167
Steele, J., 81
Steinheim, 135, 150
Sterkfontein Valley, 47, 59, 61, 65, 69, 88, 91, 112, 125, 193
Stern, J.T., 58
Stoke Newington, 140
Stoneking, M., 176
storage, 119, 120, 121, 123, 138, 139, 142, 190
Stránská Skála, 136
Straus, L., 195
Straus, W.E., 145
Stringer, C., 150, 174
Strum, S.C., 114
subsistence, 190; foragers, 123
Suggs, 235
Sunda Shelf, 91, 133, *153*, *156*

Sunghir, 187
Susman, R.L., 58, 70
Svalbard-Spitzbergen, 214
Swan River, 215
Swanscombe, 135, 138
Swartkrans cave, 61, 63, 65, 69, 70
Symons, D., 104

Tabun, 154
Taimyr peninsula, 214
Tasmania, 16, 180, 215, 216, 220, 223, 224, 225, 234, 236
Tasmanian rain forests, *224*
Tattesall, I., 52, 155
Taubach, 136
Taung, 53, 61; child, 130
Taylor, G., 31
Taylor, O., 178
tectonics, 49, 92, *92*
Tenerife, 238
Ternifine, 127
Terra Amata, 138
Terra Nullius Doctrine, 14
Terrell, J., 230, 235
Teshik Tash, 193–4, 194
thermoregulation, 151
Tethys Sea, 49
Thiselton-Dyer, W., 32
Thomsen, C.J., 21
Thorne, A., 154
three age system, 21
Tierra del Fuego, 17, 33
Tischofer cave, 194
Tito Bustillo, 192
Tobias, P., 53
Todd, L., 210
tools, 23, 101, 119, 163; Acheulean, 71, 123–5, *124*, 127, 128, 136; bone, 164–5; chalcedony, 132; Folsom, 208; manufacture, 137–8, 140, 158; Mousterian, 161, 163, 181, 193; Oldowan, 65–66, *65*, 70, 101, 111, 125; stone, 64–6, 70, 101, 123–5, 127–8, 130, 132, 135, 136, 137–8, 139, 140, 141, 158–62, 164, 169, 179–80, 180, 181, 184, 189, 190, 191, 193, 195, 201, 203, 210, 213, 214, 216–17, 218, 220, 222, 228–30; types, *161*; use of, 125, *165*; utilization, 123
Torralba, 138
Torrence, R., 119
Trinidade, 238
Trinil, 129, 133, 155
Trinkaus, E., 151
Truganini, 180
Tseitlin, S.M., 203
Turkana, 54, 62; East 47; Lake 47, 61; West, 61, 71
Turner, A., 62, 89, 90, 94, 135, 156
Tuva 136, 193, 202
Tylor, E.B., 20–1, 26, 39, 124

'Ubeidiya, 127, 128
Ushki Lake, 211
Ust-Mil, 205

Van Dieman's Land, 180
Verberie, 188
Vereschagin, N.K., 207
Verkhoyansk mountains, 205
Vértesszöllös, 135, 136, 138
vicariance, 77–9, 92, 93, 156
Vico, G., 242–3, 245
Victoria cave, 199
Vilas Ruivas, *153*
Voidimatis river, 194, 195
Volp caverns, 192
von Koenigswald, G.H.R., 130
Vonnegut, K., 100, 101
Vore, I. de, 113
Vrba, E.S., 5, 76, 82, 83, 85, 86, 87, 88, 89, 91

Waal, F. de, 103, 111
Wadi Arid, 189
Wainscoat, J., 176
Wajak, *153*, 155
Walker, A.C., 54
Wallace, A.R., 2, 26, 27, 29, 30, 32, 40, 41, 77–8, 82, 92, 175
Wang Chiyen, 130
Ward, G., 231
Warreen cave, 224
Washburn, S.L., 74, 113
watercraft, 216
Webb, J., 231
Webb, S., 226
Weidenreich, F., 145, 155
Weinberg caves, 159, *153*

Weiner, J.S., 145
Wells, H.G. 144, 145
Wenban-Smith, F., 128
Wendorf, F. 129, 190
Wernicke, 172
Western rift, 49
Whallon, R., 173–4
Wheeler, P.E., 81
White, M.J.D., 75
White, P., 216, 230
White, R., 173
White, T.D., 58–9, *60*
Wickler, S., 229
Wiessner, P., 102–3, 168–9
Willandra Lakes, 149, *153*, 215, 217, 218, 220, 225
Willendorf, 186, 187
Willner, L.A., 113–14
Wilson, A., 176
Wilson, E.O., 37, 50, 233
Wobst, H.M., 169, 187, 188
Wolpoff, M.H., 150, 154, 157
Wood, B.A., *55*, 56, *57*, 62
Wood, Reverend J.G., 21, 22
Worsaae, J.J.A., 18
Wu Rukang, 130–1
Wynne-Edwards, V.C., 85, 97, 106, 107
Wyrie Swamp, 217

Xian, 134

Yamashita-Cho cave, 228
Yayo, 127
Yesner, D., 206
Yuanmou, 130, 133
Yunxian, 131

Zeeman, 76
Zhimin, A., 196
Zhokhov Island, 214
Zhoukoudian, 53, 130, 131, 132, 154
Zimmermann, E.A.W., 34
Zinjanthropus boisei (Zinj), 52, 53, 61, 67, 70